Springer Tracts in Advanced Robotics

Volume 49

Editors: Bruno Siciliano · Oussama Khatib · Frans Groen

Christian Ott

Cartesian Impedance Control of Redundant and Flexible-Joint Robots

Springer

Professor Bruno Siciliano, Dipartimento di Informatica e Sistemistica, Università di Napoli Federico II, Via Claudio 21, 80125 Napoli, Italy, E-mail: siciliano@unina.it

Professor Oussama Khatib, Robotics Laboratory, Department of Computer Science, Stanford University, Stanford, CA 94305-9010, USA, E-mail: khatib@cs.stanford.edu

Professor Frans Groen, Department of Computer Science, Universiteit van Amsterdam, Kruislaan 403, 1098 SJ Amsterdam, The Netherlands, E-mail: groen@science.uva.nl

Author

Christian Ott
Nakamura & Yamane Laboratory
Department of Mechano-Informatics
The University of Tokyo
7-3-1 Hongo, Bunkyo-ku
113-8656 Tokyo
Japan
E-Mail: ott@ynl.t.u-tokyo.ac.jp

ISBN 978-3-642-08873-5 e-ISBN 978-3-540-69255-3

DOI 10.1007/978-3-540-69255-3

Springer Tracts in Advanced Robotics ISSN 1610-7438

Printed in acid-free paper

5 4 3 2 1 0

springer.com

STAR (Springer Tracts in Advanced Robotics) has been promoted under the auspices of EURON (European Robotics Research Network)

Foreword

By the dawn of the new millennium, robotics has undergone a major transformation in scope and dimensions. This expansion has been brought about by the maturity of the field and the advances in its related technologies. From a largely dominant industrial focus, robotics has been rapidly expanding into the challenges of the human world. The new generation of robots is expected to safely and dependably co-habitat with humans in homes, workplaces, and communities, providing support in services, entertainment, education, healthcare, manufacturing, and assistance.

Beyond its impact on physical robots, the body of knowledge robotics has produced is revealing a much wider range of applications reaching across diverse research areas and scientific disciplines, such as: biomechanics, haptics, neurosciences, virtual simulation, animation, surgery, and sensor networks among others. In return, the challenges of the new emerging areas are proving an abundant source of stimulation and insights for the field of robotics. It is indeed at the intersection of disciplines that the most striking advances happen.

The goal of the series of Springer Tracts in Advanced Robotics (STAR) is to bring, in a timely fashion, the latest advances and developments in robotics on the basis of their significance and quality. It is our hope that the wider dissemination of research developments will stimulate more exchanges and collaborations among the research community and contribute to further advancement of this rapidly growing field.

The monograph written by Christian Ott is devoted to the classical research topic of impedance control which has recently found new interest after the progress in the mechanical design of lightweight robotic systems with improved actuation and sensing principles. The contents expand the author's doctoral dissertation and are focused on two key issues, namely joint flexibility and kinematic redundancy. A number of effective controllers are developed in theory, based on consolidated approaches such as singular perturbation and passivity, and are tested in extensive experiments on the DLR humanoid manipulator 'Justin', one of the most advanced robotic systems available up to date from a technology and mechatronics standpoint. No doubt, a very fine addition to the STAR series!

Naples, Italy,
April 2008

Bruno Siciliano
STAR Editor

Preface

This monograph is based on my PhD thesis, which was written at the Institute of Robotics and Mechatronics at the German Aerospace Center (DLR e.V.), and which was defended at Saarland University in November 2005. Its main topic is the Cartesian impedance control problem. While the basic concept of impedance control is well known since several years, its implementation in a wide range of applications nowadays comes into focus of robotics research again due to recent progress in the mechanical design of robotic systems with new and improved actuation principles. Moreover, the availability of more and more complex robot mechanisms allows us also to refine the control theory by considering more detailed and thus more precise models. In the present work, in particular the effect of joint elasticity on the design of impedance controllers is analyzed in detail. The consideration of joint elasticity is particularly relevant for lightweight robots in service robotics scenarios as well as for industrial robots when high accuracy is required also for very fast movements. Another topic treated in this work is the extension to redundant systems in which the robot has more degrees of freedom than necessary to fulfill a given task. This is especially important for more complex robotic systems like robots with multiple arms and hands.

The book also contains an applications chapter with several experiments in which the control methods have been applied. This includes also some more recent experiments with DLR's upper-body manipulator "Justin", which goes beyond what was originally included in the PhD thesis.

During the the work on the thesis a lot of people gave me support and advice in many different ways. First of all I want to thank Prof. Gerd Hirzinger, Head of the Institute of Robotics and Mechatronics, for he gave me the possibility to conduct research in the exciting field of robot control.

I am very grateful to Prof. Andreas Kugi, my Ph.D. supervisor from the Chair of System Theory and Automatic Control at Saarland University, for support and invaluable advice. Whenever I visited his lab he spared much time for discussing my work.

I would also like to express my gratitude to the colleagues at DLR. In particular I want to thank Dr. Alin Albu-Schäffer for the enjoyable cooperation and the stimulating conversations over the years. Moreover, I would also like to thank Dr. Alin Albu-Schäffer, Dr. Udo Frese, and Dr. Tobias Ortmaier for proofreading the thesis on which this book is based on.

During my Ph.D. research I had the opportunity to spend a period of three month at the University of Twente, The Netherlands, where I visited the groups of Prof. Stefano Stramigioli and Prof. Arjan van der Schaft. This visit significantly influenced my understandings of passivity based control and its use for redundant and flexible robots.

Finally, I would like to thank Prof. Bruno Siciliano and Dr. Thomas Ditzinger for their patience in waiting for the final modifications of the book.

Tokyo, April 2008 Christian Ott

Contents

1 Introduction

Robotic systems are a well studied field of application for control theory. Robot arms designed for industrial applications traditionally are treated as mechanical multi-body systems, where several rigid bodies are connected by actuated joints. In the past few years the robotics community evolved growing interest in robots which are designed for the use in service robotics scenarios, in which the robots will have to work close to humans. For this kind of applications the robotic systems are desired to be built up very light. Consequently, these systems usually have considerable elasticity. Therefore, the need for control strategies arises which take account not only of the rigid body dynamics but also of the joint flexibility.

Section 1.1 gives a short overview of typical service robotics applications and their peculiarities. In Section 1.2 the DLR lightweight robots are described, on which the controllers reported in this book were implemented and tested.

The state of the art of compliant motion control schemes and of control methods for flexible joint robots are reviewed in Section 1.3 and Section 1.5, respectively. Section 1.6 finally gives an overview of this book and summarizes the contributions.

1.1 Service Robotics Applications

While for most industrial applications the environment of the robots can be considered as known in advance, the situation is quite different in fields as service robotics and medical robotics. In these applications the robots share their environment with humans. Consequently, direct interaction will be desired not only with a known environment but also with a human user.

While mobile robots are used out of laboratory environments in *real-world applications* for quite some time (e.g. house floor cleaning robots, pool cleaning robots, etc ...), the situation is different for systems which must possess also manipulation capabilities. Typical tasks for such a general service robot are for instance: wiping tasks, different pick and place operations, or the opening of doors and drawers. From a conceptual point of view different approaches ranging from combinations of

C. Ott: Cartesian Imped. Cntrl. of Redund. & Flexible-Joint Robots, STAR 49, pp. 1–11, 2008.
springerlink.com

mobile robots with anthropomorphic arms and hands to more advanced humanoid robots have been proposed for such applications. They all have in common that they contain one or two robotic arms and hands, which must be built very light in order to be compatible with the locomotion part of the system.

The need for lightweight arms (and hands) arises not only from such practical requirements, but also from safety requirements. In a typical service robotics scenario, the robot will have to work close to humans or even physically interact with them. One approach to decrease possible injuries in case of an undesired impact is the use of a compliant covering for the robot structure. In [ZKRS02] it is shown that this solution is in many cases impractical. Another approach, which was followed in the construction of the DLR lightweight robots, is to reduce the effective inertia of the arm.

After all, the design and control of modern lightweight arms clearly is a crucial point for the development of robotic systems suitable for the emerging field of service robotics. The lightweight construction of such systems naturally leads to elasticities which must be considered within the controller design.

Besides service robotics applications the need for controllers which are designed for robot models with non-negligible joint elasticities also arises from industrial robots when high performance is required for very fast movements.

1.2 The DLR Lightweight Robots

The seven-joint DLR lightweight robots are optimized especially for the use in service robotics applications. In contrast to typical industrial robots, all electronics is included in the arm structure. This is also important in order to allow for the combination of these arms with mobile systems for locomotion. The Lightweight-Robot-II (Figure 1.1 (left)) was first presented in 2001. Its main design objective was the achievement of a load to weight ratio comparable to that of the human arm. All critical components of the robot therefore have been optimized for low weight. Light *Harmonic Drive*®[1] gears with high transmission ratios and brushless DC motors were used in the construction. The Harmonic Drive® gears in the joints are the main source of elasticity. While these gears are in principle backdrivable, the external forces needed for moving a joint are very high due to the high gear ratio. In order to cope with this situation, strain gauge based joint torque sensors have been included in the joints. These sensors allow to measure also very low forces at the link side, which hardly could be measured at the motor side due to stiction. The stiffness of the torque sensors is about a factor 10 higher than the stiffness of the gears, so the overall stiffness of the joints is dominated by the gears. From a control point of view, the use of joint torque sensors, together with the measurement of the motor positions[2], allows for the implementation of a full state feedback, when a flexible joint robot model is considered (see Chapter 2).

[1] Harmonic Drive® is a trademark of the Harmonic Drive AG.

[2] For the DLR lightweight robots the measurement of the motor side position is based on the measurement of the magnetic field of the rotor of the brushless DC motor.

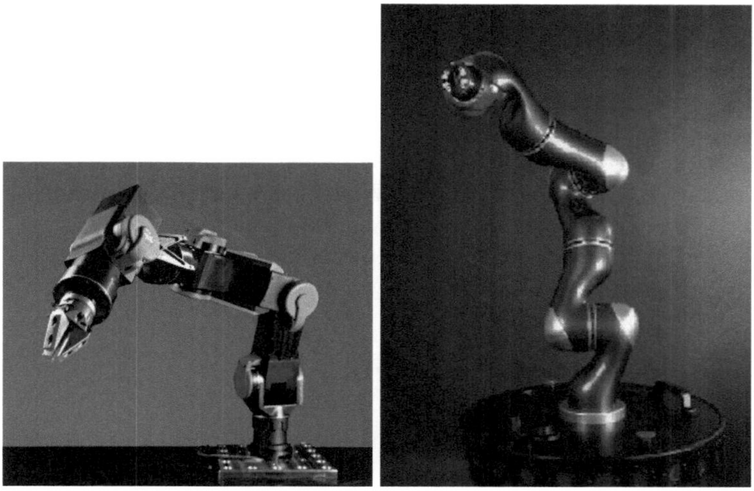

Fig. 1.1. DLR Lightweight Robots: Version II (left) and III (right)

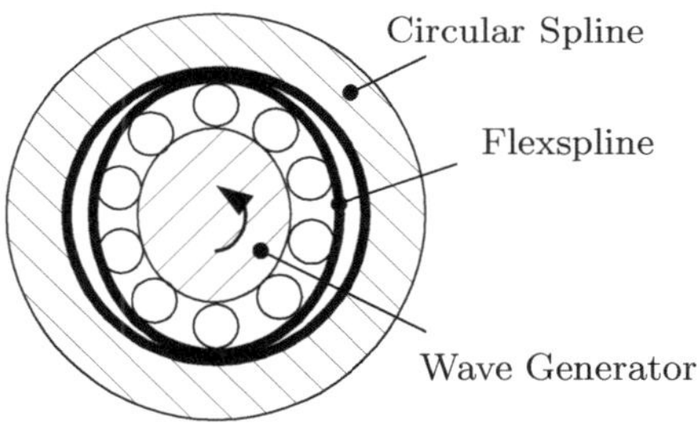

Fig. 1.2. Harmonic Drive® gear concept

As a result of the above mentioned considerations the Lightweight-Robot-II has a weight of approximately 18 kg and can carry loads up to 7 kg. Further developments of the components and the quest for a modular structure led to the design of the Lightweight-Robot-III (Figure 1.1 (right)), which has an overall weight of 14 kg and a load-to-weight ratio of about 1:1.

Since the Harmonic Drive® gears are, as mentioned above, the main source of flexibility in the DLR lightweight robots, the operation principle of this gear type will be explained shortly[3]. The gears basically consist of three parts, the *Wave Generator*, the *Flexspline*, and the *Circular Spline* (Figure 1.2). The flexspline and the circular spline are the parts having teeth. The flexspline in particular has two teeth less than the circular spline and is deformable. The wave generator, which is the driving element, has elliptic shape. Assume that the circular spline

[3] See `http://www.harmonicdrive.de` for more details on the Harmonic Drive® gear concept.

is fixed. By turning the wave generator the flexspline is deformed, and the main axis of the elliptic shape of the wave generator determines the area in which the teeth of the flexspline and the teeth of the circular spline are in contact. Consider, for instance, the case that the wave generator turns about one revolution. Since the flexspline has two teeth less than the circular spline, the flexspline will then turn about the angle of two teeth with respect to the circular spline during each revolution of the wave generator. With this a very high transmission ratio between the wave generator and the flexspline is realized.

In this work only the elasticity of the harmonic drive gears will be considered. More complete models can be found, e.g., in [TS96, KG97].

1.3 Compliant Motion Control

The realization of compliant behavior is a classical problem in robotics. This is relevant whenever the robot comes into contact with its environment and especially if the environment is only partly known.

Three different approaches are discussed in the following. The difference between these approaches and the classical position or force control concepts is that in addition to the setpoint control task also a desired *disturbance response* with respect to external forces has to be ensured. In the impedance control literature the desired setpoint $\boldsymbol{x}_d$ usually is called a *virtual* desired setpoint (or equilibrium point), since it should only be reached in case of free motion, i.e. when no external forces act on the robot.

1.3.1 Impedance Control

One possible approach to achieve compliant behavior is given by impedance control (Figure 1.3). The general concept of impedance control was introduced by Hogan in the seminal work [Hog85a, Hog85b, Hog85c] and can nowadays be considered as a classical control approach in robotics. One of the core statements of the impedance control methodology is that the controller should modulate the mechanical impedance[4] of the manipulator. This is justified by the observation that for manipulation tasks the environment of the robot is properly described by a mechanical admittance[5]. In a stronger definition of *impedance control*, for which sometimes also the term *force based impedance control* is used, it is furthermore required that not only the controlled manipulator, but also the controller itself should have impedance causality. Throughout this work the term *impedance control* refers to this stronger definition.

From a practical point of view the desired dynamic behavior of the robot, i.e. the *disturbance response* (regarding the external forces as a *disturbance*), usually

---

[4] A mechanical impedance is a mapping from (generalized) velocities to (generalized) forces.

[5] A mechanical admittance is, in contrast to a mechanical impedance, a mapping from (generalized) forces to (generalized) velocities.

is defined in terms of *Cartesian coordinates*, describing the pose of the end-effector[6]. The *Operational Space Formulation* [Kha87] provides a methodology for the controller design with respect to the Cartesian coordinates. In typical implementations, the controller has the Cartesian position $\boldsymbol{x}$ and velocity $\dot{\boldsymbol{x}}$ of the end-effector as inputs and gives the motor torques $\boldsymbol{\tau}_m$ as outputs.

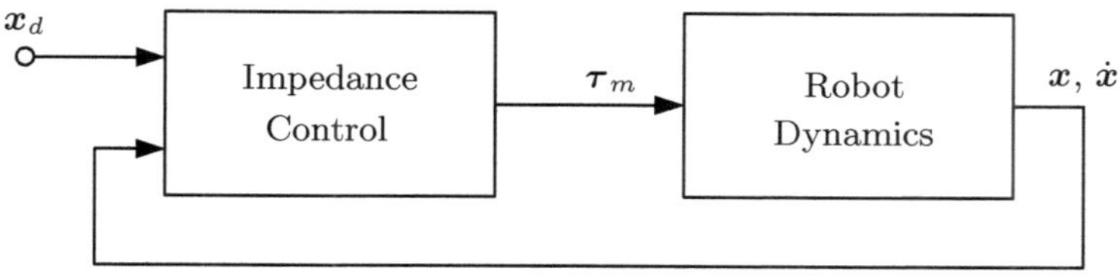

Fig. 1.3. Concept of impedance control

One relevant problem in the design of Cartesian impedance controllers is the choice of an appropriate representation of the end-effector orientation. Detailed comparisons of different orientation representations in the context of Cartesian impedance control were given by Caccavale et al. in [CNSV99] and by Natale in [Nat03]. The problem of constructing a *spatial stiffness*, suitable for the use in an impedance controller, was addressed by Fasse in [FB97, FH95, Fas97]. Further extensions of these works were presented by Stramigioli in [SD01, Str01].

1.3.2 Admittance Control

Another possible approach, which is typically followed in industrial robotics, is the concept of admittance control (Figure 1.4), sometimes also called *position based impedance control*. Here the robot is controlled by a position controller[7]. The desired compliant behavior is realized by an outer control loop. Based on measurements of the generalized environmental forces $\boldsymbol{F}_{ext}$ a desired setpoint $\boldsymbol{x}_0$ for the position control loop is generated. The admittance controller therefore has the causality of a mechanical admittance, i.e. of a mapping $\boldsymbol{F}_{ext} \rightarrow \dot{\boldsymbol{x}}_0$. The concept of admittance control is very well suited for industrial robots. These robots generally are equipped with position or velocity controllers off the shelf, while no joint torque or motor current interfaces are provided for the user. Furthermore, in usual industrial scenarios the environmental forces and torques can be measured by a six-degrees-of-freedom force-torque-sensor mounted at the tip of the robot. It is well known that the feedback of *non-collocated* forces, measured at the tip of the robot, leads to (challenging) stability problems [CH89]. This is especially relevant if a stiff force-torque-sensor, like the widely used JR3[8], is applied. Alternatively, also a compliant force-torque-sensor can be used.

[6] Hence the term *Cartesian impedance control*.

[7] Or alternatively by a velocity controller.

[8] See `http://www.jr3.com/` for detailed informations of the sensors provided by JR3 Inc.

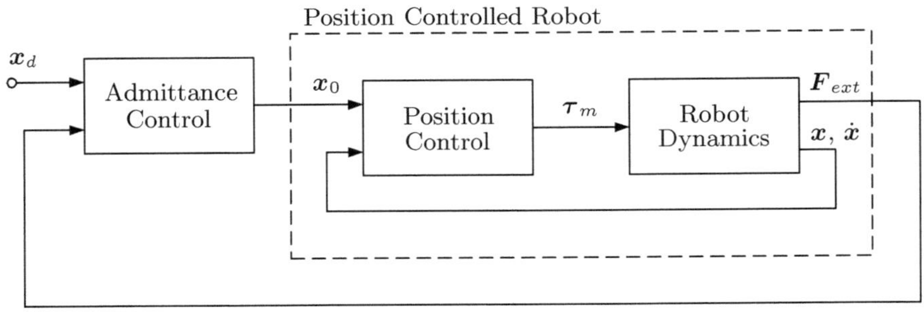

Fig. 1.4. Concept of admittance control

Furthermore, *hybrid position/force control*, which was introduced by Mason in [Mas81] and refined by Raibert and Craig in [RC81], is also closely related to the admittance control concept. The problem of coordinate invariance in hybrid position/force control schemes was discussed in [LD88] and [Duf90].

1.3.3 Conservative Congruence Transformation

In addition to the two above mentioned methods the projection of Cartesian impedance characteristics like the stiffness matrix or the damping matrix to the relevant quantities in joint coordinates represents a third possibility to realize a compliant motion controller [CK02]. The transformation law for the stiffness matrix is also called *Conservative Congruence Transformation*. This is of interest especially if the robot is equipped with a joint level impedance controller.

An interesting related concept, which currently is lively discussed in the robotics community, is to use an antagonistic arrangement of two actuators in each joint in order to realize a variable joint stiffness [LKCC91, BTBP03, BRT01]. This concept allows to implement a joint level stiffness matrix without off-diagonal elements. The possibility to achieve a desired Cartesian stiffness, however, is limited [ASFS+04].

1.4 Control of Redundant Robots

When the number m of task coordinates is smaller than the number n of configuration coordinates, the question arises, how to control the remaining $n - m$ *redundant* degrees-of-freedom. In this context a joint motion which keeps the end-effector fixed is called a *nullspace motion*. The control of such *redundant robots* is a topic widely discussed in the robotics literature. The following short literature overview refers only to those works which are related closely to the topics of this book.

A good overview of different redundancy resolution techniques is given by [Sic90]. A projection based method to optimize a given cost function via nullspace motions was presented in [HHS89]. An inverse kinematics method which uses an extended Jacobian was introduced in [Bai85]. The computation of nullspace base matrices and their utilization for the inverse kinematics of redundant robots was analyzed in [HV91] and in [CW93].

In the *Operational Space Formulation* [Kha87] a nullspace projection technique is used in order to treat redundant robots.

The impedance control problem for redundant manipulators was considered in [NSV99] in combination with a quaternion based spatial stiffness. The *Joint Space Decomposition Method* was formulated in [PCY99, PCY00] and applied to the impedance control problem in [OCY97, OCY98, OCYS98].

1.5 Control of Robots with Flexible Joints

Robotic systems with flexible joints have been treated in the robotics literature for more than twenty years. In contrast to robot models with elastic links the joint elasticity usually is modeled in form of a lumped linear spring which is located between the motor and the following link [Spo87a].

Tomei [Tom91] showed that such a system can be stabilized by a simple PD controller for the motor angle together with a compensation of the gravity torque at the desired equilibrium point. In [KS98, SK97a] similar controllers were presented based on passivity arguments. Furthermore, in [GFG99] a stability analysis for a hybrid position/force controller for a flexible joint robot without gravitational effects was given.

In practice it turned out that only quite unsatisfactory results can be achieved by restricting to purely motor position (and velocity) based feedback controllers (without additional non-collocated feedback) for the case of a flexible joint robot. In some works a controller structure based on a feedback of the joint torques as well as the link side positions was considered and it was shown that this can indeed lead to better results (see e.g. [Spo89]). From a theoretical point of view this approach usually is justified (for a sufficiently high joint stiffness) by an approximate analysis based on the singular perturbation theory [Spo89, Spo95, GHS89, Ge96].

In [ASH01a, AS01] a controller with a complete static state feedback (position and torque as well as their first derivatives) was introduced, for which (analogously to [Tom91]) asymptotic stability was shown based on the passivity properties of the controller. Contrary to the classical PD-controller, the motor inertia and the joint stiffness are included in the same passive block as the state feedback controller such that an effective damping of the joint oscillations could be achieved.

An extension of the simple PD controller by a suitable feedforward term was proposed in [DL00] in order to treat the tracking case. In [DLSZ05, ZSL+03, ZLS04, ZSL+05] the gravity compensation problem for a PD-like regulation controller was addressed.

The output feedback control problem was treated in [OKL95, NT95]. In [DLL98] an algorithm for dynamic feedback linearization of the *complete* flexible joint robot model was presented.

Another approach, consisting of a partial feedback linearization for the torque dynamics and an additional outer control loop, was presented in [LG95, LG96].

Also the tracking control problem was considered in many works, see [BOL95] for an excellent comparison. A passivity based approach was given in [LB92] for the adaptive case. A *backstepping* approach was used in [OL97] and an adaptive tracking controller was presented in [NT92].

Furthermore, variable structure controllers were analyzed in [SRS88] and [HG93].

1.6 Overview

While most of the works on control of flexible joint robots deal with the position control problem, only little works have been published for the case of the impedance control problem. As outlined in the last section the solutions for the position control problem range from simple solutions based on the singular perturbation approach to more complex ones based on modern approaches like feedback linearization or backstepping. Notice that all these results readily could be used for the design of an admittance controller. For the stability analysis of the robot in contact with its environment the situation is then similar to the case of a rigid body robot. The admittance approach, however, basically relies on the measurement of the external forces. In practice these forces usually are measured by a force-torque-sensor mounted at the end-effector. External forces, which instead act on the robot structure rather than on the end-effector, usually cannot be measured. Based on the admittance concept, therefore, a compliant behavior can only be achieved with respect to the forces acting on the end-effector. But in many applications, like for instance in service robotics where human users come in contact with the robot, this might not be acceptable. Moreover, in case of a redundant robot, like the DLR lightweight robots, additionally to the Cartesian impedance also nullspace stiffness is of interest.

This monograph concentrates on the impedance control method. Here the measurement of the external forces is not a priori necessary. Therefore, also the implementation of a nullspace stiffness will be possible.

As a desired behavior of the robot a mass-spring-damper like system will be considered. Furthermore, since inertia shaping requires the feedback of external forces even in the rigid body case (see Chapter 3), the particular case of a desired impedance behavior is analyzed in detail, where a shaping of the inertia matrix is not required. This also leads to problems concerning the damping design and the control of the nullspace motion in the redundant case. The organization of this book is summarized in the following. The rigid body case is treated in the Chapters 3 and 4. Different approaches for designing a Cartesian impedance controller for the flexible joint robot model are then analyzed in Chapter 5 to 7.

Chapter 2 treats the modeling of flexible joint robots. Herein the modeling approach from [MLS94] is applied to the particular case of a robot with flexible joints.

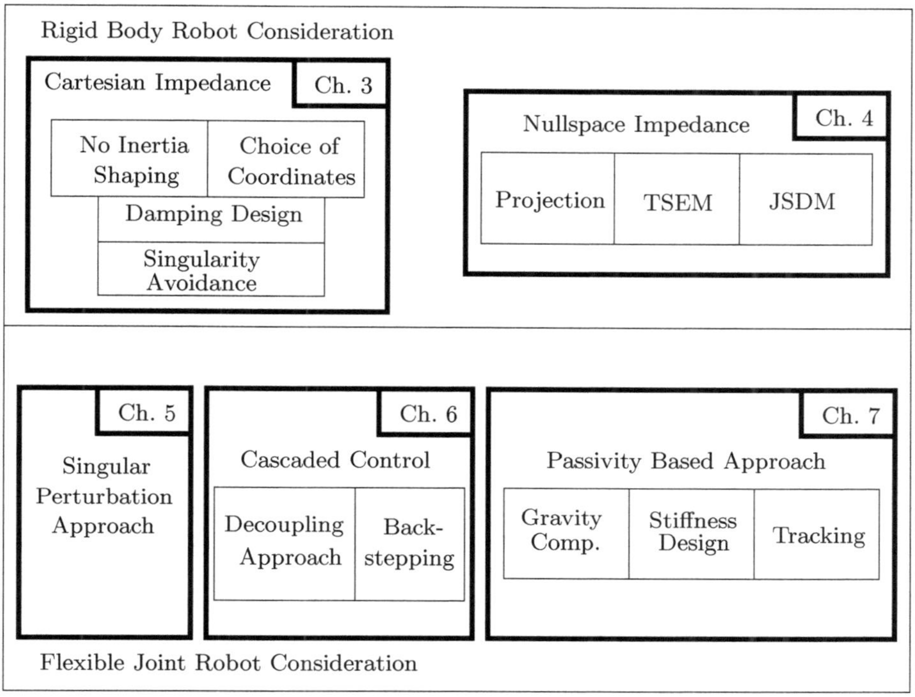

Fig. 1.5. Relation between the chapters (TSEM ... Task Space Augmentation Method, JSDM ... Joint Space Decomposition Method)

Chapter 3 deals with the impedance control problem for a rigid body robot without considering the joint flexibility. The situation of a desired impedance behavior, in which the inertia shaping is not desired, is analyzed in more detail. For this case a solution of the damping design problem is given. Additionally, some practically relevant issues, namely the singularity avoidance and the choice of Cartesian coordinates, are discussed. The results of this chapter also form the basis for the construction of Cartesian impedance controllers in the flexible joint case in Chapter 5 - Chapter 7.

Chapter 4 treats the problems of designing a nullspace stiffness for a redundant robot. In addition to the Cartesian impedance controller from Chapter 3 a nullspace impedance control law is designed. Similar to the previous chapter the case of an impedance controller without inertia shaping is analyzed. First the use of appropriate projection matrices for the design of a nullspace stiffness and damping is analyzed. Then the realization of a nullspace impedance controller based on the *Task Space Augmentation Method* as well as on the *Joint Space Decomposition Method* is treated in more detail. Within the latter approach a controller is proposed, for which a proof of the asymptotic stability is given based on the stability theory with semi-definite Lyapunov functions.

In **Chapter 5** the singular perturbation theory is applied to the flexible joint robot model from Chapter 2. Instead of the common composite controller design method used in [Spo89, Spo95, GHS89, Ge96], a modified control law is proposed, which allows to enhance the performance of the boundary layer system without computing the steady state torque explicitly. Based on this singular perturbation controller a Cartesian impedance controller is designed. While the controller can be easily implemented, the stability analysis is restricted to the singular perturbation approximation.

In the first part of **Chapter 6** the cascaded control theory is used for designing a Cartesian impedance controller. The controller is based on an inner feedback loop which decouples the torque dynamics from the link dynamics and therefore transforms the system into a triangular form. Global uniform asymptotic stability is shown for the case of free motion. Also a useful passivity property of the closed loop system is shown for the regulation case. The design idea for this controller is related to the work of Lin and Goldenberg [LG95, LG96], but it should be mentioned that the stability analysis of [LG95, LG96] cannot be extended to the considered impedance control problem[9]. Instead, an approach based on the theory of cascaded systems is used for the stability analysis.

Additionally, in the second part of this chapter a backstepping approach is applied to design a Cartesian impedance controller. The resulting controller, however, is slightly more complex than the decoupling based controller.

Chapter 7 presents an approach based on passivity theory. The design idea for this controller relies on a physical interpretation of the torque feedback. An online gravity compensation term is proposed, which, in contrast to all related previous solutions in the literature, does not lead to any constraints on the controller gains, such that it is also suitable for redundant robots. It is shown that the closed loop system can be seen as a feedback interconnection of passive subsystems. Therefore, this controller has very advantageous robustness properties. Moreover, its complexity is comparable to the simple singular perturbation controller.

It should also be mentioned that the resulting controller can be seen as a generalization of the state feedback controller presented in [ASH01a, AS01]. The controller from [ASH01a, AS01] was designed for the position control problem with a constant gravity compensation term. The physical interpretation of torque feedback together with the online gravity compensation term basically allows to formulate the controller in Cartesian coordinates and to implement also very low controller gains as desired for impedance controllers with stiffness values down to zero.

Furthermore, it is shown how the joint stiffness can be taken into account for the design of the impedance. It is also shown how the controller must be extended when non-negligible joint damping[10] is present. Finally, also an extension to the tracking case is discussed.

---

[9] This is mainly due to the fact that the considered impedance behavior is time-varying.

[10] In contrast to motor side friction the term *joint damping* refers here to a damping in parallel to the joint stiffness (see Chapter 2).

Chapter 8 presents the simulation studies and the experimental validation of the proposed controllers with the DLR lightweight robots.

In **Chapter 9** some typical applications are presented for which the Cartesian impedance controllers were used.

Chapter 10 finally gives a short summary and conclusions. It also contains a short comparison of the different presented control approaches.

2 Modeling of Flexible Joint Robots

In this chapter the dynamical model of a robot with flexible joints will be derived. The presentation is restricted to *open chain manipulators* in which the links form a serial kinematic chain without loops. Furthermore, the links of the robot are assumed to be connected either by revolute or prismatic joints[1].

When deriving a dynamical model of elastic robots basically two different sources of elasticity can be distinguished. The *distributed* elasticity caused by the robot links and a *concentrated* elasticity in the joints which is usually caused by the gears of the robot. For the former an infinite dimensional model, called *elastic link robot model*, can be derived based on the usual assumptions of the Euler-Bernoulli beam [dWSB96]. The latter, instead, leads to a finite dimensional model, called *flexible joint robot model*.

In case of the DLR lightweight robots the main sources of flexibility are the Harmonic Drive® gears. For these robots the stiffness of the links is significantly higher than the stiffness of the gears. Therefore, a flexible joint robot model will be considered in the following.

First, Section 2.1 treats the robot kinematics. In Section 2.1.1 the foundations of the description of rigid body motions are shortly reviewed. Based on that, Section 2.1 describes the kinematic modeling of serial kinematic chains. These kinematic relations are of interest not only for the derivation of the dynamical model in Section 2.2 but also for the design of Cartesian controllers in the following chapters. The *complete* and the *reduced* dynamical model of a flexible joint robot are derived in Section 2.2 and their control properties are discussed.

2.1 Robot Kinematics in a Nutshell

The following treatment of robot kinematics is based on [MLS94] and [Sel96]. Instead of giving detailed explanations of the basic kinematic relationships in robotics, the following two sections are merely intended to clarify the notation.

[1] Other types, like helical, cylindrical, spherical, or planar joints are not considered.

C. Ott: Cartesian Imped. Cntrl. of Redund. & Flexible-Joint Robots, STAR 49, pp. 13–27, 2008.
springerlink.com

2.1.1 Rigid Body Motion

The motion of a single rigid body can be represented by the relative motion of a body fixed coordinate frame[2] $\mathcal{B}$ with respect to an inertial frame $\mathcal{S}$ (see Figure 2.1). In robotic applications the inertial frame $\mathcal{S}$ usually is called *base frame* or *world frame*. The translational motion of the rigid body can be described by the vector $\boldsymbol{p}_{sb} \in \mathbb{R}^3$ from the origin of $\mathcal{S}$ to the origin of $\mathcal{B}$. The notation $\boldsymbol{p}_{sb}$ means that the vector is represented in the coordinates of $\mathcal{S}$, while ${}_b\boldsymbol{p}_{sb}$ denotes the same vector represented in the frame $\mathcal{B}$.

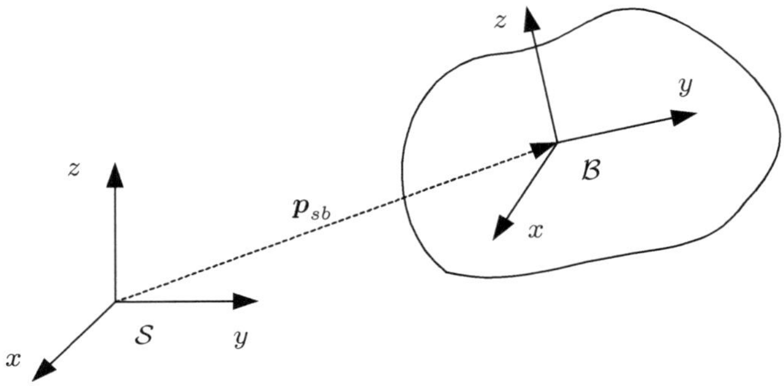

Fig. 2.1. Coordinate frames of a single rigid body

The rotation of frame $\mathcal{B}$ with respect to the world frame $\mathcal{S}$ can be described by an orthogonal matrix $\boldsymbol{R}_{sb} = [\boldsymbol{e}_x, \boldsymbol{e}_y, \boldsymbol{e}_z]$, where $\boldsymbol{e}_i$ is a unit vector, which is collinear to the ith coordinate direction of frame $\mathcal{B}$, represented in frame $\mathcal{S}$. Since all considered coordinate frames are right-handed by convention, the rotation matrices have determinant $+1$. The set of all such rotation matrices is denoted by $SO(3)$, the special orthogonal group[3] (i.e. the rotation group of $\mathbb{R}^3$). Notice that the inverse of $\boldsymbol{R}_{sb}$ can be computed by $\boldsymbol{R}_{bs} = \boldsymbol{R}_{sb}^{-1} = \boldsymbol{R}_{sb}^T$.

The spatial configuration of the rigid body is a combination of translation and rotation and can be described by the special Euclidean group $SE(3)$. The group $SE(3)$ is the semi-direct[4] product of $SO(3)$ and $\mathbb{R}^3$:

$$SE(3) = SO(3) \rtimes \mathbb{R}^3 \ . \tag{2.1}$$

---

[2] Notice that, by adopting the usual convention, all coordinate frames herein are considered as right-handed.

[3] $SO(3) \in \mathbb{R}^{3\times 3}$ is a *group* under the operation of matrix multiplication.

[4] In the semi-direct product $\mathcal{G} \rtimes \mathcal{H}$ of a group $\mathcal{G}$ and a commutative group $\mathcal{H}$ the multiplication operation is given by $(g_1, h_1)(g_2, h_2) = (g_1 g_2, h_1 + g_1(h_2))$, where $g(h)$ is a linear action of $g \in \mathcal{G}$ on $h \in \mathcal{H}$ and (g, h) is an element of $\mathcal{G} \rtimes \mathcal{H}$. In contrast to that, the direct (or Cartesian) product $\mathcal{G} \times \mathcal{H}$ has the multiplication operation $(g_1, h_1)(g_2, h_2) = (g_1 g_2, h_1 h_2)$.

An element $\boldsymbol{h}_{sb} \in SE(3)$ (i.e. the configuration of a rigid body) and its inverse $\boldsymbol{h}_{bs} = \boldsymbol{h}_{sb}^{-1}$ can be represented by the *homogeneous* matrices[5]

$$\boldsymbol{h}_{sb} = \begin{bmatrix} \boldsymbol{R}_{sb} & \boldsymbol{p}_{sb} \\ \boldsymbol{0} & 1 \end{bmatrix} \quad , \quad \boldsymbol{h}_{bs} = \begin{bmatrix} \boldsymbol{R}_{sb}^T & -\boldsymbol{R}_{sb}^T \boldsymbol{p}_{sb} \\ \boldsymbol{0} & 1 \end{bmatrix} . \tag{2.2}$$

While $\boldsymbol{h}_{sb} \in SE(3)$ represents the configuration of a rigid body, its velocity can be described by the Lie[6] algebra $se(3)$ which is the infinitesimal generator of the Lie group $SE(3)$. One common representation of a rigid body velocity is the *body twist* $\boldsymbol{V}^b \in se(3)$. The components of the body twist

$$\boldsymbol{V}^b = \begin{pmatrix} {}_b\boldsymbol{v}_{sb} \\ {}_b\boldsymbol{\omega}_{sb} \end{pmatrix}$$

have the following physical meaning: The vector ${}_b\boldsymbol{v}_{sb} \in \mathbb{R}^3$ is the velocity of the origin of frame $\mathcal{B}$ with respect to $\mathcal{S}$, represented in $\mathcal{B}$. The vector ${}_b\boldsymbol{\omega}_{sb} \in \mathbb{R}^3$ is the angular velocity of $\mathcal{B}$ with respect to $\mathcal{S}$, represented in $\mathcal{B}$.

Let $\boldsymbol{h}_{sb}(t) \in SE(3)$ be a one-parameter curve (parameterized by time) representing the trajectory of a rigid body. The relationship between the body twist $\boldsymbol{V}^b$ and the motion $\boldsymbol{h}_{sb}$ is given by $\tilde{\boldsymbol{V}}^b = \boldsymbol{h}_{sb}^{-1}\dot{\boldsymbol{h}}_{sb}$, where the application of the operator $\tilde{(\cdot)}$ to the twist is defined as

$$\tilde{\boldsymbol{V}}^b := \begin{bmatrix} {}_b\tilde{\boldsymbol{\omega}}_{sb} & {}_b\boldsymbol{v}_{sb} \\ \boldsymbol{0} & 0 \end{bmatrix} ,$$

$${}_b\tilde{\boldsymbol{\omega}}_{sb} := \begin{bmatrix} 0 & -\omega_3 & \omega_2 \\ \omega_3 & 0 & -\omega_1 \\ -\omega_2 & \omega_1 & 0 \end{bmatrix} \quad , \quad {}_b\boldsymbol{\omega}_{sb} = \begin{pmatrix} \omega_1 \\ \omega_2 \\ \omega_3 \end{pmatrix} .$$

Another common representation of a rigid body velocity is the *spatial twist* $\boldsymbol{V}^s$ which is related to the body twist $\boldsymbol{V}^b$ via $\boldsymbol{V}^s = \mathrm{Ad}_{\boldsymbol{h}_{sb}} \boldsymbol{V}^b$. Herein $\mathrm{Ad}_{\boldsymbol{h}_{sb}}$ is the *adjoint transformation*[7] associated to $\boldsymbol{h}_{sb}$. This transformation and its inverse $\mathrm{Ad}_{\boldsymbol{h}_{bs}} = \mathrm{Ad}_{\boldsymbol{h}_{sb}}^{-1}$ are given by

$$\mathrm{Ad}_{\boldsymbol{h}_{sb}} := \begin{bmatrix} \boldsymbol{R}_{sb} & \tilde{\boldsymbol{p}}_{sb}\boldsymbol{R}_{sb} \\ \boldsymbol{0} & \boldsymbol{R}_{sb} \end{bmatrix} \quad , \quad \mathrm{Ad}_{\boldsymbol{h}_{bs}} = \begin{bmatrix} \boldsymbol{R}_{sb}^T & -\boldsymbol{R}_{sb}^T \tilde{\boldsymbol{p}}_{sb} \\ \boldsymbol{0} & \boldsymbol{R}_{sb}^T \end{bmatrix} .$$

---

5 With this representation the group operation of $SE(3)$ is given by matrix multiplication.

6 Notice that $SE(3)$ and $SO(3)$ both are *Lie groups*.

7 See [MLS94] for a detailed description.

2.1.2 Forward Kinematics

In this section the forward kinematics map of an open chain manipulator with n joints is considered. For each joint a *joint variable* $q_i \in \mathbb{R}$ is introduced. In case of a revolute joint this variable represents the angle between two adjacent links. Notice that, alternatively, this angle can also be thought as an element of the set S^1, the unit circle, instead of an element of $\mathbb{R}$. In this case the angles q_i and $q_i + 2\pi$ would not be distinguished. A prismatic joint is simply described by the linear displacement between the two adjacent links.

The configuration space $\mathcal{Q}$ is the Cartesian product of the joint spaces, i.e $\mathcal{Q} := \mathbb{R}^n$. The individual joint variables q_i are summarized in the vector

$$\boldsymbol{q} = (q_1, \cdots, q_n)^T \in \mathcal{Q} .$$

In order to simplify the distinction between prismatic and rotational joints the set $\mathcal{P}$ shall be defined such that it contains all the indices i which refer to the joint variables q_i of prismatic joints.

The task will be formulated with respect to the *tool frame* which is a frame attached to the robot end-effector. The forward kinematics map $\boldsymbol{h}_{st} : \mathcal{Q} \to SE(3)$ gives the configuration of the tool frame for a given joint configuration. The computation of the forward kinematics is relevant for the implementation of Cartesian controllers as well as for the derivation of the dynamical model in the next section.

Notice first that the motion of an individual joint i (with all other joint variables fixed at $q_j = 0,\ j \neq i$) can be represented by the *joint twist* $\boldsymbol{\xi}_i$ which has the form

$$\boldsymbol{\xi}_i = \begin{pmatrix} \boldsymbol{v}_i \\ \boldsymbol{\omega}_i \end{pmatrix} .$$

In case of a revolute joint the first part of $\boldsymbol{\xi}_i$ is given by $\boldsymbol{v}_i = -\boldsymbol{\omega}_i \times \boldsymbol{l}_i$, where $\boldsymbol{\omega}_i$ is a unit vector aligned to the joint axis and $\boldsymbol{l}_i$ is an arbitrary point on the joint axis [MLS94], both represented in $\mathcal{S}$. For a prismatic joint one has $||\boldsymbol{v}_i|| = 1$ and $\boldsymbol{\omega}_i = \boldsymbol{0}$, where the vector $\boldsymbol{v}_i$ points in the direction of translation.

Given the individual joint twists as well as the reference position $\boldsymbol{h}_{st}(\boldsymbol{0})$ of the tool frame for $\boldsymbol{q} = \boldsymbol{0}$, the forward kinematics map $\boldsymbol{h}_{st}(\boldsymbol{q})$ can be computed via the well known *product of exponentials formula*

$$\boldsymbol{h}_{st}(\boldsymbol{q}) = e^{\tilde{\boldsymbol{\xi}}_1 q_1} e^{\tilde{\boldsymbol{\xi}}_2 q_2} \ldots e^{\tilde{\boldsymbol{\xi}}_n q_n} \boldsymbol{h}_{st}(\boldsymbol{0}) , \tag{2.3}$$

where the exponential terms $e^{\tilde{\boldsymbol{\xi}}_i q_i} \in SE(3)$ of a twist $\boldsymbol{\xi}_i$ are defined as

$$e^{\tilde{\boldsymbol{\xi}}_i q_i} = \begin{cases} \begin{bmatrix} e^{\tilde{\boldsymbol{\omega}}_i q_i} & \left(\boldsymbol{I} - e^{\tilde{\boldsymbol{\omega}}_i q_i}\right)(\boldsymbol{\omega}_i \times \boldsymbol{v}_i) + \boldsymbol{\omega}_i \boldsymbol{\omega}_i^T \boldsymbol{v}_i q_i \\ \boldsymbol{0} & 1 \end{bmatrix} & \text{for} \quad ||\boldsymbol{\omega}_i|| = 1 \\ \begin{bmatrix} \boldsymbol{I} & \boldsymbol{v}_i q_i \\ \boldsymbol{0} & 1 \end{bmatrix} & \text{for} \quad \boldsymbol{\omega}_i = \boldsymbol{0} \end{cases} .$$

Furthermore, the exponential terms $e^{\tilde{\omega}_i q_i} \in SO(3)$ of the unit vectors $\boldsymbol{\omega}_i$ are given by *Rodrigues' formula*

$$e^{\tilde{\boldsymbol{\omega}}_i q_i} = \boldsymbol{I} + \tilde{\boldsymbol{\omega}}_i \sin(q_i) + \tilde{\boldsymbol{\omega}}_i^2 (1 - \cos(q_i)) \ .$$

Equation (2.3) gives the configuration of the tool frame. The actual velocity of the tool frame is given by the body twist $\boldsymbol{V}^b$. As described in [MLS94] the body twist $\boldsymbol{V}^b$ can also be computed based on the joint twists. It is given by

$$\boldsymbol{V}^b = \boldsymbol{J}^b(\boldsymbol{q})\dot{\boldsymbol{q}} \ , \tag{2.4}$$

$$\boldsymbol{J}^b(\boldsymbol{q}) = \left[\boldsymbol{\xi}_1^\dagger(\boldsymbol{q}) \cdots \boldsymbol{\xi}_n^\dagger(\boldsymbol{q})\right] \ , \tag{2.5}$$

$$\boldsymbol{\xi}_i^\dagger(\boldsymbol{q}) = \mathrm{Ad}^{-1}_{e^{\tilde{\boldsymbol{\xi}}_i q_i} \cdots e^{\tilde{\boldsymbol{\xi}}_n q_n} \boldsymbol{h}_{st}(\boldsymbol{0})} \boldsymbol{\xi}_i \ ,$$

where the matrix $\boldsymbol{J}^b(\boldsymbol{q})$ is usually called *body Jacobian* or *geometrical Jacobian*[8].

2.2 Dynamical Model of a Flexible Joint Robot

Two different models are used in the literature for the modeling of flexible joint robots, known as the *complete* and the *reduced* model[9]. The *reduced* model was proposed by Spong in [Spo87b]. It differs from the *complete* model in the assumption that the kinetic energy of the rotors is determined only by their own rotation. All the modeling assumptions which lead to the *complete* and the *reduced* model are given in the following together with a detailed derivation of the models based on the Lagrangian equations of motion [Arn89, MLS94] and on the formulation of the kinematics as presented in the previous section.

2.2.1 Modeling Assumptions for the *Complete Model*

For the derivation of the complete dynamical model the following three assumptions are made:

Assumption 2.1. *The rotors of the motors are rotational symmetric rigid bodies, and the rotation axis and the symmetry axis of each joint coincide.*

The rotor angle of the ith motor which describes the rotation around the symmetry axis will be denoted by $\theta_{\mathrm{motor},i}$.

Assumption 2.2. *The electrical part of the dynamics (i.e. the inner current control loop of the motors) is sufficiently fast such that it can be neglected.*

The motors are thus modeled as ideal torque sources and the torque of the motor i acting on the ith rotor inertia is given by $\tau_{\mathrm{motor},i}$.

8 Notice that the term *Jacobian* is a bit misleading here because $\boldsymbol{J}^b(\boldsymbol{q})$ is in general not the differential of any mapping.

9 The terminology *complete* and *reduced* model is taken from [DLT96].

In order to obtain a description which is independent of the transmission ratios N_i of the gears, the *motor torque* variables $\tau_{m,i}$ and the *motor angle* variables θ_i are introduced. These variables are related to the physical values $\tau_{\text{motor},i}$ and $\theta_{\text{motor},i}$ by the scaling values N_i and $1/N_i$, i.e.

$$\tau_{m,i} = N_i \tau_{\text{motor},i} \qquad i = 1, \cdots, n \; , \tag{2.6}$$

$$\theta_i = \frac{1}{N_i} \theta_{\text{motor},i} \qquad i = 1, \cdots, n \; . \tag{2.7}$$

Similarly to $\boldsymbol{q} \in \mathcal{Q}$, the motor angles θ_i and the motor torques $\tau_{m,i}$ are summarized in the vectors $\boldsymbol{\theta} := (\theta_1, \cdots, \theta_n)^T \in \mathbb{R}^n$ and $\boldsymbol{\tau}_m := (\tau_{m,1}, \cdots, \tau_{m,n})^T \in \mathbb{R}^n$.

Assumption 2.3. *For each joint the elasticity of the gear is modeled as a linear spring located between the rotor and the subsequent link.*

Assumption 2.3 is illustrated in Figure 2.2 for the second joint of the DLR-Lightweight-Robot-III. The joint angle which determines the location of the next link is given by q_i. The spring between the rotor and the link is described by the linear relationship $\tau_i = K_i(\theta_i - q_i)$, where k_i is the stiffness coefficient of the spring and τ_i is the *joint torque*. The stiffness potential of the spring in the ith joint is then given by $V_{k,i}(q_i, \theta_i) := 1/2 K_i(\theta_i - q_i)^2$.

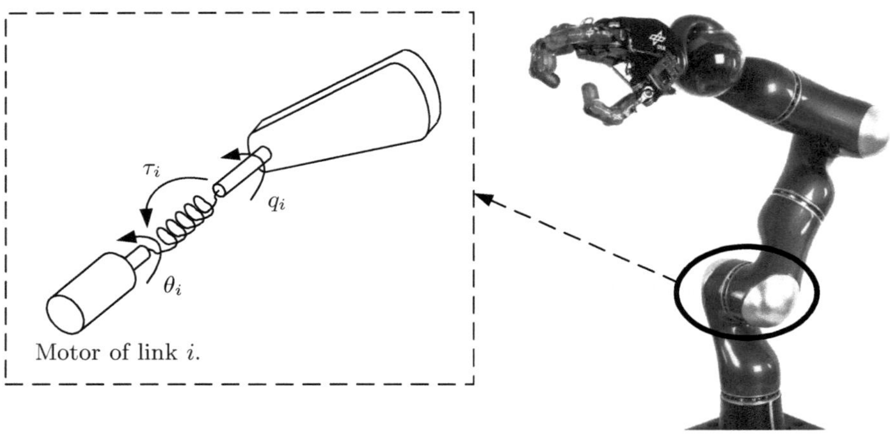

Fig. 2.2. Model of a flexible joint robot

2.2.2 Derivation of the *Complete Model*

For a robot with n flexible joints one has to consider the dynamics of $2n$ rigid bodies:

- The n links[10], building up the serial kinematic chain and
- the n rotors of the motors.

[10] It is assumed that the distal link contains also the tool and the load of the robot.

The following derivation is based on the Lagrangian equations of motion [Arn89, MLS94]. Therefore, the kinetic energy[11] $T(\boldsymbol{q}, \dot{\boldsymbol{q}}, \dot{\boldsymbol{\theta}})$ and the potential energy $V_{pot}(\boldsymbol{q}, \boldsymbol{\theta})$ of the system have to be computed.

Kinetic Energy of Rigid Bodies

Consider first the kinetic energy of an arbitrary rigid body. Let m_b be the mass and $\boldsymbol{\mathcal{I}}_b$ the inertia tensor expressed in the body fixed frame $\mathcal{B}$ which is located in the center of gravity of the considered body. Then the kinetic energy T_b of this rigid body can be written as the sum of a translational and a rotational component. It is given by

$$T_b = \frac{1}{2} m_b \left({}_b\boldsymbol{v}_{sb}\right)^T \left({}_b\boldsymbol{v}_{sb}\right) + \frac{1}{2} ({}_b\boldsymbol{\omega}_{sb})^T \boldsymbol{\mathcal{I}}_b \left({}_b\boldsymbol{\omega}_{sb}\right) \tag{2.8}$$

$$= \frac{1}{2} (\boldsymbol{V}^b)^T \boldsymbol{\mathcal{M}}_b \boldsymbol{V}^b \ , \tag{2.9}$$

with the generalized inertia matrix

$$\boldsymbol{\mathcal{M}}_b = \begin{bmatrix} m_b\boldsymbol{I} & \boldsymbol{0} \\ \boldsymbol{0} & \boldsymbol{\mathcal{I}}_b \end{bmatrix} , \tag{2.10}$$

where $\boldsymbol{I}$ is a 3×3 unit matrix. The vectors ${}_b\boldsymbol{v}_{sb}$ and ${}_b\boldsymbol{\omega}_{sb}$ are the translational and angular velocity vectors of the frame $\mathcal{B}$ with respect to frame $\mathcal{S}$ represented in frame $\mathcal{B}$ and, consequently, $\boldsymbol{V}^b$ is the body twist of frame $\mathcal{B}$.

Computing the kinetic energy thus can be reduced to computing the body twists of the rotors and the links. Therefore, the following coordinate frames are assigned: For each of the n links a frame $\mathcal{L}_i$ is assigned, fixed to the link i, such that its origin coincides with the center of gravity of the link. For the n rotors the frame $\mathcal{R}_i$ is chosen in such a way that it is attached not to the rotor itself but to the previous link and the z-coordinate axis of $\mathcal{R}_i$ is identical to the ith joint axis (see Figure 2.3). From these coordinate frames, the forward kinematics mappings and the body twists of the links and the rotors can be computed via equations (2.3) and (2.4).

The Kinetic Energy of the Links

It can easily be observed that the body twists of the links $\boldsymbol{V}^b_{l,i}$ do not depend on the motor angles and can be written as

$$\boldsymbol{V}^b_{l,i} = \boldsymbol{J}^b_{l,i}(\boldsymbol{q})\dot{\boldsymbol{q}} \quad i = 1, \cdots, n \ , \tag{2.11}$$

[11] During the derivation it will become clear that the kinetic energy is independent of the motor angles $\boldsymbol{\theta}$.

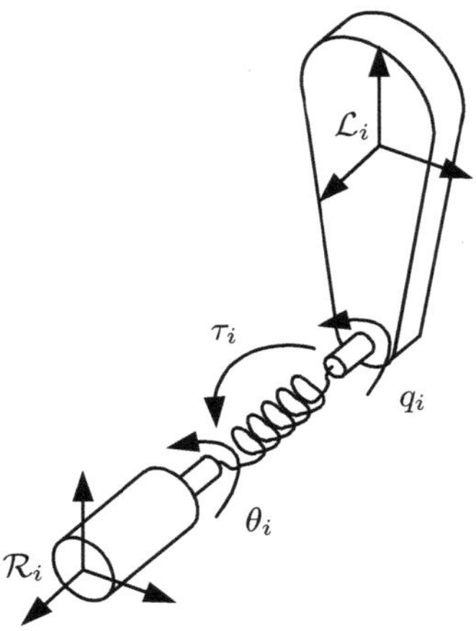

Fig. 2.3. Assigned coordinate frames in the flexible joint robot model

where $\boldsymbol{J}^b_{l,i}(\boldsymbol{q})$ is the body Jacobian[12] of the frame $\mathcal{L}_i$. Therefore, the kinetic energy of the links is given by

$$T_l(\boldsymbol{q},\dot{\boldsymbol{q}}) = \frac{1}{2}\sum_{i=1}^{n}(\boldsymbol{V}^b_{l,i})^T\boldsymbol{\mathcal{M}}_{l,i}\boldsymbol{V}^b_{l,i} ,$$

where $\boldsymbol{\mathcal{M}}_{l,i}$ is the generalized inertia matrix of the link i. By introducing the inertia matrix of the links $\boldsymbol{M}_l(\boldsymbol{q})$ via

$$\boldsymbol{M}_l(\boldsymbol{q}) = \sum_{i=1}^{n}\boldsymbol{J}^b_{l,i}(\boldsymbol{q})^T\boldsymbol{\mathcal{M}}_{l,i}\boldsymbol{J}^b_{l,i}(\boldsymbol{q}) ,$$

the kinetic energy $T_l(\boldsymbol{q},\dot{\boldsymbol{q}})$ can simply be written as

$$T_l(\boldsymbol{q},\dot{\boldsymbol{q}}) = \frac{1}{2}\dot{\boldsymbol{q}}^T\boldsymbol{M}_l(\boldsymbol{q})\dot{\boldsymbol{q}} .$$

The Kinetic Energy of the Rotors

The body twists of the rotors have the form

$$\boldsymbol{V}^b_{r,i} = \begin{pmatrix} {}_r\boldsymbol{v}_{sr} \\ \\ {}_r\boldsymbol{\omega}_{sr} \end{pmatrix} = \boldsymbol{J}^b_{r,i}(\boldsymbol{q})\dot{\boldsymbol{q}} + \begin{pmatrix} 0\\0\\0\\0\\0\\1 \end{pmatrix}\dot{\theta}_i \quad i = 1,\cdots,n , \tag{2.12}$$

[12] Notice that the last $n-i$ columns of $\boldsymbol{J}^b_{l,i}(\boldsymbol{q})$ have only zero elements because the motion of the link i does only depend on the joints 1 to i. Equation (2.5) is therefore actually used to compute the i nonzero columns of $\boldsymbol{J}^b_{l,i}(\boldsymbol{q})$.

where $\boldsymbol{J}^b_{r,i}(\boldsymbol{q})$ is the body Jacobian[13] of the frame $\mathcal{R}_i$. Therefore, the kinetic energy of the rotors $T_r(\boldsymbol{q}, \dot{\boldsymbol{q}}, \dot{\boldsymbol{\theta}})$ can be computed as

$$T_r(\boldsymbol{q}, \dot{\boldsymbol{q}}, \dot{\boldsymbol{\theta}}) = \frac{1}{2} \sum_{i=1}^{n} (\boldsymbol{V}^b_{r,i})^T \boldsymbol{\mathcal{M}}_{r,i} \boldsymbol{V}^b_{r,i} ,$$

where $\boldsymbol{\mathcal{M}}_{r,i}$ is the generalized inertia matrix of the rotor i. Due to the symmetry Assumption 2.1 the inertia tensors of the rotors have the following diagonal form when represented in the coordinate frames $\mathcal{R}_i$:

$$\boldsymbol{\mathcal{I}}_{r,i} = \begin{bmatrix} A_i & 0 & 0 \\ 0 & A_i & 0 \\ 0 & 0 & B_i \end{bmatrix} \qquad i = 1, \cdots, n . \tag{2.13}$$

Thus, the kinetic energy of the rotors has the form

$$T_r(\dot{\boldsymbol{\theta}}, \boldsymbol{q}, \dot{\boldsymbol{q}}) = \frac{1}{2} \begin{pmatrix} \dot{\boldsymbol{q}} \\ \dot{\boldsymbol{\theta}} \end{pmatrix}^T \begin{bmatrix} \boldsymbol{M}_r(\boldsymbol{q}) & \boldsymbol{S}(\boldsymbol{q}) \\ \boldsymbol{S}(\boldsymbol{q})^T & \boldsymbol{B} \end{bmatrix} \begin{pmatrix} \dot{\boldsymbol{q}} \\ \dot{\boldsymbol{\theta}} \end{pmatrix} , \tag{2.14}$$

where the matrices $\boldsymbol{M}_r(\boldsymbol{q})$, $\boldsymbol{S}(\boldsymbol{q})$, and $\boldsymbol{B}$ are given by

$$\boldsymbol{M}_r(\boldsymbol{q}) := \sum_{i=1}^{n} \boldsymbol{J}^b_{r,i}(\boldsymbol{q})^T \boldsymbol{\mathcal{M}}_{r,i} \boldsymbol{J}^b_{r,i}(\boldsymbol{q}) , \tag{2.15}$$

$$\boldsymbol{S}(\boldsymbol{q}) := \begin{bmatrix} \boldsymbol{s}_1(\boldsymbol{q}) & \cdots & \boldsymbol{s}_n(\boldsymbol{q}) \end{bmatrix} , \tag{2.16}$$

$$\boldsymbol{s}_i(\boldsymbol{q}) := \boldsymbol{J}^b_{r,i}(\boldsymbol{q})^T \begin{pmatrix} 0 \\ 0 \\ 0 \\ 0 \\ 0 \\ B_i \end{pmatrix} , \tag{2.17}$$

$$\boldsymbol{B} := \mathrm{diag}(B_i) \in \mathbb{R}^{n \times n} . \tag{2.18}$$

It can be shown that $\boldsymbol{S}(\boldsymbol{q})$ is a *strictly upper triangular* matrix (see, e.g., [DLT96]). Finally, the complete kinetic energy of the $2n$ rigid bodies is given by

$$\begin{aligned} T(\boldsymbol{q}, \dot{\boldsymbol{q}}, \dot{\boldsymbol{\theta}}) &= T_l(\boldsymbol{q}, \dot{\boldsymbol{q}}) + T_r(\boldsymbol{q}, \dot{\boldsymbol{q}}, \dot{\boldsymbol{\theta}}) \\ &= \frac{1}{2} \begin{pmatrix} \dot{\boldsymbol{q}} \\ \dot{\boldsymbol{\theta}} \end{pmatrix}^T \begin{bmatrix} \boldsymbol{M}_r(\boldsymbol{q}) + \boldsymbol{M}_l(\boldsymbol{q}) & \boldsymbol{S}(\boldsymbol{q}) \\ \boldsymbol{S}(\boldsymbol{q})^T & \boldsymbol{B} \end{bmatrix} \begin{pmatrix} \dot{\boldsymbol{q}} \\ \dot{\boldsymbol{\theta}} \end{pmatrix} . \end{aligned}$$

Potential Energy

The potential energy consists of two parts:

- the stiffness potential V_k of the springs, and
- the gravity potential V_g.

[13] Notice that the last $n - i + 1$ columns of $\boldsymbol{J}^b_{r,i}(\boldsymbol{q})$ have only zero elements because the motion of the rotor i does only depend on the joints 1 to $i - 1$.

The combined stiffness potential of all joints is given by

$$V_k(\boldsymbol{q},\boldsymbol{\theta}) = \sum_{i=1}^{n} V_{k,i}(q_i,\theta_i) \tag{2.19}$$
$$= \frac{1}{2}(\boldsymbol{\theta}-\boldsymbol{q})^T \boldsymbol{K}(\boldsymbol{\theta}-\boldsymbol{q}) \ , \tag{2.20}$$

where $\boldsymbol{K}$ is a diagonal matrix containing the joint stiffness values as diagonal elements:

$$\boldsymbol{K} := \mathrm{diag}(K_i) \in \mathbb{R}^{n\times n} \ . \tag{2.21}$$

Due to Assumption 2.1 the gravitational potential is independent of $\boldsymbol{\theta}$. Let $m_{l,i}$ and $m_{r,i}$ denote the masses of the links and the rotors, respectively. The vector $\boldsymbol{e}_g \in \mathbb{R}^3$ is a unit vector pointing in the direction of gravity and $g \in \mathbb{R}$ is the gravitational acceleration. Then the gravitational potential $V_g(\boldsymbol{q})$ is given by

$$V_g(\boldsymbol{q}) = -g\boldsymbol{e}_g^T \left(\sum_{i=1}^{n} m_{l,i}\boldsymbol{p}_{sl,i}(\boldsymbol{q}) + m_{r,i}\boldsymbol{p}_{sr,i}(\boldsymbol{q}) \right) \ , \tag{2.22}$$

where $\boldsymbol{p}_{sl,i}(\boldsymbol{q}) \in \mathbb{R}^3$ and $\boldsymbol{p}_{sr,i}(\boldsymbol{q}) \in \mathbb{R}^3$ are the vectors pointing from the origin of $\mathcal{S}$ to the origins of $\mathcal{L}_i$ and $\mathcal{R}_i$, respectively, and can be computed using (2.3). The complete potential energy $V_{pot}(\boldsymbol{q},\boldsymbol{\theta})$ of the flexible joint robot model is given by the sum of the stiffness potential and the gravity potential as

$$V_{pot}(\boldsymbol{q},\boldsymbol{\theta}) = V_g(\boldsymbol{q}) + V_k(\boldsymbol{q},\boldsymbol{\theta}) \tag{2.23}$$
$$= V_g(\boldsymbol{q}) + \frac{1}{2}(\boldsymbol{\theta}-\boldsymbol{q})^T \boldsymbol{K}(\boldsymbol{\theta}-\boldsymbol{q}) \ . \tag{2.24}$$

Lagrangian Equations

In order to simplify the notation the configuration vector

$$\hat{\boldsymbol{q}} = \begin{pmatrix} \boldsymbol{q} \\ \boldsymbol{\theta} \end{pmatrix} \in \mathbb{R}^{2n} \tag{2.25}$$

is introduced. Then the Lagrangian $L(\hat{\boldsymbol{q}},\dot{\hat{\boldsymbol{q}}})$ of the flexible joint robot model is given by the difference between kinetic and potential energy:

$$L(\hat{\boldsymbol{q}},\dot{\hat{\boldsymbol{q}}}) = T(\boldsymbol{q},\dot{\boldsymbol{q}},\dot{\boldsymbol{\theta}}) - V_{pot}(\boldsymbol{q},\boldsymbol{\theta}) \ . \tag{2.26}$$

The equations of motion of the complete model are computed via

$$\frac{\mathrm{d}}{\mathrm{d}t}\left(\frac{\partial L(\hat{\boldsymbol{q}},\dot{\hat{\boldsymbol{q}}})}{\partial \dot{\hat{\boldsymbol{q}}}} \right) - \frac{\partial L(\hat{\boldsymbol{q}},\dot{\hat{\boldsymbol{q}}})}{\partial \hat{\boldsymbol{q}}} = \boldsymbol{Q} \ , \tag{2.27}$$

where $\boldsymbol{Q} \in \mathbb{R}^{2n}$ are the generalized forces corresponding to $\hat{\boldsymbol{q}}$. The motor torques $\boldsymbol{\tau}_m$ clearly correspond to the generalized forces acting on the rotors. For the design of impedance controllers also additional external torques $\boldsymbol{\tau}_{ext} \in \mathbb{R}^n$ are considered which act on the links of the robot. The generalized forces are thus given by

$$\boldsymbol{Q} = \begin{pmatrix} \boldsymbol{\tau}_{ext} \\ \boldsymbol{\tau}_m \end{pmatrix} .$$

By evaluating (2.27) one gets the equations of motion of the form

$$\boxed{\boldsymbol{H}(\boldsymbol{q}) \begin{pmatrix} \ddot{\boldsymbol{q}} \\ \ddot{\boldsymbol{\theta}} \end{pmatrix} + \boldsymbol{\Gamma}(\boldsymbol{q}, \dot{\boldsymbol{q}}) \begin{pmatrix} \dot{\boldsymbol{q}} \\ \dot{\boldsymbol{\theta}} \end{pmatrix} + \begin{pmatrix} \boldsymbol{g}(\boldsymbol{q}) - \boldsymbol{K}(\boldsymbol{\theta} - \boldsymbol{q}) \\ \boldsymbol{K}(\boldsymbol{\theta} - \boldsymbol{q}) \end{pmatrix} = \begin{pmatrix} \boldsymbol{\tau}_{ext} \\ \boldsymbol{\tau}_m \end{pmatrix}} , \tag{2.28}$$

where $\boldsymbol{H}(\boldsymbol{q}) = (H_{ij}(\boldsymbol{q}))$ is the *complete inertia matrix*

$$\boldsymbol{H}(\boldsymbol{q}) = \begin{bmatrix} \boldsymbol{M}_l(\boldsymbol{q}) + \boldsymbol{M}_r(\boldsymbol{q}) & \boldsymbol{S}(\boldsymbol{q}) \\ \boldsymbol{S}(\boldsymbol{q})^T & \boldsymbol{B} \end{bmatrix} ,$$

and the gravity torques $\boldsymbol{g}(\boldsymbol{q})$ are given by the gravity potential

$$\boldsymbol{g}(\boldsymbol{q}) = \left(\frac{\partial V_g(\boldsymbol{q})}{\partial \boldsymbol{q}} \right)^T . \tag{2.29}$$

The term $\boldsymbol{\Gamma}(\boldsymbol{q}, \dot{\boldsymbol{q}})\dot{\hat{\boldsymbol{q}}}$ consists of the centrifugal and the Coriolis forces. The matrix $\boldsymbol{\Gamma}(\boldsymbol{q}, \dot{\boldsymbol{q}}) = (\Gamma_{ij}(\boldsymbol{q}, \dot{\boldsymbol{q}}))$, which is sometimes called *Coriolis/centrifugal matrix*, is not unique in general. But it can be chosen such that its elements are related to the *Christoffel symbols*[14] Γ_{ijk} [MLS94, SS96] via

$$\Gamma_{ij}(\boldsymbol{q}, \dot{\boldsymbol{q}}) = \sum_{k=1}^{2n} \Gamma_{ijk}(\boldsymbol{q})\dot{\hat{q}}_k , \tag{2.30}$$

$$\Gamma_{ijk}(\boldsymbol{q}) = \frac{1}{2} \left(\frac{\partial H_{ij}(\boldsymbol{q})}{\partial \hat{q}_k} + \frac{\partial H_{ik}(\boldsymbol{q})}{\partial \hat{q}_j} - \frac{\partial H_{kj}(\boldsymbol{q})}{\partial \hat{q}_i} \right) . \tag{2.31}$$

Moreover, it should be mentioned that the lower right $n \times n$ block of $\boldsymbol{\Gamma}(\boldsymbol{q}, \dot{\boldsymbol{q}})$ is structurally zero.

Equation (2.28) represents the so-called *complete model* of a flexible joint robot. One can see that the equations for the joint angles $\boldsymbol{q}$ and the equations for the motor angles $\boldsymbol{\theta}$ are coupled by the stiffness term $\boldsymbol{K}(\boldsymbol{\theta} - \boldsymbol{q})$ as well as by the outer-diagonal part $\boldsymbol{S}(\boldsymbol{q})$ of the inertia matrix $\boldsymbol{H}(\boldsymbol{q})$. This outer-diagonal part is often neglected in the literature and leads to the *reduced model* considered next.

[14] Of the first kind.

2.2.3 The *Reduced Model*

In addition to the previous modeling assumptions, which led to the *complete model*, the following assumption is made.

Assumption 2.4. *The rotational part of the kinetic energy of the rotors is determined only by the rotors' relative movement with respect to the previous link.*

This assumption is justified for the DLR lightweight robots by the fact that the gear ratios N_i are quite high. For the DLR-Lightweight-Robot-II they have values in the range 100-160. Therefore, the rotors will turn much faster than the links of the robot and hence the kinetic energy of the rotors will not be affected considerably by the motion of the links. Notice also that this modeling assumption is automatically fulfilled in a robot construction in which the motors are all located in the base of the robot[15]. This is the case for cable actuated robots, such as for instance the Dexter arm [ZSL+03].

From Assumption 2.4 it follows that the angular velocity of the rotors from (2.12) is given by

$$ {}_r\boldsymbol{\omega}_{sr,i} \approx \begin{pmatrix} 0 \\ 0 \\ 1 \end{pmatrix} \dot{\theta}_i \qquad i = 1, \cdots, n \ , \tag{2.32} $$

while the translational velocity ${}_r\boldsymbol{v}_{sr,i}$ is the same as in (2.12). By computing the new kinetic energy of the rotors, one can see that the coupling part $\boldsymbol{S}(\boldsymbol{q})$ in the inertia matrix vanishes[16], i.e. $\boldsymbol{S}(\boldsymbol{q}) = \boldsymbol{0}$. Thus, the kinetic energy of the rotors (2.14) simplifies to

$$ T_r(\boldsymbol{q}, \dot{\boldsymbol{q}}, \dot{\boldsymbol{\theta}}) = \frac{1}{2} \begin{pmatrix} \dot{\boldsymbol{q}} \\ \dot{\boldsymbol{\theta}} \end{pmatrix}^T \begin{bmatrix} \bar{\boldsymbol{M}}_r(\boldsymbol{q}) & \boldsymbol{0} \\ \boldsymbol{0} & \boldsymbol{B} \end{bmatrix} \begin{pmatrix} \dot{\boldsymbol{q}} \\ \dot{\boldsymbol{\theta}} \end{pmatrix} , $$

with the link side inertia matrix of the rotors $\bar{\boldsymbol{M}}_r(\boldsymbol{q})$ given by

$$ \bar{\boldsymbol{M}}_r(\boldsymbol{q}) = \sum_{i=1}^{n} m_{r,i} \bar{\boldsymbol{J}}^b_{r,i}(\boldsymbol{q})^T \bar{\boldsymbol{J}}^b_{r,i}(\boldsymbol{q}) \ , \tag{2.33} $$

$$ \bar{\boldsymbol{J}}^b_{r,i}(\boldsymbol{q}) = \begin{bmatrix} \boldsymbol{I}_{3\times 3} & \boldsymbol{0}_{3\times 3} \end{bmatrix} \boldsymbol{J}^b_{r,i}(\boldsymbol{q}) \ . \tag{2.34} $$

The *reduced model* can then be derived by following the same steps as in case of the complete model. This leads to the equations of motion:

$$ \boxed{\begin{aligned} \boldsymbol{M}(\boldsymbol{q})\ddot{q} + \boldsymbol{C}(\boldsymbol{q}, \dot{q})\dot{q} + \boldsymbol{g}(\boldsymbol{q}) &= \boldsymbol{K}(\boldsymbol{\theta} - \boldsymbol{q}) + \boldsymbol{\tau}_{ext} \\ \boldsymbol{B}\ddot{\boldsymbol{\theta}} + \boldsymbol{K}(\boldsymbol{\theta} - \boldsymbol{q}) &= \boldsymbol{\tau}_m \end{aligned}} \tag{2.35} $$

[15] In this case the translational part of the kinetic energy of the rotors would vanish, too.

[16] It should be noted that there exist some simple kinematic arrangements of joint/motor axes that lead to $\boldsymbol{S}(\boldsymbol{q}) = \boldsymbol{0}$, even without the simplifying Assumption 2.4.

Herein the link side inertia matrix $\boldsymbol{M}(\boldsymbol{q})$ is given by $\boldsymbol{M}(\boldsymbol{q}) = \boldsymbol{M}_l(\boldsymbol{q}) + \bar{\boldsymbol{M}}_r(\boldsymbol{q})$ and $\boldsymbol{C}(\boldsymbol{q}, \dot{\boldsymbol{q}})$ is the Coriolis/centrifugal matrix with respect to[17] $\boldsymbol{M}(\boldsymbol{q})$.

The *reduced model* (2.35) will be considered in the following chapters for the controller design. In contrast to the complete model, the only coupling between the dynamics for $\boldsymbol{q}$ and the dynamics for $\boldsymbol{\theta}$ in the reduced model is the joint stiffness. This special structure can be exploited for the controller design.

2.2.4 Properties of the Dynamical Model

In this section some interesting properties of the dynamical model are given. Unless otherwise mentioned all the properties hold for both the complete and the reduced model.

Clearly both of the models are underactuated. There are $2n$ configuration variables and only n control inputs, the motor torques. Furthermore, both models are differentially flat [FLMR95] and the reduced model can even be linearized by static state feedback [Spo87b, DLT96]. The complete model can be linearized by dynamic state feedback [DLL98].

From a control point of view, also the following two well known properties are important.

Property 2.5. *The inertia matrix*[18] $\boldsymbol{M}(\boldsymbol{q}) \in \mathbb{R}^{n \times n}$ *is symmetric and positive definite:*

$$\boldsymbol{M}(\boldsymbol{q}) = \boldsymbol{M}(\boldsymbol{q})^T \ , \ \boldsymbol{y}^T \boldsymbol{M}(\boldsymbol{q}) \boldsymbol{y} > 0 \quad \forall \boldsymbol{q}, \boldsymbol{y} \neq \boldsymbol{0} \in \mathbb{R}^n \ .$$

Property 2.6. *The matrix*[19] $\dot{\boldsymbol{M}}(\boldsymbol{q}) - 2\boldsymbol{C}(\boldsymbol{q}, \dot{\boldsymbol{q}}) \in \mathbb{R}^{n \times n}$ *is skew symmetric, if* $\boldsymbol{C}(\boldsymbol{q}, \dot{\boldsymbol{q}})$ *is chosen via the Christoffel symbols as in equation (2.30):*

$$\boldsymbol{y}^T (\dot{\boldsymbol{M}}(\boldsymbol{q}) - 2\boldsymbol{C}(\boldsymbol{q}, \dot{\boldsymbol{q}})) \boldsymbol{y} = 0 \quad \forall \boldsymbol{y}, \boldsymbol{q}, \dot{\boldsymbol{q}} \in \mathbb{R}^n \ .$$

The proofs for these two properties are standard in the context of rigid body robots (see, e.g., [SS96]) and will therefore be omitted here. It should only be mentioned that Property 2.6 is strongly related to the passivity property of the system and depends on the special choice of $\boldsymbol{C}(\boldsymbol{q}, \dot{\boldsymbol{q}})$ via the Christoffel symbols.

In case that the robot contains not only revolute but also prismatic joints, it is useful to consider a subset $\mathcal{Q}^p \subseteq \mathcal{Q}$ in which all[20] the prismatic joints $i \in \mathcal{P}$ keep bounded by some lower and upper bounds $q_{i,min}$ and $q_{i,max}$, respectively. This subset shall be denoted by

$$\mathcal{Q}^p := \{\boldsymbol{q} \in \mathcal{Q} | \ q_{i,min} \leq q_i \leq q_{i,max} \ \ \forall i \in \mathcal{P}\} \ . \tag{2.36}$$

---

[17] The matrix $\boldsymbol{C}(\boldsymbol{q}, \dot{\boldsymbol{q}})$ thus corresponds to the upper left part of $\boldsymbol{\Gamma}(\boldsymbol{q}, \dot{\boldsymbol{q}})$ when $\boldsymbol{H}(\boldsymbol{q})$ is replaced by $\boldsymbol{M}(\boldsymbol{q})$ in equation (2.31).

[18] Respectively $\boldsymbol{H}(\boldsymbol{q}) \in \mathbb{R}^{2n \times 2n}$ for the complete model.

[19] Respectively $\dot{\boldsymbol{H}}(\boldsymbol{q}) - 2\boldsymbol{\Gamma}(\boldsymbol{q}, \dot{\boldsymbol{q}}) \in \mathbb{R}^{2n \times 2n}$ for the complete model.

[20] Notice that one could of course exclude those prismatic joints from the boundedness assumption which generate only horizontal displacements.

Then the following property, which will be useful in the analysis of the desired impedance behavior in Section 3.2 and for the controller design in Chapter 6, holds for all $\boldsymbol{q} \in \mathcal{Q}^p$.

Property 2.7. *Within $\mathcal{Q}^p$ the eigenvalues $\lambda_i(\boldsymbol{M}(\boldsymbol{q}))$ of the inertia matrix keep bounded, i.e.*

$$0 < \lambda_{m,M} \leq \lambda_i(\boldsymbol{M}(\boldsymbol{q})) \leq \lambda_{M,M} < \infty \quad \forall i = 1, \cdots, n, \ \forall \boldsymbol{q} \in \mathcal{Q}^p \ ,$$

where

$$\lambda_{m,M} := \inf_{\boldsymbol{q} \in \mathcal{Q}^p} \lambda_{min}(\boldsymbol{M}(\boldsymbol{q})) \tag{2.37}$$

and

$$\lambda_{M,M} := \sup_{\boldsymbol{q} \in \mathcal{Q}^p} \lambda_{max}(\boldsymbol{M}(\boldsymbol{q})) \tag{2.38}$$

are the strict positive minimum and maximum eigenvalues of $\boldsymbol{M}(\boldsymbol{q})$ in $\mathcal{Q}^p$.

Notice that (for a robot with both rotational and prismatic joints) the particular values of the eigenvalues of the inertia matrix depend on the choice of the physical units for rotational and prismatic joints. However, for the analysis in this book the particular values of the bounds $\lambda_{m,M}$ and $\lambda_{M,M}$ are not important. Only the existence of such bounds will be important. This is, of course, ensured for any choice of physical units.

Finally, a property of the gravity model is given, which will be of interest for the passivity based controller in Chapter 7.

Property 2.8. *The gravity torques $\boldsymbol{g}(\boldsymbol{q})$ are given by the potential function $V_g(\boldsymbol{q})$ in the form $\boldsymbol{g}(\boldsymbol{q}) = (\partial V_g(\boldsymbol{q})/\partial \boldsymbol{q})^T$, and for every matrix norm $||\cdot||$ there exists an $\alpha > 0$ such that*

$$||\partial \boldsymbol{g}(\boldsymbol{q})/\partial \boldsymbol{q}|| < \alpha \qquad \forall \boldsymbol{q} \in \mathcal{Q}^p$$

holds.

The existence of an upper bound for the Hessian of the gravity potential follows from the fact that $V_g(\boldsymbol{q})$, as defined in (2.22), consists of terms which are trigonometric functions (for rotational joints) and affine functions (for prismatic joints) of the joint variables q_i [dWSB96]. Notice that the existence of such a bound is ensured for an arbitrary matrix norm. The particular value of α, however, will be affected by the choice of this norm. This will be discussed in more detail in Chapter 7.

2.2.5 Some Remarks Concerning Friction

For the derivation of the models in Section 2.2.2 and Section 2.2.3 no friction is considered. Modeling and compensation of motor side friction for the DLR lightweight robots was treated in detail in [AS01]. It is interesting to note that viscous friction, on both link and motor sides, does not destroy the feedback linearization properties, neither of the reduced nor of the complete model.

The control algorithms derived in this book deal exclusively with the above frictionless models. In practice they clearly should be used in combination with a friction compensation as for instance the one proposed in [AS01]. Notice that this friction compensation was also used for the experiments presented in Chapter 8 and Chapter 9.

A further generalization of the robot model is the introduction of joint damping, i.e. damping in parallel to the joint spring. In the Lagrangian $L(\hat{\boldsymbol{q}}, \dot{\hat{\boldsymbol{q}}})$ the joint damping can formally be included by the use of a Rayleigh dissipation function $1/2(\dot{\boldsymbol{\theta}} - \dot{\boldsymbol{q}})^T \boldsymbol{D}(\dot{\boldsymbol{\theta}} - \dot{\boldsymbol{q}})$, where $\boldsymbol{D}$ is a diagonal and positive semi-definite damping matrix. The system equations for the reduced model with joint damping are given by

$$\boldsymbol{M}(\boldsymbol{q})\ddot{\boldsymbol{q}} + \boldsymbol{C}(\boldsymbol{q}, \dot{\boldsymbol{q}})\dot{\boldsymbol{q}} + \boldsymbol{g}(\boldsymbol{q}) = \boldsymbol{K}(\boldsymbol{\theta} - \boldsymbol{q}) + \boldsymbol{D}(\dot{\boldsymbol{\theta}} - \dot{\boldsymbol{q}}) + \boldsymbol{\tau}_{ext} \tag{2.39}$$

$$\boldsymbol{B}\ddot{\boldsymbol{\theta}} + \boldsymbol{K}(\boldsymbol{\theta} - \boldsymbol{q}) + \boldsymbol{D}(\dot{\boldsymbol{\theta}} - \dot{\boldsymbol{q}}) = \boldsymbol{\tau}_m \ . \tag{2.40}$$

This slightly more general model will only be considered in Chapter 7 in which a passivity based controller is presented. The joint damping though is usually[21] very low such that it can be neglected for the controller design.

2.3 Summary

In this chapter the kinematic and dynamical modeling of flexible joint robots was discussed. At first prerequisites from robot kinematics were reported which are common in the robotics community. Based on these kinematic relationships two dynamical models for a flexible joint robot were derived, namely the *complete* and the *reduced* model. All the assumptions which lead to these models were described in detail. The modeling approach from [MLS94] was applied to derive the mathematical model of a robot with flexible joints. Finally, some properties of the dynamical models which are of great interest for the controller design were shortly discussed.

[21] An identification of the damping parameters of the DLR lightweight robots can be found in [AS01].

3 Cartesian Impedance Control: The Rigid Body Case

In this chapter the classical theory of impedance control for rigid body robots is described. The presentation in Section 3.1 is based on the seminal work of Hogan about the concept of impedance control [Hog85a, Hog85b, Hog85c] and on the *Operational Space Formulation* by Khatib [Kha87]. In Section 3.2 the case of a desired impedance is analyzed, in which the inertial behavior must not be shaped explicitly. This brings about the problem of designing the damping matrix in an appropriate way. Furthermore, some additional aspects concerning singularity avoidance, the choice of coordinates, and the stiffness design are discussed.

This chapter refers to the rigid body part of the robot model without considering joint elasticities. The presented topics serve as a prerequisite for the design of Cartesian impedance controllers for the flexible joint robot model from Section 2.2.3 and can readily be combined with the controllers from Chapter 5, 6, and 7.

3.1 Complete Decoupling

Throughout this chapter the flexibility of the joints is neglected. While in the previous chapter the joint torques $\boldsymbol{\tau} \in \mathbb{R}^n$ were related to the motor side positions $\boldsymbol{\theta}$ and the link angles $\boldsymbol{q}$ via $\boldsymbol{\tau} = \boldsymbol{K}(\boldsymbol{\theta} - \boldsymbol{q})$, they are now considered as the control inputs. The joint angles are summarized in the vector $\boldsymbol{q} \in \mathcal{Q}^p$. By using the notation of Chapter 2 the considered dynamical model of the robot is given by[1]

$$\boldsymbol{M}(\boldsymbol{q})\ddot{\boldsymbol{q}} + \boldsymbol{C}(\boldsymbol{q}, \dot{\boldsymbol{q}})\dot{\boldsymbol{q}} + \boldsymbol{g}(\boldsymbol{q}) = \boldsymbol{\tau} + \boldsymbol{\tau}_{ext} \ , \tag{3.1}$$

where $\boldsymbol{M}(\boldsymbol{q})$ is the inertia matrix, $\boldsymbol{C}(\boldsymbol{q}, \dot{\boldsymbol{q}})$ is the Coriolis/centrifugal matrix, $\boldsymbol{g}(\boldsymbol{q})$ is the vector of gravity torques, and $\boldsymbol{\tau}_{ext}$ is the vector of external torques.

[1] Notice that the joint torques $\boldsymbol{\tau}$, not the motor torques $\boldsymbol{\tau}_m$, are considered here as the control inputs for the rigid body model. Therefore, the respective inertia matrix is $\boldsymbol{M}(\boldsymbol{q})$ instead of $\boldsymbol{M}(\boldsymbol{q}) + \boldsymbol{B}$.

C. Ott: Cartesian Imped. Cntrl. of Redund. & Flexible-Joint Robots, STAR 49, pp. 29–44, 2008.
springerlink.com

3.1.1 Task Formulation

The goal of impedance control is to realize a particular desired dynamical relationship between the robot motion and the external torques. In case of the Cartesian impedance control problem, this relationship is specified in terms of coordinates which describe the motion of the end-effector. In general, the actual configuration of the end-effector can be represented in special coordinates by a homogeneous matrix (see Section 2.1), i.e. as an element of $SE(3)$, which can be computed based on the product of exponentials formula (2.3). For the purpose of controller design, instead, a minimal representation in terms of m end-effector coordinates $\boldsymbol{x} \in \mathbb{R}^m$ is often preferred. In case that all degrees-of-freedom of the end-effector motion are considered in the task, one has $m = 6$. In the following it is assumed that the relationship between these Cartesian coordinates $\boldsymbol{x}$ and the configuration coordinates $\boldsymbol{q} \in \mathcal{Q}$ is given by a known function $\boldsymbol{f} : \mathcal{Q} \rightarrow \mathbb{R}^m$, i.e. $\boldsymbol{x} = \boldsymbol{f}(\boldsymbol{q})$. In Section 3.5 some more details about possible choices for the Cartesian coordinates $\boldsymbol{x}$ are given.

Throughout this chapter only the non-redundant case is considered, for which the number of joint angles n and the number of Cartesian coordinates m are the same $m = n$. Possible extensions to the redundant case will be treated in the next chapter.

For the formulation of the desired dynamic behavior in terms of the Cartesian coordinates $\boldsymbol{x}$, also the first and the second time derivatives, $\dot{\boldsymbol{x}}$ and $\ddot{\boldsymbol{x}}$, are considered. With the Jacobian[2] $\boldsymbol{J}(\boldsymbol{q}) = \frac{\partial \boldsymbol{f}(\boldsymbol{q})}{\partial \boldsymbol{q}}$ these derivatives can be written as

$$\dot{\boldsymbol{x}} = \boldsymbol{J}(\boldsymbol{q})\dot{\boldsymbol{q}} \, , \tag{3.2}$$

$$\ddot{\boldsymbol{x}} = \boldsymbol{J}(\boldsymbol{q})\ddot{\boldsymbol{q}} + \dot{\boldsymbol{J}}(\boldsymbol{q})\dot{\boldsymbol{q}} \, . \tag{3.3}$$

Considering (2.3), one can see that the Cartesian coordinates $\boldsymbol{x} = \boldsymbol{f}(\boldsymbol{q})$ depend on the joint variables q_i by trigonometric (for revolute joints) and affine (for prismatic joints) terms. Therefore, one can conclude that the maximum possible singular value $\sigma_{M,J}$ of the Jacobian keeps bounded[3] within $\mathcal{Q}^p$, i.e.

$$\sigma_{M,J} := \sup_{\boldsymbol{q} \in \mathcal{Q}^p} \sigma_{max}(\boldsymbol{J}(\boldsymbol{q})) < \infty \, . \tag{3.4}$$

Besides the restriction to the non-redundant case, which was already mentioned before, it will, furthermore, be assumed that the Jacobian $\boldsymbol{J}(\boldsymbol{q})$ is non-singular[4]. In general $\boldsymbol{J}(\boldsymbol{q})$ will, of course, not be non-singular in the whole configuration space $\mathcal{Q}^p$. Then the analysis of this chapter is restricted to an area $\bar{\mathcal{Q}}^p$ in which the invertibility is ensured. This is established herein by requiring that the minimum possible singular value of the Jacobian

$$\sigma_{m,J} := \inf_{\boldsymbol{q} \in \bar{\mathcal{Q}}^p} \sigma_{min}(\boldsymbol{J}(\boldsymbol{q})) \tag{3.5}$$

---

[2] In the robotics literature this matrix is sometimes called *analytical Jacobian*, in contrast to the geometrical Jacobian, or the body Jacobian, respectively.

[3] Note that $\mathcal{Q}^p$, as defined in (2.36), describes a subset of the configuration space $\mathcal{Q}$ in which the prismatic joint variables keep bounded.

[4] An appropriate singularity treatment will be briefly described in Section 3.4.

must be bigger than some value $\sigma_0 > 0$. Moreover, it is assumed that the mapping $\boldsymbol{f}(\boldsymbol{q})$ is one-to-one in $\bar{\mathcal{Q}}^p$. The corresponding area in Cartesian coordinates, i.e. the image of $\bar{\mathcal{Q}}^p$ through $\boldsymbol{f}$, is denoted by $\bar{\mathcal{Q}}^p_c$

$$\bar{\mathcal{Q}}^p := \{\boldsymbol{q} \in \mathcal{Q}^p \mid \sigma_{min}(\boldsymbol{J}(\boldsymbol{q})) > \sigma_0 \text{ and } \boldsymbol{f}(\boldsymbol{q}) \text{ is one-to-one}\} , \tag{3.6}$$

$$\bar{\mathcal{Q}}^p_c := \boldsymbol{f}(\bar{\mathcal{Q}}^p) = \{\boldsymbol{x} \in \mathbb{R}^n \mid \exists \boldsymbol{q} \in \bar{\mathcal{Q}}^p \ , \ \boldsymbol{f}(\boldsymbol{q}) = \boldsymbol{x}\} . \tag{3.7}$$

In the set $\bar{\mathcal{Q}}^p_c$ the Cartesian coordinates $\boldsymbol{x}$ can be used as generalized coordinates of the system (3.1). In general it is, of course, not possible to find Cartesian coordinates, for which $\bar{\mathcal{Q}}^p_c$ corresponds to the complete state space $\mathbb{R}^n$. Still, for the analysis of a Cartesian controller it is often interesting to investigate if a (local) stability statement holds also globally under the assumption $\bar{\mathcal{Q}}^p_c = \mathbb{R}^n$. If this is true, the stability region is only restricted by the particular choice of coordinates.

The external torques $\boldsymbol{\tau}_{ext}$ shall be related to the vector of generalized external forces $\boldsymbol{F}_{ext}$ via $\boldsymbol{\tau}_{ext} = \boldsymbol{J}(\boldsymbol{q})^T \boldsymbol{F}_{ext}$. Therefore, $\boldsymbol{F}_{ext}$ is the dual variable to $\dot{\boldsymbol{x}}$ and hence the power which is exchanged between the robot and its environment is given by $\boldsymbol{\tau}^T_{ext}\dot{\boldsymbol{q}} = \boldsymbol{F}^T_{ext}\dot{\boldsymbol{x}}$.

In order to specify the desired impedance behavior the position error $\tilde{\boldsymbol{x}} = \boldsymbol{x} - \boldsymbol{x}_d$ between $\boldsymbol{x}$ and a (possibly time-varying) *virtual equilibrium position*[5] $\boldsymbol{x}_d$ is introduced. Then the control objective is to achieve a dynamical relationship of the form

$$\boldsymbol{\Lambda}_d\ddot{\tilde{\boldsymbol{x}}} + \boldsymbol{D}_d\dot{\tilde{\boldsymbol{x}}} + \boldsymbol{K}_d\tilde{\boldsymbol{x}} = \boldsymbol{F}_{ext} \tag{3.8}$$

between $\tilde{\boldsymbol{x}}$ and $\boldsymbol{F}_{ext}$, where $\boldsymbol{K}_d$, $\boldsymbol{D}_d$, and $\boldsymbol{\Lambda}_d$ are the symmetric and positive definite matrices of the desired stiffness, damping, and inertia, respectively. In principle one could of course also consider a more general impedance behavior. But in most robotic applications the restriction of the impedance controller to a desired behavior in form of such a *mass-spring-damper-system* is sufficient.

3.1.2 Robot Model in Task Coordinates

The robot model (3.1) is written in joint coordinates $\boldsymbol{q}$, while the desired behavior (3.8) is defined in task coordinates $\boldsymbol{x}$. For the controller design it is easier to rewrite the model (3.1) also in task coordinates as it is done in the operational space formulation [Kha87]. Substituting $\ddot{\boldsymbol{q}} = \boldsymbol{J}(\boldsymbol{q})^{-1}(\ddot{\boldsymbol{x}} - \dot{\boldsymbol{J}}(\boldsymbol{q})\dot{\boldsymbol{q}})$ from (3.3) and $\boldsymbol{\tau}_{ext} = \boldsymbol{J}(\boldsymbol{q})^T\boldsymbol{F}_{ext}$ into equation (3.1) leads to

$$\boldsymbol{M}(\boldsymbol{q})\boldsymbol{J}(\boldsymbol{q})^{-1}\left(\ddot{\boldsymbol{x}} - \dot{\boldsymbol{J}}(\boldsymbol{q})\dot{\boldsymbol{q}}\right) + \boldsymbol{C}(\boldsymbol{q},\dot{\boldsymbol{q}})\dot{\boldsymbol{q}} + \boldsymbol{g}(\boldsymbol{q}) = \boldsymbol{\tau} + \boldsymbol{J}(\boldsymbol{q})^T\boldsymbol{F}_{ext} \ .$$

With $\dot{\boldsymbol{q}} = \boldsymbol{J}(\boldsymbol{q})^{-1}\dot{\boldsymbol{x}}$ from (3.2) substituted in the second and the third term and by pre-multiplying the resulting equation by $\boldsymbol{J}(\boldsymbol{q})^{-T}$, one gets

$$\begin{aligned}&\boldsymbol{J}(\boldsymbol{q})^{-T}\boldsymbol{M}(\boldsymbol{q})\boldsymbol{J}(\boldsymbol{q})^{-1}\ddot{\boldsymbol{x}} + \boldsymbol{J}(\boldsymbol{q})^{-T}\boldsymbol{C}(\boldsymbol{q},\dot{\boldsymbol{q}})\boldsymbol{J}(\boldsymbol{q})^{-1}\dot{\boldsymbol{x}} - \\ &\quad \boldsymbol{J}(\boldsymbol{q})^{-T}\boldsymbol{M}(\boldsymbol{q})\boldsymbol{J}(\boldsymbol{q})^{-1}\dot{\boldsymbol{J}}(\boldsymbol{q})\boldsymbol{J}(\boldsymbol{q})^{-1}\dot{\boldsymbol{x}} + \boldsymbol{J}(\boldsymbol{q})^{-T}\boldsymbol{g}(\boldsymbol{q}) = \boldsymbol{J}(\boldsymbol{q})^{-T}\boldsymbol{\tau} + \boldsymbol{F}_{ext} \ .\end{aligned}$$

[5] As already mentioned in the introduction the set-point in impedance control usually is called a *virtual* equilibrium point, because it actually should only be reached in case that no external forces act on the robot.

This equation can now be written in the form

$$\boldsymbol{\Lambda}(\boldsymbol{x})\ddot{\boldsymbol{x}} + \boldsymbol{\mu}(\boldsymbol{x},\dot{\boldsymbol{x}})\dot{\boldsymbol{x}} + \boldsymbol{J}(\boldsymbol{q})^{-T}\boldsymbol{g}(\boldsymbol{q}) = \boldsymbol{J}(\boldsymbol{q})^{-T}\boldsymbol{\tau} + \boldsymbol{F}_{ext} \ , \tag{3.9}$$

where the matrices $\boldsymbol{\Lambda}(\boldsymbol{x})$ and $\boldsymbol{\mu}(\boldsymbol{x},\dot{\boldsymbol{x}})$ are given by

$$\boldsymbol{\Lambda}(\boldsymbol{x}) = \boldsymbol{J}(\boldsymbol{q})^{-T}\boldsymbol{M}(\boldsymbol{q})\boldsymbol{J}(\boldsymbol{q})^{-1} \ , \tag{3.10}$$

$$\boldsymbol{\mu}(\boldsymbol{x},\dot{\boldsymbol{x}}) = \boldsymbol{J}(\boldsymbol{q})^{-T}\left(\boldsymbol{C}(\boldsymbol{q},\dot{\boldsymbol{q}}) - \boldsymbol{M}(\boldsymbol{q})\boldsymbol{J}(\boldsymbol{q})^{-1}\dot{\boldsymbol{J}}(\boldsymbol{q})\right)\boldsymbol{J}(\boldsymbol{q})^{-1} \ , \tag{3.11}$$

with $\boldsymbol{q} = \boldsymbol{f}^{-1}(\boldsymbol{x})$ and $\dot{\boldsymbol{q}} = \boldsymbol{J}(\boldsymbol{f}^{-1}(\boldsymbol{x}))\dot{\boldsymbol{x}}$. From a control point of view the use of both the joint variables $\boldsymbol{q}$ and the Cartesian coordinates $\boldsymbol{x}$ simultaneously in (3.10) and (3.11) is a little bit misleading, because the considered state variables are only $\boldsymbol{x}$ and $\dot{\boldsymbol{x}}$ throughout this chapter. A *direct* representation of some components in terms of $\boldsymbol{q}$ on the other hand is sometimes much clearer and simpler[6]. Whenever the two representations are mixed, $\boldsymbol{q}$ and $\dot{\boldsymbol{q}}$ can be thought of being replaced by $\boldsymbol{q} = \boldsymbol{f}^{-1}(\boldsymbol{x})$ and $\dot{\boldsymbol{q}} = \boldsymbol{J}(\boldsymbol{f}^{-1}(\boldsymbol{x}))\dot{\boldsymbol{x}}$, respectively.

In analogy to the external torques also the gravity torques $\boldsymbol{g}(\boldsymbol{q})$ and the joint torques $\boldsymbol{\tau}$ can be rewritten in form of the equivalent task space gravity forces $\boldsymbol{F}_g(\boldsymbol{x}) = \boldsymbol{J}(\boldsymbol{q})^{-T}\boldsymbol{g}(\boldsymbol{q})$ and the new input vector $\boldsymbol{F}_\tau$, which is related to $\boldsymbol{\tau}$ via $\boldsymbol{\tau} = \boldsymbol{J}(\boldsymbol{q})^T\boldsymbol{F}_\tau$. Therefore, the system equations finally have the form

$$\boldsymbol{\Lambda}(\boldsymbol{x})\ddot{\boldsymbol{x}} + \boldsymbol{\mu}(\boldsymbol{x},\dot{\boldsymbol{x}})\dot{\boldsymbol{x}} + \boldsymbol{F}_g(\boldsymbol{x}) = \boldsymbol{F}_\tau + \boldsymbol{F}_{ext} \ . \tag{3.12}$$

The matrices $\boldsymbol{\Lambda}(\boldsymbol{x})$ and $\boldsymbol{\mu}(\boldsymbol{x},\dot{\boldsymbol{x}})$ are the inertia matrix and the *Coriolis/centrifugal matrix* with respect to the coordinates $\boldsymbol{x}$.

Before the classical impedance control law for the model (3.12) is formulated, two important lemmata of the model (3.12) are presented, which follow directly from Property 2.5 and Property 2.6 (see [dWSB96]).

Lemma 3.1. *The matrix $\boldsymbol{\Lambda}(\boldsymbol{x})$, as defined in (3.10), is symmetric and positive definite for all positions $\boldsymbol{x} \in \bar{\mathcal{Q}}_c^p$.*

Proof. Due to the restriction of $\boldsymbol{x}$ to $\bar{\mathcal{Q}}_c^p$, the matrix $\boldsymbol{\Lambda}(\boldsymbol{x})$ is well defined. The symmetry and the positive definiteness of $\boldsymbol{\Lambda}(\boldsymbol{x})$ follow directly from its definition in (3.10) together with Property 2.5.

Lemma 3.2. *The matrix $\dot{\boldsymbol{\Lambda}}(\boldsymbol{x}) - 2\boldsymbol{\mu}(\boldsymbol{x},\dot{\boldsymbol{x}})$, with $\boldsymbol{\Lambda}(\boldsymbol{x})$ and $\boldsymbol{\mu}(\boldsymbol{x},\dot{\boldsymbol{x}})$ as defined in (3.10) and (3.11), is skew symmetric for all $\boldsymbol{x} \in \bar{\mathcal{Q}}_c^p$ and all $\dot{\boldsymbol{x}} \in \mathbb{R}^m$.*

Proof. Due to the restriction of $\boldsymbol{x}$ to $\bar{\mathcal{Q}}_c^p$ both matrices $\boldsymbol{\Lambda}(\boldsymbol{x})$ and $\boldsymbol{\mu}(\boldsymbol{x},\dot{\boldsymbol{x}})$ are well defined. By Lemma A.22, one has to show that the equality $\dot{\boldsymbol{\Lambda}}(\boldsymbol{x}) = \boldsymbol{\mu}(\boldsymbol{x},\dot{\boldsymbol{x}}) + \boldsymbol{\mu}(\boldsymbol{x},\dot{\boldsymbol{x}})^T$ holds. In the remaining part of the proof the arguments of the matrices will be dropped in order to simplify the notation. From (3.10) and (3.11) and

[6] Since the joint angles $\boldsymbol{q}$ and not the Cartesian coordinates $\boldsymbol{x}$ are the measured quantities a representation in terms of $\boldsymbol{q}$ is also required for the actual implementation.

the symmetry of $\boldsymbol{M}$ (Property 2.5) it follows that the matrices $\dot{\boldsymbol{\Lambda}}$ and $\boldsymbol{\mu}+\boldsymbol{\mu}^T$ are given by

$$\begin{aligned}\dot{\boldsymbol{\Lambda}} &= \frac{\mathrm{d}}{\mathrm{d}t}(\boldsymbol{J}^{-T}\boldsymbol{M}\boldsymbol{J}^{-1}) \\ &= \frac{\mathrm{d}}{\mathrm{d}t}\boldsymbol{J}^{-T}\boldsymbol{M}\boldsymbol{J}^{-1} + \boldsymbol{J}^{-T}\dot{\boldsymbol{M}}\boldsymbol{J}^{-1} + \boldsymbol{J}^{-T}\boldsymbol{M}\frac{\mathrm{d}}{\mathrm{d}t}\boldsymbol{J}^{-1}\ , \\ \boldsymbol{\mu}+\boldsymbol{\mu}^T &= \boldsymbol{J}^{-T}(\boldsymbol{C}+\boldsymbol{C}^T)\boldsymbol{J}^{-1} - \boldsymbol{J}^{-T}\boldsymbol{M}\boldsymbol{J}^{-1}\dot{\boldsymbol{J}}\boldsymbol{J}^{-1} \\ &\quad -\boldsymbol{J}^{-T}\dot{\boldsymbol{J}}^T\boldsymbol{J}^{-T}\boldsymbol{M}\boldsymbol{J}^{-1}\ .\end{aligned}$$

Considering the equality $\dot{\boldsymbol{M}} = \boldsymbol{C}+\boldsymbol{C}^T$, which follows from Property 2.6 together with Lemma A.22, the term $\dot{\boldsymbol{\Lambda}}-\boldsymbol{\mu}-\boldsymbol{\mu}^T$ results in

$$\begin{aligned}\dot{\boldsymbol{\Lambda}}-\boldsymbol{\mu}-\boldsymbol{\mu}^T &= \frac{\mathrm{d}}{\mathrm{d}t}\boldsymbol{J}^{-T}\boldsymbol{M}\boldsymbol{J}^{-1} + \boldsymbol{J}^{-T}\boldsymbol{M}\frac{\mathrm{d}}{\mathrm{d}t}\boldsymbol{J}^{-1} \\ &\quad + \boldsymbol{J}^{-T}\boldsymbol{M}\boldsymbol{J}^{-1}\dot{\boldsymbol{J}}\boldsymbol{J}^{-1} + \boldsymbol{J}^{-T}\dot{\boldsymbol{J}}^T\boldsymbol{J}^{-T}\boldsymbol{M}\boldsymbol{J}^{-1} \\ &= (\frac{\mathrm{d}}{\mathrm{d}t}\boldsymbol{J}^{-T} + \boldsymbol{J}^{-T}\dot{\boldsymbol{J}}^T\boldsymbol{J}^{-T})\boldsymbol{M}\boldsymbol{J}^{-1} + \boldsymbol{J}^{-T}\boldsymbol{M}(\frac{\mathrm{d}}{\mathrm{d}t}\boldsymbol{J}^{-1} + \boldsymbol{J}^{-1}\dot{\boldsymbol{J}}\boldsymbol{J}^{-1})\ .\end{aligned}$$

From the equality

$$\begin{aligned}\boldsymbol{0} &= \frac{\mathrm{d}}{\mathrm{d}t}(\boldsymbol{I})\boldsymbol{J}^{-1} = \frac{\mathrm{d}}{\mathrm{d}t}(\boldsymbol{J}^{-1}\boldsymbol{J})\boldsymbol{J}^{-1} \\ &= \frac{\mathrm{d}}{\mathrm{d}t}\boldsymbol{J}^{-1} + \boldsymbol{J}^{-1}\dot{\boldsymbol{J}}\boldsymbol{J}^{-1}\end{aligned}$$

one can conclude $\dot{\boldsymbol{\Lambda}}-\boldsymbol{\mu}-\boldsymbol{\mu}^T = \boldsymbol{0}$ which completes the proof.

Notice that these properties are not surprising at all, since (3.12) is nothing else than the model (3.1) written in another set of coordinates. The matrix $\boldsymbol{\mu}(\boldsymbol{x},\dot{\boldsymbol{x}})$ in (3.12), however, could also have been chosen differently from (3.11). The special form in (3.11) has the advantage that it ensures the validity of Lemma 3.2. Equation (3.11) corresponds to the transformation of the Christoffel symbols [Fra97, Boo03] from (2.30) into Cartesian coordinates, which means that the same form of $\boldsymbol{\mu}(\boldsymbol{x},\dot{\boldsymbol{x}})$ as in (3.11) is obtained, if it is computed via the Christoffel symbols of $\boldsymbol{\Lambda}(\boldsymbol{x})$. Equation (3.10) clearly is the classical coordinate transformation for a covariant tensor of rank 2 [Fra97, Boo03].

The restriction to $\bar{\mathcal{Q}}_c^p$ ensures also that, similar to Property 2.7, the eigenvalues of the Cartesian inertia matrix keep bounded from above and below by some non-zero bounds.

Property 3.3. *Within $\bar{\mathcal{Q}}_c^p$ the eigenvalues $\lambda_i(\boldsymbol{\Lambda}(\boldsymbol{x}))$ of the Cartesian inertia matrix keep bounded, i.e.*

$$0 < \frac{\lambda_{m,M}}{\sigma_{M,J}^2} \leq \lambda_i(\boldsymbol{\Lambda}(\boldsymbol{x})) \leq \frac{\lambda_{M,M}}{\sigma_{m,J}^2} < \infty \quad i=1,\cdots,n,\ \forall \boldsymbol{x}\in\bar{\mathcal{Q}}_c^p\ ,$$

with $\lambda_{m,M}$ and $\lambda_{M,M}$ as the minimum and maximum possible eigenvalues of the inertia matrix as defined in (2.37) and (2.38), and $\sigma_{m,J}$ and $\sigma_{M,J}$ as the *minimum and maximum possible singular value of the Jacobian as defined in (3.4) and (3.5).*

Proof. This property follows directly from the definition of $\boldsymbol{\Lambda}(\boldsymbol{x})$ in (3.10) together with Property 2.7 and the definition of $\bar{\mathcal{Q}}_c^p$.

This property will be of interest for the analysis of the desired impedance behavior in Section 3.2 as well as for the controller design in Chapter 6.

3.1.3 Classical Impedance Controller

The classical impedance control law [Hog85a, Hog85b] can be directly computed from equation (3.12) [Kha87]. The control input $\boldsymbol{F}_\tau$ which leads to the desired closed loop system (3.8) is given by

$$\begin{aligned}\boldsymbol{F}_\tau = &\boldsymbol{F}_g(\boldsymbol{x}) + \boldsymbol{\Lambda}(\boldsymbol{x})\ddot{\boldsymbol{x}}_d + \boldsymbol{\mu}(\boldsymbol{x},\dot{\boldsymbol{x}})\dot{\boldsymbol{x}} - \boldsymbol{\Lambda}(\boldsymbol{x})\boldsymbol{\Lambda}_d^{-1}\left(\boldsymbol{K}_d\tilde{\boldsymbol{x}} + \boldsymbol{D}_d\dot{\tilde{\boldsymbol{x}}}\right) + \\ &\left(\boldsymbol{\Lambda}(\boldsymbol{x})\boldsymbol{\Lambda}_d^{-1} - \boldsymbol{I}\right)\boldsymbol{F}_{ext}\ .\end{aligned}$$

This Cartesian impedance controller is then actually implemented via the joint torques $\boldsymbol{\tau}$ as follows

$$\begin{aligned}\boldsymbol{\tau} &= \boldsymbol{J}(\boldsymbol{q})^T\boldsymbol{F}_\tau && (3.13)\\ &= \boldsymbol{g}(\boldsymbol{q}) + \boldsymbol{J}(\boldsymbol{q})^T\left(\boldsymbol{\Lambda}(\boldsymbol{x})\ddot{\boldsymbol{x}}_d + \boldsymbol{\mu}(\boldsymbol{x},\dot{\boldsymbol{x}})\dot{\boldsymbol{x}}\right) - \\ &\quad \boldsymbol{J}(\boldsymbol{q})^T\boldsymbol{\Lambda}(\boldsymbol{x})\boldsymbol{\Lambda}_d^{-1}\left(\boldsymbol{K}_d\tilde{\boldsymbol{x}} + \boldsymbol{D}_d\dot{\tilde{\boldsymbol{x}}}\right) + \\ &\quad \boldsymbol{J}(\boldsymbol{q})^T\left(\boldsymbol{\Lambda}(\boldsymbol{x})\boldsymbol{\Lambda}_d^{-1} - \boldsymbol{I}\right)\boldsymbol{F}_{ext}\ . && (3.14)\end{aligned}$$

One can see that the shaping of the desired impedance in this case also contains a feedback of the external forces $\boldsymbol{F}_{ext}$. These forces are usually measured by means of a force-torque-sensor mounted at the end-effector. But in general there might also be external forces which do not act on the tool but directly on the robot structure and thus cannot be measured. Since these forces are not included in the measurement of $\boldsymbol{F}_{ext}$, it is clear that the closed loop impedance behavior with respect to these forces will be quite different. Notice that, from a practical point of view, this is typically much more relevant in service robotics than in industrial applications. Clearly, when the robot is interacting with humans this interaction is not necessarily restricted to the tool of the robot.

The need for feedback of the external forces follows from the requirement that not only the stiffness and damping behavior of the robot should be shaped but also the inertial behavior. However, in many applications this is not necessary. Therefore, a simplified control objective is considered in the next section, in which an explicit shaping of the apparent inertia is not considered and the emphasis is laid on the shaping of the stiffness and the damping.

3.2 Avoidance of Inertia Shaping

The feedback of external forces $\boldsymbol{F}_{ext}$ can be avoided when the desired inertia $\boldsymbol{\Lambda}_d$ is identical to the robot inertia $\boldsymbol{\Lambda}(\boldsymbol{x})$

$$\boldsymbol{\Lambda}_d = \boldsymbol{\Lambda}(\boldsymbol{x})\ . \tag{3.15}$$

Since the desired inertia depends on the position $\boldsymbol{x}$, also the relevant centrifugal and Coriolis terms should be considered in the specification of the desired closed loop behavior. This is necessary in order to fulfill Lemma 3.2 and thus to ensure the passivity of the system for the regulation case (see also Proposition 3.5 below). The desired dynamic relationship between $\tilde{\boldsymbol{x}}$ and $\boldsymbol{F}_{ext}$ is given as

$$\boldsymbol{\Lambda}(\boldsymbol{x})\ddot{\tilde{\boldsymbol{x}}} + (\boldsymbol{\mu}(\boldsymbol{x}, \dot{\boldsymbol{x}}) + \boldsymbol{D}_d)\,\dot{\tilde{\boldsymbol{x}}} + \boldsymbol{K}_d\tilde{\boldsymbol{x}} = \boldsymbol{F}_{ext} \ , \tag{3.16}$$

where $\boldsymbol{K}_d$ and $\boldsymbol{D}_d$ are again the symmetric and positive definite matrices of desired stiffness and desired damping, respectively. In the regulation case (i.e. for $\dot{\boldsymbol{x}}_d = \boldsymbol{0}$) this control objective is often called *compliance control* problem.

Before presenting the controller which leads to the closed loop system (3.16), a short comparison to the original desired dynamics (3.8) shall be given in order to justify the choice of (3.16). Consider first the case of free motion, i.e. $\boldsymbol{F}_{ext} = \boldsymbol{0}$. The asymptotic stability of the original desired dynamics (3.8) is ensured for this case simply by the fact that the desired stiffness, damping, and inertia matrices are positive definite. The original desired dynamics is even linear and time-invariant. The stability properties of the new dynamics (3.16) are not that obvious. Clearly, the system (3.16) is nonlinear due to the use of a position dependent inertia matrix. Furthermore, it is time-varying[7], since the virtual equilibrium position $\boldsymbol{x}_d(t)$ may be a two-times continuously differentiable function of the time t. Some relevant definitions and lemmata regarding the stability analysis of time-varying systems can be found in Appendix A.1.

Without going into the details, it should be mentioned that the stability of (3.16) for the case of free motion can be shown by considering the (time-varying) Lyapunov function

$$V(\tilde{\boldsymbol{x}}, \dot{\tilde{\boldsymbol{x}}}, t) = \frac{1}{2}\dot{\tilde{\boldsymbol{x}}}^T\boldsymbol{\Lambda}(\boldsymbol{x})\dot{\tilde{\boldsymbol{x}}} + \frac{1}{2}\tilde{\boldsymbol{x}}^T\boldsymbol{K}_d\tilde{\boldsymbol{x}} \ . \tag{3.17}$$

The change of this Lyapunov function along the solutions of (3.16) is negative semi-definite, which implies stability but not asymptotic stability. However, for free motion, the dynamics in (3.16) is equivalent to the closed loop system of the well known *PD+ controller* [PP88] written in task coordinates. Asymptotic stability of the PD+ controller in configuration space was shown in [PP88]. Furthermore, a strict Lyapunov function for this system was presented in [SK97b]. In both works the boundedness of the inertia matrix $\boldsymbol{M}(\boldsymbol{q})$, as formulated in Property 2.7, was required. The equivalent statement for the Cartesian inertia matrix $\boldsymbol{\Lambda}(\boldsymbol{x})$ is ensured in Property 3.3 due to the restriction to $\bar{\mathcal{Q}}_c^p$. According to the proof in [SK97b] (for the same controller in configuration space) the stability statement formulated in the following proposition holds even globally, if $\bar{\mathcal{Q}}_c^p$ corresponds to the complete state space $\mathbb{R}^n$.

Proposition 3.4. *Let the desired trajectory $\boldsymbol{x}_d(t)$ be continuously differentiable twice. Assume further that the Cartesian coordinates are valid globally, i.e. $\bar{\mathcal{Q}}_c^p = \mathbb{R}^n$. Then for $\boldsymbol{F}_{ext} = \boldsymbol{0}$ the system (3.16) with symmetric and positive definite matrices $\boldsymbol{K}_d$ and $\boldsymbol{D}_d$ is uniformly globally asymptotically stable.*

[7] Notice the occurrence of both $\boldsymbol{x}$ and $\tilde{\boldsymbol{x}} = \boldsymbol{x} - \boldsymbol{x}_d(t)$ in (3.16).

Another important feature of the original desired dynamics (3.8) is that in the regulation case (i.e. $\dot{\boldsymbol{x}}_d = \boldsymbol{0}$) it represents a passive mapping from external generalized forces $\boldsymbol{F}_{ext}$ to the velocity $\dot{\boldsymbol{x}}$. This property is especially important if the interaction of the robot with passive[8] environments is considered. By considering (3.17) as a storage function it can be shown for the regulation case that the system (3.16) also represents a passive mapping from the external force $\boldsymbol{F}_{ext}$ to the velocity $\dot{\boldsymbol{x}}$.

Proposition 3.5. *For $\dot{\boldsymbol{x}}_d(t) = \boldsymbol{0}$, the system (3.16) with symmetric and positive definite matrices $\boldsymbol{K}_d$ and $\boldsymbol{D}_d$ is time-invariant and represents a passive mapping from the external force $\boldsymbol{F}_{ext}$ to the velocity $\dot{\boldsymbol{x}}$.*

By following the same argumentation as for the classical impedance controller from the last section, one gets the control law

$$\boldsymbol{F}_\tau = \boldsymbol{F}_g(\boldsymbol{x}) + \boldsymbol{\Lambda}(\boldsymbol{x})\ddot{\boldsymbol{x}}_d + \boldsymbol{\mu}(\boldsymbol{x},\dot{\boldsymbol{x}})\dot{\boldsymbol{x}}_d - \boldsymbol{K}_d\tilde{\boldsymbol{x}} - \boldsymbol{D}_d\dot{\tilde{\boldsymbol{x}}} \;,$$

to achieve the dynamics (3.16). For the implementation of this controller a formulation at the joint level

$$\begin{aligned}\boldsymbol{\tau} &= \boldsymbol{J}(\boldsymbol{q})^T\boldsymbol{F}_\tau \\ &= \boldsymbol{g}(\boldsymbol{q}) + \boldsymbol{J}(\boldsymbol{q})^T\left(\boldsymbol{\Lambda}(\boldsymbol{x})\ddot{\boldsymbol{x}}_d + \boldsymbol{\mu}(\boldsymbol{x},\dot{\boldsymbol{x}})\dot{\boldsymbol{x}}_d - \boldsymbol{K}_d\tilde{\boldsymbol{x}} - \boldsymbol{D}_d\dot{\tilde{\boldsymbol{x}}}\right)\end{aligned} \tag{3.18}$$

is more convenient.

The desired stiffness usually is given by the application. The new desired impedance behavior (3.16) then brings up the problem of how to choose the damping matrix. This problem is treated in the next section.

3.3 Design of the Damping Matrix

In case of the original desired dynamics (3.8) the desired inertia matrix $\boldsymbol{\Lambda}_d$ is constant and can be designed in principle such that its eigenvectors coincide with the eigenvectors of $\boldsymbol{K}_d$. The design of the damping matrix can then be reduced to the design of the damping coefficients for n decoupled linear and time-invariant second order systems, each describing the dynamics of the system along one of the eigenvectors.

For the new desired dynamics (3.16) the situation is different. The desired stiffness matrix must be constant and is usually defined by the application being considered. The design of the desired damping matrix, instead, is not so clear. Notice that a constant and diagonal matrix $\boldsymbol{D}_d$, in general, is not a good choice, since the inertia matrix is non-diagonal and time-varying. Instead, one should also take into account the particular structure and the change of $\boldsymbol{\Lambda}(\boldsymbol{x})$ during movement.

For the design of the damping matrix it should be mentioned that all the stability and passivity properties from above will also hold, when the constant

[8] Here the velocity $\dot{\boldsymbol{x}}$ is considered as the input and the negative external force $-\boldsymbol{F}_{ext}$ as the output of the environment.

matrix $\boldsymbol{D}_d$ in (3.16) is replaced by an arbitrary, but positive definite, position dependent[9] matrix $\boldsymbol{D}_d(\boldsymbol{x})$.

In [ASOFH03] two different methods, how to choose the matrix $\boldsymbol{D}_d(\boldsymbol{x})$ for a given symmetric and positive definite stiffness matrix $\boldsymbol{K}_d$, were proposed and evaluated for the DLR lightweight robots. In the following only the method based on the generalized eigenvalue decomposition of symmetric matrices will be presented in more detail. For this, the following lemma is formulated[10] [Har97].

Lemma 3.6. *Given a symmetric and positive definite matrix $\boldsymbol{A} \in \mathbb{R}^{n\times n}$ and a symmetric matrix $\boldsymbol{B} \in \mathbb{R}^{n\times n}$. Then one can find a non-singular matrix $\boldsymbol{Q} \in \mathbb{R}^{n\times n}$ and a diagonal matrix $\boldsymbol{B}_0 \in \mathbb{R}^{n\times n}$, such that $\boldsymbol{Q}^T\boldsymbol{Q} = \boldsymbol{A}$ and $\boldsymbol{B} = \boldsymbol{Q}^T\boldsymbol{B}_0\boldsymbol{Q}$.*

The elements of this diagonal matrix $\boldsymbol{B}_0$ are called the *generalized eigenvalues* of $\boldsymbol{B}$ with respect to $\boldsymbol{A}$. If the matrix $\boldsymbol{B}$ is positive definite, the generalized eigenvalues will be positive. In order to design the matrix $\boldsymbol{D}_d(\boldsymbol{x})$ in (3.16) a *quasi-static* analysis is performed, which means that in each position $\boldsymbol{x}_0$ the system is approximated by the following linear time-invariant system

$$\boldsymbol{\Lambda}(\boldsymbol{x}_0)\ddot{\tilde{\boldsymbol{x}}} + \boldsymbol{D}_d(\boldsymbol{x}_0)\dot{\tilde{\boldsymbol{x}}} + \boldsymbol{K}_d\tilde{\boldsymbol{x}} = \boldsymbol{F}_{ext} \ , \tag{3.19}$$

wherein the Coriolis/centrifugal matrix is neglected and the matrices $\boldsymbol{\Lambda}(\boldsymbol{x}_0)$ and $\boldsymbol{D}_d(\boldsymbol{x}_0)$ are considered as constant. By applying Lemma 3.6 (with $\boldsymbol{\Lambda}(\boldsymbol{x}_0)$ and $\boldsymbol{K}_d$ corresponding to $\boldsymbol{A}$ and $\boldsymbol{B}$ respectively) the matrices $\boldsymbol{K}_d$ and $\boldsymbol{\Lambda}(\boldsymbol{x}_0)$ can be diagonalized simultaneously by a non-singular matrix $\boldsymbol{Q}(\boldsymbol{x}_0)$, such that the system (3.19) can be written in the form

$$\boldsymbol{Q}(\boldsymbol{x}_0)^T\boldsymbol{Q}(\boldsymbol{x}_0)\ddot{\tilde{\boldsymbol{x}}} + \boldsymbol{D}_d(\boldsymbol{x}_0)\dot{\tilde{\boldsymbol{x}}} + \boldsymbol{Q}(\boldsymbol{x}_0)^T\boldsymbol{B}_0(\boldsymbol{x}_0)\boldsymbol{Q}(\boldsymbol{x}_0)\tilde{\boldsymbol{x}} = \boldsymbol{F}_{ext} \ , \tag{3.20}$$

with a positive definite diagonal matrix $\boldsymbol{B}_0(\boldsymbol{x}_0)$.

Let $\lambda^{\Lambda}_{K,i}$ be the ith diagonal element of $\boldsymbol{B}_0$. This is the ith generalized eigenvalue of $\boldsymbol{K}_d$ with respect to $\boldsymbol{\Lambda}(\boldsymbol{x}_0)$. Then the matrix $\boldsymbol{D}_d(\boldsymbol{x}_0)$ can be chosen as $\boldsymbol{D}_d(\boldsymbol{x}_0) = 2\boldsymbol{Q}(\boldsymbol{x}_0)^T \operatorname{diag}(\xi_i\sqrt{\lambda^{\Lambda}_{K,i}})\boldsymbol{Q}(\boldsymbol{x}_0)$, where ξ_i is a damping factor[11] to be chosen in the range $[0,1]$.

Clearly, by this choice the system can be written in the state variable $\boldsymbol{z} = \boldsymbol{Q}(\boldsymbol{x}_0)\tilde{\boldsymbol{x}}$ in the following form

$$\ddot{\boldsymbol{z}} + 2\operatorname{diag}(\xi_i\sqrt{\lambda^{\Lambda}_{K,i}})\dot{\boldsymbol{z}} + \operatorname{diag}(\lambda^{\Lambda}_{K,i})\boldsymbol{z} = \boldsymbol{Q}(\boldsymbol{x}_0)^{-T}\boldsymbol{F}_{ext} \ . \tag{3.21}$$

It should be mentioned again that the above approximations are only used for the design of the damping matrix and do not affect the stability properties of the system, because it is ensured that the damping matrix is always positive definite.

---

[9] It can even be chosen time-varying.

[10] The author would like to thank Udo Frese for pointing out this feature of positive definite matrices.

[11] Corresponding to the ith generalized eigenvector.

3.4 Singularity Treatment

In the previous sections it was assumed that $\boldsymbol{J}(\boldsymbol{q})$ is non-singular in the considered workspace. Generally, two different types of singularities are to be distinguished: representation singularities and kinematic singularities. The first ones are the result of the particular choice of local coordinates $\boldsymbol{x}$ which describe the configuration $\boldsymbol{h}_{st}(\boldsymbol{q}) \in SE(3)$ of the tool-frame. It is well known that every minimal representation of $SO(3)$, and hence also of $SE(3)$, will have such singularities. They can in practice be handled by using different representations for different regions of the workspace. Switching between these different representations, however, is a critical issue. The only way to avoid such singularities completely is to use a non-minimal parametrization instead. A method how to construct a singularity-free Cartesian stiffness based on such a non-minimal parametrization will be reviewed in Section 3.5.2.

The second type of singularities, the kinematic singularities, however, cannot be avoided. They are inherently connected to the kinematic structure of the manipulator, representing the inability to produce an arbitrary movement of the end-effector at certain joint configurations. From a mathematical point of view the kinematic singularities are the singularities of the body Jacobian. It is clear that near a singular configuration the desired impedance behavior (3.16) in general cannot be achieved exactly but will be *distorted*. By using a control law like (3.18) the robot simply might *get stuck* when moving through a singular configuration because the second term in (3.18) might vanish.

One method for the control of manipulators at kinematic singularities was proposed in [CK95]. In this work a factorization of the determinant of the body Jacobian is required, in which each term corresponds to one of the different singular configurations of the robot. When approaching such a singular configuration the controller splits up into two parts. One part in the controller is used to control the *distance* of the joint configuration from the next singular configuration via the relevant term in the factorization. The other part basically controls the movement *orthogonally*[12] to the singular direction by the classical operational space formulation [Kha87]. While this is a quite effective approach, especially (but not only) for the use in a position controller, in case of an impedance controller one sometimes[13] prefers simply to avoid the singular configurations.

In this section a possible solution for avoiding the singularities is introduced. This will implicitly restrict the manipulators workspace, but this restriction will be done in a generic way without the need to know all the singular configurations in advance.

In order to implement the singularity avoidance, the Cartesian impedance controller from (3.18) is combined with a second impedance controller which forces the manipulator to move away from singular configurations. The combination of the two controllers can be done by considering the superposition

[12] Notice that the term *orthogonality*, as it was used in [CK95], actually requires the choice of a particular metric.

[13] It clearly depends on the application whether the avoidance of singularities is admissible or not.

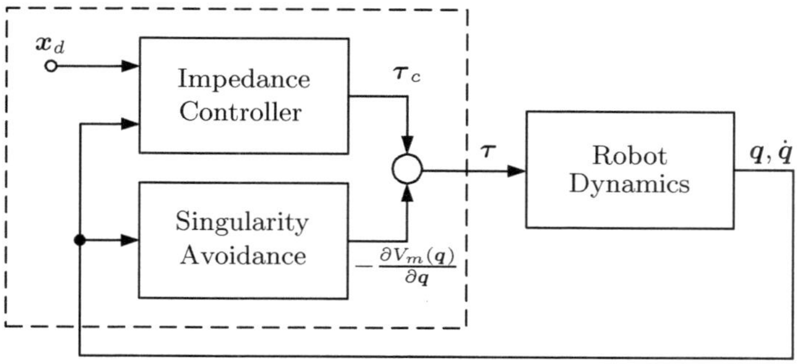

Fig. 3.1. Superposition of the Cartesian impedance controller and the singularity avoidance

principle for impedances [Hog85a, BS98]. According to this multiple impedance components coupled to an admittance may be assembled simply by adding their output torques even if the behavior of the impedances is nonlinear.

For the construction of a singularity avoidance potential, the kinematic manipulability measure [Yos90]

$$m_{kin}(\boldsymbol{q}) = \sqrt{\det(\boldsymbol{J}(\boldsymbol{q})\boldsymbol{J}(\boldsymbol{q})^T)} \tag{3.22}$$

is used. The singularity avoidance potential is chosen as

$$V_m(\boldsymbol{q}) = \begin{cases} k_s(m_{kin}(\boldsymbol{q}) - m_0)^2 & m_{kin}(\boldsymbol{q}) \leq m_0 \\ 0 & m_{kin}(\boldsymbol{q}) > m_0 \end{cases} , \tag{3.23}$$

where $k_s \in \mathbb{R}$ is a positive scalar factor controlling the gain of the singularity avoidance. The upper bound $m_0 > 0 \in \mathbb{R}$ for $m_{kin}(\boldsymbol{q})$ determines the area around a singular configuration, in which the singularity avoidance will be active. These two parameters are the design parameters.

Let $\boldsymbol{\tau}_c$ denote the output torque of the Cartesian impedance controller from the last section, then the complete control law with singularity avoidance is given by

$$\boldsymbol{\tau} = \boldsymbol{\tau}_c - \frac{\partial V_m(\boldsymbol{q})}{\partial \boldsymbol{q}} . \tag{3.24}$$

Notice that instead of $m_{kin}(\boldsymbol{q})$ one could also use the dynamic manipulability measure $m_{dyn}(\boldsymbol{q}) = \sqrt{\det(\boldsymbol{J}(\boldsymbol{q})\boldsymbol{M}(\boldsymbol{q})^{-1}\boldsymbol{M}(\boldsymbol{q})^{-T}\boldsymbol{J}(\boldsymbol{q})^T)}$ [Yos90]. With regard to the implementation the use of the kinematic measure has the advantage of consuming less computation power.

Some additional remarks concerning the effects of the singularity avoidance on the stability properties of the closed loop system are in order. Notice that the singularity avoidance will only be active when the manipulator configuration is near to a singularity. Far[14] from the singularities, only the Cartesian impedance

[14] Measured by the condition $m_{kin}(\boldsymbol{q}) > m_0$.

controller is active. If one assumes that the virtual equilibrium position is far away from the singularities and that the desired stiffness is not *too low*, it is clear that the singularity avoidance will not affect the (local) stability of the system. On the other hand it should be mentioned that it indeed could happen that, near to a singularity, the torques due to the Cartesian stiffness potential and the singularity avoidance potential counterbalance each other. Thereby, the complete potential function may have a local minimum other than the desired equilibrium position. Then the robot clearly could get stuck while moving through this local minimum, which, however, will be close to the singular configuration.

3.5 Remarks on the Stiffness Implementation

In the previous sections, it was assumed that the end-effector pose can be described by some end-effector coordinates $\boldsymbol{x} = \boldsymbol{f}(\boldsymbol{q})$ and that the Cartesian error $\tilde{\boldsymbol{x}}$ reads accordingly as $\tilde{\boldsymbol{x}} = \boldsymbol{x} - \boldsymbol{x}_d$. In contrast to the damping matrix discussed in Section 3.3, the desired stiffness matrix $\boldsymbol{K}_d$ is required to be a constant matrix. Otherwise the stability statement from Proposition 3.4 would not be valid any more.

In this section it will be explained how the particular choice of the used Cartesian coordinates $\boldsymbol{f}(\boldsymbol{q})$ in the control law (3.18) affects the resulting stiffness behavior. For this it is assumed that the Cartesian coordinates $\boldsymbol{x} \in \mathbb{R}^6$ can be split up into two components $\boldsymbol{x}_t \in \mathbb{R}^3$ and $\boldsymbol{x}_r \in \mathbb{R}^3$ describing the end-effector position and orientation, respectively. Analogously, the translational error is denoted by $\tilde{\boldsymbol{x}}_t \in \mathbb{R}^3$ and the rotational error by $\tilde{\boldsymbol{x}}_r \in \mathbb{R}^3$. In accordance with the translational and rotational error, the stiffness matrix $\boldsymbol{K}_d$ is partitioned into a translational stiffness $\boldsymbol{K}_t$, a rotational stiffness $\boldsymbol{K}_r$, and a coupling stiffness $\boldsymbol{K}_c$, i.e.

$$\boldsymbol{K}_d = \begin{bmatrix} \boldsymbol{K}_t & \boldsymbol{K}_c \\ \boldsymbol{K}_c^T & \boldsymbol{K}_r \end{bmatrix} .$$

As described in Section 2.1 the pose of the end-effector, i.e. the pose of the tool frame $\mathcal{T}$, is given by the forward kinematics map $\boldsymbol{h}_{st}(\boldsymbol{q}) : \mathcal{Q} \to SE(3)$, which can be computed by the product of exponentials formula (2.3). According to the notation of Chapter 2 the rotational part of $\boldsymbol{h}_{st}(\boldsymbol{q})$ is denoted by $\boldsymbol{R}_{st}(\boldsymbol{q}) \in SO(3)$ and the translational part by $\boldsymbol{p}_{st}(\boldsymbol{q}) \in \mathbb{R}^3$. Given a (time-varying) *virtual* pose $\boldsymbol{h}_{sd}(t) \in SE(3)$, corresponding to a (time-varying) desired frame $\mathcal{D}$, the deviation of $\boldsymbol{h}_{st}(\boldsymbol{q})$ from $\boldsymbol{h}_{sd}(t)$ can be described by

$$\boldsymbol{h}_{dt}(\boldsymbol{q},t) = \boldsymbol{h}_{sd}^{-1}(t)\boldsymbol{h}_{st}(\boldsymbol{q}) = \begin{bmatrix} \boldsymbol{R}_{dt}(\boldsymbol{q},t) & {}_d\boldsymbol{p}_{dt}(\boldsymbol{q},t) \\ \boldsymbol{0} & 1 \end{bmatrix} . \quad (3.25)$$

The following discussion is split up into two parts. Section 3.5.1 clarifies some aspects concerning the implementation of the translational stiffness $\boldsymbol{K}_t$, and

Section 3.5.2 describes different orientation representations according to the rotational stiffness $\boldsymbol{K}_r$. The coupling stiffness $\boldsymbol{K}_c$ can be designed similarly. Since it is not used in the applications of Chapter 9 it is not considered here in detail, i.e. $\boldsymbol{K}_c = \boldsymbol{0}$.

3.5.1 Translational Stiffness

The vector ${}_d\boldsymbol{p}_{dt}(\boldsymbol{q}, t)$ in (3.25) corresponds to a vector from the origin of the desired frame $\mathcal{D}$ to the origin of the tool frame $\mathcal{T}$, represented in frame $\mathcal{D}$. From this one can get an appropriate set of translational end-effector coordinates by rotating this vector into the base frame $\mathcal{S}$

$$\boldsymbol{p}_{dt}(\boldsymbol{q}, t) = \boldsymbol{R}_{sd}(t)\,{}_d\boldsymbol{p}_{dt}(\boldsymbol{q}, t) = \boldsymbol{p}_{st}(\boldsymbol{q}) - \boldsymbol{p}_{sd}(t) \ . \tag{3.26}$$

Then, one can choose $\boldsymbol{x}_t = \boldsymbol{p}_{st}(\boldsymbol{q})$ as translational coordinates and $\boldsymbol{p}_{sd}(t)$ as their corresponding virtual equilibrium position. By this choice the translational part of the stiffness matrix corresponds to a stiffness, which is represented in the base frame $\mathcal{S}$, and consequently the eigenvectors of $\boldsymbol{K}_t$, corresponding to the principal axes of the stiffness matrix, are constant vectors expressed in the base frame.

For some applications it is instead desired to use a stiffness implementation in which the principal axes of the stiffness matrix are defined (as constant vectors) in the desired frame $\mathcal{D}$ or the tool frame $\mathcal{T}$. Then, one must refer to a different set of coordinates. If for instance the vector ${}_d\boldsymbol{p}_{dt}(\boldsymbol{q}, t)$ is used as the translational part $\tilde{\boldsymbol{x}}_t$ of the Cartesian error $\tilde{\boldsymbol{x}}$, this corresponds to a stiffness representation in $\mathcal{D}$. For a stiffness representation in $\mathcal{T}$ one can accordingly use translational coordinates of the form ${}_t\boldsymbol{p}_{dt}(\boldsymbol{q}, t) = \boldsymbol{R}_{td}(\boldsymbol{q}, t)\,{}_d\boldsymbol{p}_{dt}(\boldsymbol{q}, t)$. Notice that (except for the choice $\boldsymbol{x}_t = \boldsymbol{p}_{st}(\boldsymbol{q})$) one cannot write $\tilde{\boldsymbol{x}}_t$ as the difference between a configuration dependent coordinate $\boldsymbol{x}_t(\boldsymbol{q})$ and a time-varying (but configuration-independent) virtual equilibrium position $\boldsymbol{x}_d(t)$ any more. However, the impedance controllers of the last sections can of course be easily adapted to this case. This will be treated in more detail in Section 3.5.3.

3.5.2 Rotational Stiffness

While the procedure for the translational part of the stiffness was quite straightforward, the situation is more complex for the orientation representation. It is well known that no global minimal representation of $SO(3)$ exists. Different choices of orientation coordinates for the use in Cartesian controllers were analyzed in detail in [Nat03]. In this section two different approaches for the implementation of a rotational stiffness are shortly discussed, namely *Euler angles* and *unit quaternions*.

Consider a set of Euler angles $\boldsymbol{\phi}_{st}(\boldsymbol{q}) = \boldsymbol{\phi}(\boldsymbol{R}_{st}(\boldsymbol{q})) \in \mathbb{R}^3$ computed directly from the rotation matrix $\boldsymbol{R}_{st}(\boldsymbol{q})$. One can build the *classical* Euler angle based orientation error $\boldsymbol{\phi}^*_{dt}(\boldsymbol{q}, t) = \boldsymbol{\phi}_{st}(\boldsymbol{q}) - \boldsymbol{\phi}_{sd}(t)$. But this approach to formulate the orientation error has a major drawback: Singularities are encountered whenever either the robot orientation $\boldsymbol{\phi}_{st}(\boldsymbol{q})$ or the desired equilibrium orientation $\boldsymbol{\phi}_{sd}(t)$ is singular. This can happen for arbitrary small orientation errors. A much better approach is to compute the Euler angles from the difference frame $\boldsymbol{h}_{td}(\boldsymbol{q},t)$ via ${}_t\boldsymbol{\phi}_{td}(\boldsymbol{q},t) = \boldsymbol{\phi}(\boldsymbol{R}_{td}(\boldsymbol{q},t))$. This representation, sometimes called *modified Euler angle representation*, is much more robust against singularities. If, for instance, the well known *roll-pitch-yaw* representation is used, then the representation does not contain any singularities up to an orientation error of $\pi/2$.

The main disadvantage of an Euler angle based stiffness is the occurrence of representation singularities. These can only be avoided if one refers to a non-minimal representation of the end-effector orientation. Herein, only the use of the so-called *unit quaternions*[15] will be shortly discussed. A detailed exposition of unit quaternions is beyond the scope of this section, only some relevant properties are reported. More details on the use of unit quaternions can be found e.g. in [CNSV99, Nat03] and the references cited therein.

A unit quaternion $(\eta, \boldsymbol{\epsilon})$ consists of a scalar part $\eta \in \mathbb{R}$ and a vector part $\boldsymbol{\epsilon} \in \mathbb{R}^3$, which fulfill the condition $\eta^2 + \boldsymbol{\epsilon}^T\boldsymbol{\epsilon} = 1$. The relation between a rotation matrix and unit quaternions is given as follows. Suppose that a rotation matrix is specified by a rotation of an angle α about an axis $\boldsymbol{r}$ (with $||\boldsymbol{r}||_2 = 1$), then the corresponding unit quaternion is given by $\eta = \cos(\alpha/2)$, $\boldsymbol{\epsilon} = \boldsymbol{r}\sin(\alpha/2)$. With the restriction of α to the interval $[-\pi, \pi]$ the set of unit quaternions is a one-to-one covering of $SO(3)$. In contrast to Euler angles, unit quaternions give a singularity-free representation of $SO(3)$.

Let $\boldsymbol{\epsilon}_{dt}(\boldsymbol{q},t)$ be the vector part of the unit quaternion according to the rotation matrix $\boldsymbol{R}_{dt}(\boldsymbol{q},t)$. For the implementation of a rotational stiffness, one can then refer to the following orientation error $\tilde{\boldsymbol{x}}_r(\boldsymbol{q},t) := 2\boldsymbol{\epsilon}_{dt}(\boldsymbol{q},t)$. Notice that this quantity (and thus also the resulting stiffness term) is periodic with respect to α.

Herein, only the most commonly used types of stiffness implementations were discussed. Their relation to the controllers from this book will be clarified in the next section. For another interesting class of stiffness implementations the reader is referred to [FH95, FB97, Fas97, ZF00, SD01, Str01], in which the so-called *spatial* stiffness is discussed.

3.5.3 Consequences for the Closed Loop Dynamics

The previous two sections led to a Cartesian error $\tilde{\boldsymbol{x}}(\boldsymbol{q},t)$ which cannot be written as the difference between an actual and a desired quantity. In the following it is shown how the model from Section 3.1.2 must be adapted in order to cope with this situation.

The first and second derivatives of the error vector $\tilde{\boldsymbol{x}}(\boldsymbol{q},t)$ are given by

$$\dot{\tilde{\boldsymbol{x}}} = \frac{\partial \tilde{\boldsymbol{x}}(\boldsymbol{q},t)}{\partial \boldsymbol{q}}\dot{\boldsymbol{q}} + \frac{\partial \tilde{\boldsymbol{x}}(\boldsymbol{q},t)}{\partial t} ,$$

$$\ddot{\tilde{\boldsymbol{x}}} = \frac{\partial \tilde{\boldsymbol{x}}(\boldsymbol{q},t)}{\partial \boldsymbol{q}}\ddot{\boldsymbol{q}} + \frac{\mathrm{d}}{\mathrm{d}t}\left(\frac{\partial \tilde{\boldsymbol{x}}(\boldsymbol{q},t)}{\partial \boldsymbol{q}}\right)\dot{\boldsymbol{q}} + \frac{\mathrm{d}}{\mathrm{d}t}\frac{\partial \tilde{\boldsymbol{x}}(\boldsymbol{q},t)}{\partial t} .$$

[15] For unit quaternions also the term *Euler parameters* is sometimes used.

These equations replace (3.2) and (3.3). The equations of motion for the error vector $\tilde{\boldsymbol{x}}$ can now be derived by following the same steps as in Section 3.1.2. For the ease of presentation the following substitutions are made

$$\begin{aligned} \frac{\partial \tilde{\boldsymbol{x}}(\boldsymbol{q},t)}{\partial \boldsymbol{q}} &= \boldsymbol{J}_x(\boldsymbol{q},t) \ , \\ \frac{\partial \tilde{\boldsymbol{x}}(\boldsymbol{q},t)}{\partial t} &= -\boldsymbol{v}_t(\boldsymbol{q},t) \ . \end{aligned}$$

Therefore, one gets the following equations of motion

$$\begin{aligned} \boldsymbol{\Lambda}(\boldsymbol{q},t)(\ddot{\tilde{\boldsymbol{x}}} + \dot{\boldsymbol{v}}_t(\boldsymbol{q},t)) + \boldsymbol{\mu}(\boldsymbol{q},\dot{\boldsymbol{q}},t)(\dot{\tilde{\boldsymbol{x}}} + \boldsymbol{v}_t(\boldsymbol{q},t)) + \boldsymbol{J}_x(\boldsymbol{q},t)^{-T}\boldsymbol{g}(\boldsymbol{q}) - \\ \boldsymbol{J}_x(\boldsymbol{q},t)^{-T}(\boldsymbol{\tau} + \boldsymbol{\tau}_{ext}) = \boldsymbol{0} \ , \end{aligned}$$

where $\boldsymbol{\Lambda}(\boldsymbol{q},t)$ and $\boldsymbol{\mu}(\boldsymbol{q},\dot{\boldsymbol{q}},t)$ are the relevant Cartesian inertia matrix and Coriolis/centrifugal matrix

$$\begin{aligned} \boldsymbol{\Lambda}(\boldsymbol{q},t) &= \boldsymbol{J}_x(\boldsymbol{q},t)^{-T}\boldsymbol{M}(\boldsymbol{q})\boldsymbol{J}_x(\boldsymbol{q},t)^{-1} \ , \\ \boldsymbol{\mu}(\boldsymbol{q},\dot{\boldsymbol{q}},t) &= \boldsymbol{J}_x(\boldsymbol{q},t)^{-T}\left(\boldsymbol{C}(\boldsymbol{q},\dot{\boldsymbol{q}}) - \boldsymbol{M}(\boldsymbol{q})\boldsymbol{J}_x(\boldsymbol{q},t)^{-1}\dot{\boldsymbol{J}}_x(\boldsymbol{q},t)\right)\boldsymbol{J}_x(\boldsymbol{q},t)^{-1} \ , \end{aligned}$$

analogous to (3.10) and (3.11). Comparing this equation with (3.12), one can see that the same controllers as in the previous sections can be used, when the following substitutions are made

$$\begin{aligned} \boldsymbol{J}(\boldsymbol{q}) &\to \boldsymbol{J}_x(\boldsymbol{q},t) \ , \\ \dot{\boldsymbol{x}}_d &\to -\boldsymbol{v}_t(\boldsymbol{q},t) \ , \\ \ddot{\boldsymbol{x}}_d &\to -\dot{\boldsymbol{v}}_t(\boldsymbol{q},t) \ . \end{aligned}$$

Notice that, although these quantities become time-varying now, these substitutions do not alter the stability statements presented in this book.

Explicit formulas of $\boldsymbol{J}_x(\boldsymbol{q},t)$ and $\boldsymbol{v}_t(\boldsymbol{q},t)$ for the different sets of local coordinates can be given in terms of the body Jacobian $\boldsymbol{J}^b(\boldsymbol{q})$. Notice therefore that $\boldsymbol{J}^b(\boldsymbol{q})$, as defined in Section 2.1, is used to compute the translational velocity ${}_t\boldsymbol{v}_{st} = \boldsymbol{R}_{ts}\dot{\boldsymbol{p}}_{st}$ and the angular velocity ${}_t\boldsymbol{\omega}_{st} = \boldsymbol{R}_{ts}\boldsymbol{\omega}_{st}$, expressed in the tool frame

$$\begin{pmatrix} {}_t\boldsymbol{v}_{st} \\ {}_t\boldsymbol{\omega}_{st} \end{pmatrix} = \boldsymbol{J}^b(\boldsymbol{q})\dot{\boldsymbol{q}} \ . \tag{3.27}$$

3.6 Summary

In this chapter the Cartesian impedance control problem for a (conventional) robot model without joint flexibility was discussed. First the general solution according to a desired impedance in form of a generalized mass-spring-damper system was treated. Then the situation for an impedance controller without inertia shaping was analyzed in detail. In particular this brought up the problem of designing the Cartesian damping matrix appropriately. For this problem a solution based on the simultaneous diagonalization of positive definite matrices was proposed. Moreover, the singularity avoidance problem was discussed, and a solution based on the superposition of impedances [Hog85c] was proposed. Then some additional details on the choice of Cartesian coordinates were discussed which lead to stiffness representations with respect to different frames. These issues, though practically relevant, are only rarely addressed in the robotics literature. Finally, an alternative orientation stiffness implementation which avoids the analytical singularities inherent to any minimal representation of $SO(3)$ was briefly discussed for further reference.

4 Nullspace Stiffness

In the previous chapter the non-redundant case was treated in which the number m of task coordinates is equal to the number n of configuration coordinates ($m = n$). The DLR lightweight robots, instead, have $n = 7$ joints, while the end-effector motion is described only by $m = 6$ degrees-of-freedom. The remaining $n - m$ degrees-of-freedom can be used for so-called *nullspace motions* that keep the end-effector fixed. Therefore, the question arises how to control these *redundant* degrees-of-freedom.

Throughout this chapter the same notation is used as in Chapter 3. Thus it is assumed that the end-effector task can be described by m local coordinates $\boldsymbol{x} \in \mathbb{R}^m$, while the robot configuration is described by the n coordinates $\boldsymbol{q} \in \mathcal{Q}$. The forward kinematics map $\boldsymbol{f} : \mathcal{Q} \rightarrow \mathbb{R}^m$ from configuration space to task space and the relevant (analytical) Jacobian $\boldsymbol{J}(\boldsymbol{q}) = \frac{\partial \boldsymbol{f}(\boldsymbol{q})}{\partial \boldsymbol{q}} \in \mathbb{R}^{m \times n}$ are known

$$\boldsymbol{x} = \boldsymbol{f}(\boldsymbol{q}) \ , \tag{4.1}$$

$$\dot{\boldsymbol{x}} = \boldsymbol{J}(\boldsymbol{q})\dot{\boldsymbol{q}} \ . \tag{4.2}$$

In the redundant case one has $m < n$ and hence the end-effector coordinates only partly specify the robot configuration.

In this chapter only the non-singular case is treated. Thus it is assumed that the Jacobian $\boldsymbol{J}(\boldsymbol{q})$, which is now a rectangular matrix, has full rank, i.e. $\mathrm{rank}(\boldsymbol{J}(\boldsymbol{q})) = m$, in the considered workspace. An appropriate way how to deal with kinematic singularities is described in Section 3.4.

Again a rigid body model of the manipulator dynamics is assumed throughout this chapter. Hence, the considered robot dynamics is given by

$$\boldsymbol{M}(\boldsymbol{q})\ddot{\boldsymbol{q}} + \boldsymbol{C}(\boldsymbol{q}, \dot{\boldsymbol{q}})\dot{\boldsymbol{q}} + \boldsymbol{g}(\boldsymbol{q}) = \boldsymbol{\tau} + \boldsymbol{\tau}_{ext} \ , \tag{4.3}$$

with the same notation as used in Chapter 3.

One approach to treat a redundant robot is to augment the task coordinates $\boldsymbol{x}$ by $r = n - m$ auxiliary variables, the nullspace coordinates. This augmentation can be done either for the generalized positions (via nullspace position coordinates $\boldsymbol{n} \in \mathbb{R}^r$), as it is done in the *Task Space Augmentation Method*

C. Ott: Cartesian Imped. Cntrl. of Redund. & Flexible-Joint Robots, STAR 49, pp. 45–63, 2008.
springerlink.com

[Bai85, Sic90], or for the generalized velocities (via nullspace velocity coordinates $\boldsymbol{v}_n \in \mathbb{R}^r$), as in the *Joint Space Decomposition Method* [Par99, OCY98]. The problem of finding appropriate coordinates and the decoupling of the task dynamics from the nullspace dynamics will be discussed for both methods.

The additional nullspace coordinates will be constructed using a nullspace base matrix $\boldsymbol{Z}(\boldsymbol{q}) \in \mathbb{R}^{r\times n}$ which is composed of r row vectors that are linearly independent and span the (right) nullspace of the Jacobian $\boldsymbol{J}(\boldsymbol{q})$. Section 4.1 treats the problem how to compute such a matrix. Thereafter, first a simple approach to control the nullspace motion is described in Section 4.2 based on the superposition principle for impedances. This approach is easy to implement, since it does not rely on an explicit construction of nullspace coordinates. However, it does not enable a complete stability analysis. Then the two above mentioned methods for the generation of nullspace coordinates are described in Section 4.3 and 4.4 and their use for the impedance control problem is analyzed.

4.1 Computation of the Nullspace Base Matrix

In this section it is shown, how to construct a nullspace base matrix $\boldsymbol{Z}(\boldsymbol{q}) \in \mathbb{R}^{r\times n}$ for the Jacobian $\boldsymbol{J}(\boldsymbol{q}) \in \mathbb{R}^{m\times n}$. The presentation focuses on an approach which gives the nullspace base matrix in a symbolic form.

4.1.1 The General Case

As already mentioned in the introduction of this chapter it is assumed that the matrix $\boldsymbol{J}(\boldsymbol{q})$ has full row rank in the considered workspace, i.e. singularities are avoided. By reordering the columns of $\boldsymbol{J}(\boldsymbol{q})$, one can always write $\boldsymbol{J}(\boldsymbol{q})$ in the partitioned form

$$\boldsymbol{J}(\boldsymbol{q}) = \begin{bmatrix}\boldsymbol{J}_m(\boldsymbol{q}) \ \boldsymbol{J}_r(\boldsymbol{q})\end{bmatrix} , \tag{4.4}$$

such that the left part, i.e. the quadratic matrix $\boldsymbol{J}_m(\boldsymbol{q}) \in \mathbb{R}^{m\times m}$, is (at least locally) invertible. Let further the desired nullspace base matrix be partitioned correspondingly as $\boldsymbol{Z}(\boldsymbol{q}) = \begin{bmatrix}\boldsymbol{Z}_m(\boldsymbol{q}) \ \boldsymbol{Z}_r(\boldsymbol{q})\end{bmatrix}$, where the matrices $\boldsymbol{Z}_m(\boldsymbol{q}) \in \mathbb{R}^{r\times m}$ and $\boldsymbol{Z}_r(\boldsymbol{q}) \in \mathbb{R}^{r\times r}$ are to be determined, such that $\boldsymbol{J}(\boldsymbol{q})\boldsymbol{Z}(\boldsymbol{q})^T = \boldsymbol{0}$ holds and such that $\boldsymbol{Z}(\boldsymbol{q})$ has full row rank.

With the chosen partitioning the condition $\boldsymbol{J}(\boldsymbol{q})\boldsymbol{Z}(\boldsymbol{q})^T = \boldsymbol{0}$ can be written as

$$\boldsymbol{J}_m(\boldsymbol{q})\boldsymbol{Z}_m(\boldsymbol{q})^T + \boldsymbol{J}_r(\boldsymbol{q})\boldsymbol{Z}_r(\boldsymbol{q})^T = \boldsymbol{0} . \tag{4.5}$$

One possible solution of this equation is given by the particular choice

$$\boldsymbol{Z}_m(\boldsymbol{q})^T = -\boldsymbol{J}_m(\boldsymbol{q})^{-1}\boldsymbol{J}_r(\boldsymbol{q}) , \tag{4.6}$$

$$\boldsymbol{Z}_r(\boldsymbol{q})^T = \boldsymbol{I} , \tag{4.7}$$

which by construction gives a full rank nullspace base matrix[1]. The matrix $\boldsymbol{Z}(\boldsymbol{q})$ is then given by

$$\boldsymbol{Z}(\boldsymbol{q}) = \begin{bmatrix} -\boldsymbol{J}_r(\boldsymbol{q})^T \boldsymbol{J}_m(\boldsymbol{q})^{-T} & \boldsymbol{I} \end{bmatrix} . \tag{4.8}$$

This form of $\boldsymbol{Z}(\boldsymbol{q})$ was originally proposed by Huang and Varma in [HV91] and has afterwards been used by many authors. But the matrix $\boldsymbol{Z}(\boldsymbol{q})$ obviously is not unique and could have been chosen differently. One notable modification was presented for instance by Chen and Walker in [CW93] in which a scaling of $\boldsymbol{Z}(\boldsymbol{q})$ by the factor $\det(\boldsymbol{J}_m(\boldsymbol{q}))$ was proposed, i.e. $\bar{\boldsymbol{Z}}(\boldsymbol{q}) = \det(\boldsymbol{J}_m(\boldsymbol{q}))\boldsymbol{Z}(\boldsymbol{q})$. Notice that the inverse of $\boldsymbol{J}_m(\boldsymbol{q})$ can be written as

$$\boldsymbol{J}_m(\boldsymbol{q})^{-1} = \operatorname{adj}(\boldsymbol{J}_m(\boldsymbol{q}))/\det(\boldsymbol{J}_m(\boldsymbol{q})) ,$$

where $\operatorname{adj}(\boldsymbol{J}_m(\boldsymbol{q}))$ is the adjoint matrix to $\boldsymbol{J}_m(\boldsymbol{q})$ (see, e.g., [HJ90]). Thus, the new nullspace base matrix, as proposed by Chen and Walker, is given by

$$\bar{\boldsymbol{Z}}(\boldsymbol{q}) = \begin{bmatrix} -\boldsymbol{J}_r(\boldsymbol{q})^T \operatorname{adj}(\boldsymbol{J}_m(\boldsymbol{q}))^T & \det(\boldsymbol{J}_m(\boldsymbol{q}))\boldsymbol{I} \end{bmatrix} . \tag{4.9}$$

One very interesting[2] property of this nullspace base matrix is the equality $m_{kin}(\boldsymbol{q}) = \det(\boldsymbol{J}(\boldsymbol{q})\boldsymbol{J}(\boldsymbol{q})^T) = \det(\bar{\boldsymbol{Z}}(\boldsymbol{q})\bar{\boldsymbol{Z}}(\boldsymbol{q})^T)$. The proof of this and other useful properties can be found in [CW93].

4.1.2 The Case of a One-Dimensional Nullspace

The search for an invertible submatrix $\boldsymbol{J}_m(\boldsymbol{q})$ which was required in the last section can be avoided when the degree of redundancy $r = n-m$ is one. This is for instance the case for the DLR lightweight robots. Then the matrix $\bar{\boldsymbol{Z}}(\boldsymbol{q}) \in \mathbb{R}^{r\times n}$ reduces to the row vector $\boldsymbol{z}(\boldsymbol{q}) = \begin{bmatrix} z_1(\boldsymbol{q}) & \cdots & z_n(\boldsymbol{q}) \end{bmatrix}$. Consider the matrix

$$\boldsymbol{J}_z(\boldsymbol{q}) = \begin{pmatrix} \boldsymbol{J}(\boldsymbol{q}) \\ \boldsymbol{z}(\boldsymbol{q}) \end{pmatrix} . \tag{4.10}$$

In the following the vector $\boldsymbol{z}(\boldsymbol{q})$ will be constructed such that the matrix $\boldsymbol{J}_z(\boldsymbol{q})$ is invertible. Notice that then the inverse of $\boldsymbol{J}_z(\boldsymbol{q})$ can be written in the form $\boldsymbol{J}_z(\boldsymbol{q})^{-1} = \operatorname{adj}(\boldsymbol{J}_z(\boldsymbol{q}))/\det(\boldsymbol{J}_z(\boldsymbol{q}))$. Due to $\boldsymbol{J}_z(\boldsymbol{q})\operatorname{adj}(\boldsymbol{J}_z(\boldsymbol{q})) = \det(\boldsymbol{J}_z(\boldsymbol{q}))\boldsymbol{I}$, the last column $\boldsymbol{c}_n(\boldsymbol{q})$ of $\operatorname{adj}(\boldsymbol{J}_z(\boldsymbol{q}))$ clearly fulfills the condition $\boldsymbol{J}(\boldsymbol{q})\boldsymbol{c}_n(\boldsymbol{q}) = \boldsymbol{0}$.

Notice that the adjoint is the transposed matrix of co-factors[3] and therefore the elements of the column vector $\boldsymbol{c}_n(\boldsymbol{q})$ are given by the $(n,i)^{th}$ co-factors of

[1] Notice that the rank of any matrix is equivalent to the dimension of its largest non-singular submatrix [HJ90].

[2] Notice that this property suggests an efficient method to compute the kinematic manipulability measure $m_{kin}(\boldsymbol{q}) = \sqrt{\det(\boldsymbol{J}(\boldsymbol{q})\boldsymbol{J}(\boldsymbol{q})^T)}$ from [Yos90] which was also used in (3.22).

[3] The co-factors of a matrix are the *signed* minors, see, e.g., [HJ90].

$\boldsymbol{J}_z(\boldsymbol{q})$. Since these $(n,i)^{th}$ co-factors are computed from the submatrix $\boldsymbol{J}(\boldsymbol{q})$ alone and do not depend on $\boldsymbol{z}(\boldsymbol{q})$, one can choose $\boldsymbol{z}(\boldsymbol{q}) = \boldsymbol{c}_n(\boldsymbol{q})^T$. The elements $z_i(\boldsymbol{q})$ of $\boldsymbol{z}(\boldsymbol{q})$ can thus be computed by

$$z_i(\boldsymbol{q}) = (-1)^{n+i}\det(\boldsymbol{J}_i(\boldsymbol{q})) , \tag{4.11}$$

where $\boldsymbol{J}_i(\boldsymbol{q}) \in \mathbb{R}^{m\times m}$ is the matrix $\boldsymbol{J}(\boldsymbol{q})$ with the ith column omitted.

4.2 Projection Based Approaches

The superposition principle for impedances provides a suitable framework for the combination of different impedance behaviors. As described in detail by Hogan [Hog85a, Hog85b, Hog85c], this superposition can be done very easily by adding up the outputs, i.e. joint torques, of the different impedances (see also Section 3.4). The Cartesian impedance controller from Section 3.2 gives the following joint torque as an output, cf. (3.18),

$$\boldsymbol{\tau}_c = \boldsymbol{g}(\boldsymbol{q}) + \boldsymbol{J}(\boldsymbol{q})^T \boldsymbol{F}_{imp} , \tag{4.12}$$
$$\boldsymbol{F}_{imp} = \boldsymbol{\Lambda}(\boldsymbol{x})\ddot{\boldsymbol{x}}_d + \boldsymbol{\mu}(\boldsymbol{x},\dot{\boldsymbol{x}})\dot{\boldsymbol{x}}_d - \boldsymbol{K}_d\tilde{\boldsymbol{x}} - \boldsymbol{D}_d\dot{\tilde{\boldsymbol{x}}} . \tag{4.13}$$

Hence, the combination with a nullspace impedance, which results in a joint torque output vector $\boldsymbol{\tau}_n$, can simply be done by

$$\boldsymbol{\tau} = \boldsymbol{\tau}_c + \boldsymbol{\tau}_n , \tag{4.14}$$

as it is sketched in Figure 4.1. In this section it will be shown, how such a nullspace impedance with output $\boldsymbol{\tau}_n$ can be designed in an intuitive way. First, note that for a chosen virtual equilibrium position $\boldsymbol{q}_{d,0}$, one can define a simple joint space impedance in the form

$$\boldsymbol{\tau}_0 = -\boldsymbol{D}_n\dot{\boldsymbol{q}} - \boldsymbol{K}_n(\boldsymbol{q} - \boldsymbol{q}_{d,0}) , \tag{4.15}$$

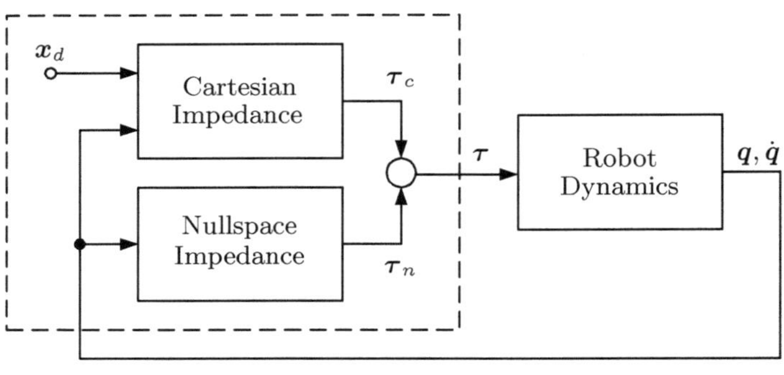

Fig. 4.1. Superposition of the outputs of the Cartesian impedance controller and the nullspace impedance controller

where the symmetric and positive definite matrices $\boldsymbol{K}_n \in \mathbb{R}^{n\times n}$ and $\boldsymbol{D}_n \in \mathbb{R}^{n\times n}$ represent a desired stiffness and damping behavior with respect to the configuration coordinates. Notice that, in view of the combination with the Cartesian impedance, $\boldsymbol{q}_{d,0}$ should be chosen *properly*, such that it is consistent with the virtual Cartesian equilibrium position $\boldsymbol{x}_d$. This means it should satisfy the condition $\boldsymbol{f}(\boldsymbol{q}_{d,0}) = \boldsymbol{x}_d$. If this is not established, it clearly is not possible to reach statically (and for the case of free motion) both the desired Cartesian position and the desired nullspace configuration simultaneously. Instead, only a (local) minimization of the constrained cost function $(\boldsymbol{q} - \boldsymbol{q}_{d,0})^T \boldsymbol{K}_n (\boldsymbol{q} - \boldsymbol{q}_{d,0})|_{\boldsymbol{f}(\boldsymbol{q})=\boldsymbol{x}_d}$ can be achieved while fulfilling the (usually more important) Cartesian equilibrium condition $\boldsymbol{x} = \boldsymbol{x}_d$.

The impedance in (4.15) cannot be used directly because it would distort the desired Cartesian impedance behavior. In order to kinematically decouple the joint space impedance from the Cartesian impedance behavior, one may project the torque $\boldsymbol{\tau}_0$ via a matrix $\boldsymbol{P}(\boldsymbol{q})$, i.e.

$$\boldsymbol{\tau}_n = \boldsymbol{P}(\boldsymbol{q})\boldsymbol{\tau}_0 \ , \tag{4.16}$$

onto the complement of the range of $\boldsymbol{J}(\boldsymbol{q})^T$. In the following, three different kinds of projection matrices $\boldsymbol{P}_i(\boldsymbol{q})$ will be examined in more detail. Consider therefore first the following proposition about the construction of projection matrices [BM92] (see also the Definitions A.18 and A.19).

Proposition 4.1. *Consider an m-dimensional subspace $\mathcal{A}$ of $\mathbb{R}^n$ described as the range of an $n \times m$ matrix $\boldsymbol{A} \in \mathbb{R}^{n\times m}$ with full rank, i.e. $\mathcal{A} := \{\boldsymbol{y} \in \mathbb{R}^n | \exists \boldsymbol{x} \in \mathbb{R}^m, \boldsymbol{y} = \boldsymbol{A}\boldsymbol{x}\}$. Let $\boldsymbol{G} \in \mathbb{R}^{n\times n}$ be a symmetric and positive definite matrix defining a metric for the co-domain space $\mathbb{R}^n$. Then the matrix*[4]

$$\boldsymbol{P} = \boldsymbol{A}\left(\boldsymbol{A}^T\boldsymbol{G}\boldsymbol{A}\right)^{-1}\boldsymbol{A}^T\boldsymbol{G} \quad \in \mathbb{R}^{n\times n} \tag{4.17}$$

is a projection matrix (i.e. it fulfills the idempotency property $\boldsymbol{P}\boldsymbol{P} = \boldsymbol{P}$*), and it projects an element of $\mathbb{R}^n$ orthogonally (with respect to $\boldsymbol{G}$) onto $\mathcal{A}$ (see Figure 4.2). It thus fulfills the condition $\boldsymbol{P}\boldsymbol{A} = \boldsymbol{A}$. Furthermore, a projection matrix which projects an element of $\mathbb{R}^n$ onto the orthogonal complement $\mathcal{A}^\perp$ of $\mathcal{A}$ is given by*

$$\boldsymbol{P}_c = \boldsymbol{I} - \boldsymbol{P} \quad \in \mathbb{R}^{n\times n}. \tag{4.18}$$

Notice that the conditions $\boldsymbol{P}_c\boldsymbol{A} = \boldsymbol{0}$ and $\boldsymbol{P}_c\boldsymbol{P}_c = \boldsymbol{P}_c$ both hold.

The equations (4.17) and (4.18) offer different ways to construct the projection matrix for the nullspace impedance. In the next two subsections a projection matrix $\boldsymbol{P}(\boldsymbol{q})$ onto the nullspace of $\boldsymbol{J}(\boldsymbol{q})^T$ is designed based on the nullspace base matrix $\boldsymbol{Z}(\boldsymbol{q})$ by using (4.17). This projection is then compared to the *dynamically consistent* nullspace projection from the operational space formulation which is based on (4.18).

[4] The factor $(\boldsymbol{A}^T\boldsymbol{G}\boldsymbol{A})^{-1}$ basically acts as a normalization herein.

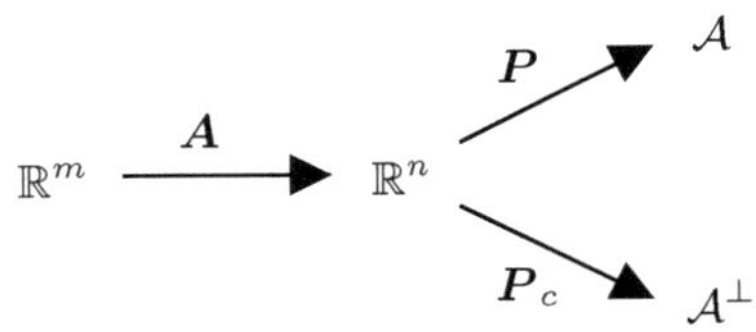

Fig. 4.2. Projections onto the range space and onto its complement

Static Nullspace Projection

The desired projection $P(q)$ should be designed such that any torque vector τ_n of the form $\tau_n = P(q)\tau_0$ is orthogonal (with respect to some, so far arbitrary, metric $G(q)$) to the Cartesian torque $\tau_c = J(q)^T F_{imp}$. This means that $\tau_c^T G(q)\tau_n = 0$ should hold and clearly results in the requirement

$$J(q)G(q)P(q) = 0 . \tag{4.19}$$

The matrix $G(q)$ is assumed to be chosen symmetric and positive definite in order to qualify as a metric. The matrix $Z(q)$ consists of r vectors $z_i(q)$ which span the (right) nullspace of the Jacobian, i.e $Z(q) = \begin{bmatrix} z_1(q) & \cdots & z_r(q) \end{bmatrix}^T$. Then a suitable choice for $P(q)$, which ensures the condition (4.19), is a projection onto the range of $G(q)^{-1}Z(q)^T$ (see Figure 4.3). By (4.17) from Proposition 4.1 (with $G(q)^{-1}Z(q)^T$ corresponding to A and $G(q)$ as the considered metric) one can compute such a projection matrix as

$$P_1(q) = G(q)^{-1}Z(q)^T \left(Z(q)G(q)^{-1}Z(q)^T\right)^{-1} Z(q) . \tag{4.20}$$

Notice that the projection $P_1(q)$ is solely based on the kinematics of the manipulator, the dynamics has not been taken into account so far and the metric $G(q)$ could therefore be chosen arbitrarily. If for instance the Euclidean metric is chosen and the vectors $z_i(q)$ in $Z(q)$ form an orthonormal basis, then the projection has the simple form $Z(q)^T Z(q)$.

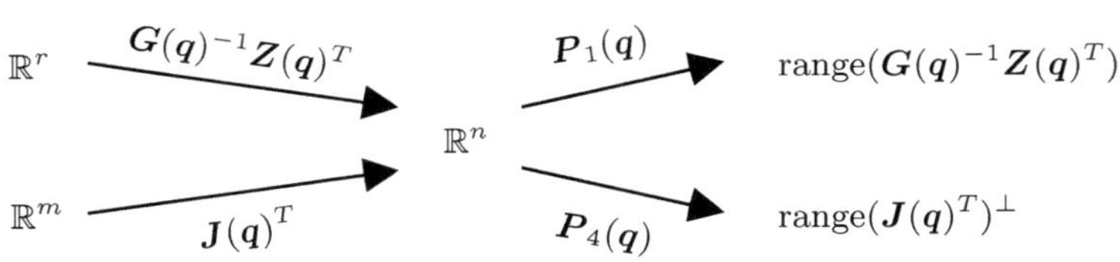

Fig. 4.3. Nullspace projections

Dynamically Consistent Projection

The fact that the metric $G(q)$ of the last section can be chosen arbitrarily is the result of considering only the kinematics, but not the dynamics. In order to take also the dynamics of the manipulator into account one may consider the relation between the Cartesian acceleration $\ddot{\boldsymbol{x}}$ and the joint torques $\boldsymbol{\tau}$. Therefore, one pre-multiplies the manipulator dynamics (4.3) by $\boldsymbol{J}(\boldsymbol{q})\boldsymbol{M}(\boldsymbol{q})^{-1}$ and gets, with $\ddot{\boldsymbol{x}} = \boldsymbol{J}(\boldsymbol{q})\ddot{\boldsymbol{q}} + \dot{\boldsymbol{J}}(\boldsymbol{q})\dot{\boldsymbol{q}}$, the equation

$$\ddot{\boldsymbol{x}} - \dot{\boldsymbol{J}}(\boldsymbol{q})\dot{\boldsymbol{q}} + \boldsymbol{J}(\boldsymbol{q})\boldsymbol{M}(\boldsymbol{q})^{-1}(\boldsymbol{C}(\boldsymbol{q},\dot{\boldsymbol{q}})\dot{\boldsymbol{q}} + \boldsymbol{g}(\boldsymbol{q})) = \boldsymbol{J}(\boldsymbol{q})\boldsymbol{M}(\boldsymbol{q})^{-1}(\boldsymbol{\tau} + \boldsymbol{\tau}_{ext}) \ .$$

From this one can easily see that a torque $\boldsymbol{\tau}_0$ which is projected by $\boldsymbol{P}_1(\boldsymbol{q})$ in general may produce also an acceleration $\ddot{\boldsymbol{x}}$ of the end-effector. But this is not desired, because the Cartesian impedance controller and the nullspace impedance controller should not interact. A projection of the torques $\boldsymbol{\tau}_0$ which does not admit a direct feed-through from $\boldsymbol{\tau}_0$ to $\ddot{\boldsymbol{x}}$ is called *dynamically consistent*[5]. This clearly is achieved, whenever

$$\boldsymbol{J}(\boldsymbol{q})\boldsymbol{M}(\boldsymbol{q})^{-1}\boldsymbol{P}(\boldsymbol{q}) = \boldsymbol{0} \tag{4.21}$$

holds. From (4.21) one can see that by pre-multiplying $\boldsymbol{P}_1(\boldsymbol{q})$ by $\boldsymbol{M}(\boldsymbol{q})$ the resulting matrix gets dynamically consistent

$$\boldsymbol{P}_2(\boldsymbol{q}) = \boldsymbol{M}(\boldsymbol{q})\boldsymbol{P}_1(\boldsymbol{q}) \ . \tag{4.22}$$

But this choice has the disadvantage that $\boldsymbol{P}_2(\boldsymbol{q})$ contains also a scaling and therefore is not a projection matrix any more, because it does not fulfill the idempotency property $\boldsymbol{P}(\boldsymbol{q})\boldsymbol{P}(\boldsymbol{q}) = \boldsymbol{P}(\boldsymbol{q})$.

A more elegant solution is to choose the inverse inertia matrix $\boldsymbol{M}(\boldsymbol{q})^{-1}$ as the metric $\boldsymbol{G}(\boldsymbol{q})$ for the joint torques. This metric is called the *natural metric* by many authors (see, e.g., [BK00]) and results in the projection

$$\boldsymbol{P}_3(\boldsymbol{q}) = \boldsymbol{M}(\boldsymbol{q})\boldsymbol{Z}(\boldsymbol{q})^T \left(\boldsymbol{Z}(\boldsymbol{q})\boldsymbol{M}(\boldsymbol{q})\boldsymbol{Z}(\boldsymbol{q})^T\right)^{-1} \boldsymbol{Z}(\boldsymbol{q}) \ . \tag{4.23}$$

The Projection in the Operational Space Formulation

Another formulation of a dynamically consistent nullspace projection matrix can be found by considering (4.18) from Proposition 4.1. According to this a projection onto the orthogonal complement of $\boldsymbol{J}(\boldsymbol{q})^T$ (see Figure 4.3), with $\boldsymbol{M}(\boldsymbol{q})^{-1}$ as the relevant metric for the joint torques, is given by

$$\boldsymbol{P}_4(\boldsymbol{q}) = \boldsymbol{I} - \boldsymbol{J}(\boldsymbol{q})^T \left(\boldsymbol{J}(\boldsymbol{q})\boldsymbol{M}(\boldsymbol{q})^{-1}\boldsymbol{J}(\boldsymbol{q})^T\right)^{-1} \boldsymbol{J}(\boldsymbol{q})\boldsymbol{M}(\boldsymbol{q})^{-1} \ . \tag{4.24}$$

This form of the projection matrix was introduced in the operational space formalism [Kha87]. Here the computation of the nullspace base matrix $\boldsymbol{Z}(\boldsymbol{q})$ is

[5] It should be noted that the dynamic consistency of the nullspace projection does not correspond to a full decoupling between the Cartesian motion and the nullspace motion. There may still be couplings in form of centrifugal or Coriolis terms. This will become more clear in Section 4.4.

avoided. On the other hand it contains the inverse[6] of the inertia matrix and therefore is from a computational point of view more difficult to be implemented in realtime. The matrix $\boldsymbol{Z}(\boldsymbol{q})\boldsymbol{M}(\boldsymbol{q})\boldsymbol{Z}(\boldsymbol{q})^T$, which must be inverted in (4.23), in contrary is only an $r \times r$ matrix[7]. The resulting matrices, however, are equivalent, i.e. $\boldsymbol{P}_3(\boldsymbol{q}) = \boldsymbol{P}_4(\boldsymbol{q})$. This can be seen by verifying the property $(\boldsymbol{P}_3(\boldsymbol{q}) - \boldsymbol{P}_4(\boldsymbol{q}))\left[\boldsymbol{J}(\boldsymbol{q})^T \ \boldsymbol{M}(\boldsymbol{q})\boldsymbol{Z}(\boldsymbol{q})^T\right] = \boldsymbol{0}$, in which the matrix $\left[\boldsymbol{J}(\boldsymbol{q})^T \ \boldsymbol{M}(\boldsymbol{q})\boldsymbol{Z}(\boldsymbol{q})^T\right]$ is invertible[8].

Discussion of the Stability Properties

At the first glance, it seems that the superposition principle for impedances could be applied to the implementation of a nullspace impedance without any problems. The impedance is parameterized by joint space stiffness and damping matrices as well as by the desired nullspace equilibrium $\boldsymbol{q}_{d,0}$ and hence allows a quite intuitive design.

If one considers the stability analysis, this design method seems to be not critical as long as all superimposed impedance components are passive mappings from $\dot{\boldsymbol{q}}$ to the output torque. Then also their superposition (which is nothing else than a parallel interconnection) gives a passive impedance, and a stability analysis can be performed for the whole system as usual by examining the sum of the individual storage functions. The problem with this argumentation is that the mapping $\dot{\boldsymbol{q}} \rightarrow \boldsymbol{\tau}_n$, which represents the nullspace impedance, is not necessarily passive (due to the projection via $\boldsymbol{P}(\boldsymbol{q})$)[9]. This conceptual problem is the main reason for considering other nullspace treatments in the next sections which are more difficult to be implemented.

4.3 Task Space Augmentation Method

The *Task Space Augmentation Method* aims at directly extending the task coordinates $\boldsymbol{x} = \boldsymbol{f}(\boldsymbol{q})$. If a suitable set of coordinates $\boldsymbol{n} = \boldsymbol{n}(\boldsymbol{q})$ can be found such that the Jacobian $\boldsymbol{J}_e(\boldsymbol{q})$ corresponding to the new coordinates

$$\boldsymbol{y} = \begin{pmatrix} \boldsymbol{x} \\ \boldsymbol{n} \end{pmatrix} , \tag{4.25}$$

$$\dot{\boldsymbol{y}} = \boldsymbol{J}_e(\boldsymbol{q})\dot{\boldsymbol{q}} = \begin{pmatrix} \boldsymbol{J}(\boldsymbol{q}) \\ \frac{\partial \boldsymbol{n}(\boldsymbol{q})}{\partial \boldsymbol{q}} \end{pmatrix} \dot{\boldsymbol{q}} \tag{4.26}$$

---

[6] Notice also that this inversion may be numerically ill-posed. While the invertibility is ensured by the fact that the inertia matrix is positive definite, its eigenvalues usually can have different orders of magnitude and $\boldsymbol{M}(\boldsymbol{q})$ may contain very small eigenvalues in certain joint configurations.

[7] Notice that this matrix reduces in case of a one-dimensional nullspace to a scalar.

[8] The invertibility of this matrix can be shown similar to Lemma A.24.

[9] Instead, the mapping $\dot{\boldsymbol{q}} \rightarrow \boldsymbol{\tau}_0$ is passive.

is invertible, then the impedance control problem can be reduced to the non-redundant case. The choice of a proper set of additional coordinates $\boldsymbol{n}(\boldsymbol{q})$, however, is problematic, since $\boldsymbol{J}_e(\boldsymbol{q})$ will in general have more singular configurations than $\boldsymbol{J}(\boldsymbol{q})$.

In the *Extended Jacobian Method* proposed by Baillieul [Bai85], for instance, the nullspace coordinates

$$\boldsymbol{n}(\boldsymbol{q}) = \boldsymbol{Z}(\boldsymbol{q}) \frac{\partial V_n(\boldsymbol{q})}{\partial \boldsymbol{q}}^T \tag{4.27}$$

were introduced, wherein the target function $V_n(\boldsymbol{q})$ is a desired function to be optimized locally via nullspace movements. The potential function $V_n(\boldsymbol{q})$ should thereby be chosen such that it has a constrained local optimum

$$\boldsymbol{q}^* = \arg_{\boldsymbol{q}} \min_{\boldsymbol{f}(\boldsymbol{q})=\boldsymbol{x}} V_n(\boldsymbol{q}) \ .$$

Optimizing the target function $V_n(\boldsymbol{q})$ then means to track a desired nullspace motion of the form $\boldsymbol{n}_d(t) = \boldsymbol{0}$, which obviously simplifies to a regulation problem.

The main disadvantage of this method is that it does in general not lead to a dynamically consistent nullspace projection, since $\boldsymbol{J}(\boldsymbol{q})\boldsymbol{M}(\boldsymbol{q})^{-1}(\partial \boldsymbol{n}(\boldsymbol{q})/\partial \boldsymbol{q})^T \neq \boldsymbol{0}$. Hence, the nullspace dynamics is fully coupled with the original task dynamics which is usually not desired. Furthermore, this method also introduces new singularities whenever $(\partial \boldsymbol{n}(\boldsymbol{q})/\partial \boldsymbol{q})\boldsymbol{Z}(\boldsymbol{q})^T = \boldsymbol{0}$ [Bai85].

In the following, one possible way is described, how a decoupling of the Cartesian impedance from the nullspace impedance can be forced even when a dynamically inconsistent differential $\partial \boldsymbol{n}(\boldsymbol{q})/\partial \boldsymbol{q}$ is used. The dynamical equations (4.3) in task space coordinates can be computed via equation (3.10) and (3.11) by replacing the Jacobian $\boldsymbol{J}(\boldsymbol{q})$ with $\boldsymbol{J}_e(\boldsymbol{q})$. Let further the joint torques $\boldsymbol{\tau}$ be expressed in terms of a Cartesian component $\boldsymbol{F}_x \in \mathbb{R}^m$ and a nullspace component $\boldsymbol{F}_n \in \mathbb{R}^r$, which are related to $\boldsymbol{\tau}$ via

$$\boldsymbol{\tau} = \boldsymbol{J}_e(\boldsymbol{q})^T \begin{pmatrix} \boldsymbol{F}_x \\ \boldsymbol{F}_n \end{pmatrix} \ . \tag{4.28}$$

The external torques $\boldsymbol{\tau}_{ext}$ are decomposed in the same way in terms of $\boldsymbol{F}_{ext,x}$ and $\boldsymbol{F}_{ext,n}$. The dynamical equations, resulting from (3.10)-(3.11), can then be written as

$$\boldsymbol{\Lambda}_e(\boldsymbol{q}) \begin{pmatrix} \ddot{\boldsymbol{x}} \\ \ddot{\boldsymbol{n}} \end{pmatrix} + \boldsymbol{\mu}_e(\boldsymbol{q}, \dot{\boldsymbol{q}}) \begin{pmatrix} \dot{\boldsymbol{x}} \\ \dot{\boldsymbol{n}} \end{pmatrix} + \boldsymbol{J}_e(\boldsymbol{q})^{-T}\boldsymbol{g}(\boldsymbol{q}) = \begin{pmatrix} \boldsymbol{F}_x + \boldsymbol{F}_{ext,x} \\ \boldsymbol{F}_n + \boldsymbol{F}_{ext,n} \end{pmatrix} \tag{4.29}$$

where the inertia matrix $\boldsymbol{\Lambda}_e(\boldsymbol{q})$ and the Coriolis/centrifugal matrix $\boldsymbol{\mu}_e(\boldsymbol{q}, \dot{\boldsymbol{q}})$ can be partitioned accordingly as

$$\boldsymbol{\Lambda}_e(\boldsymbol{q}) = \begin{bmatrix} \boldsymbol{\Lambda}_{e,x}(\boldsymbol{q}) & \boldsymbol{\Lambda}_{e,c}(\boldsymbol{q}) \\ \boldsymbol{\Lambda}_{e,c}(\boldsymbol{q})^T & \boldsymbol{\Lambda}_{e,n}(\boldsymbol{q}) \end{bmatrix} \ , \tag{4.30}$$

$$\boldsymbol{\mu}_e(\boldsymbol{q},\dot{\boldsymbol{q}}) = \begin{bmatrix} \boldsymbol{\mu}_{e,xx}(\boldsymbol{q},\dot{\boldsymbol{q}}) & \boldsymbol{\mu}_{e,xn}(\boldsymbol{q},\dot{\boldsymbol{q}}) \\ \boldsymbol{\mu}_{e,nx}(\boldsymbol{q},\dot{\boldsymbol{q}}) & \boldsymbol{\mu}_{e,nn}(\boldsymbol{q},\dot{\boldsymbol{q}}) \end{bmatrix} . \tag{4.31}$$

By using the controller from Section 3.2 one would get a closed loop dynamics with a fully occupied inertia matrix. Instead, it is desired that the dynamics for $\boldsymbol{x}$ is decoupled from the dynamics for $\boldsymbol{n}$. This can at least approximately be achieved by a pre-compensation[10] of the form

$$\begin{aligned} \begin{pmatrix} \boldsymbol{F}_x \\ \boldsymbol{F}_n \end{pmatrix} &= \boldsymbol{J}_e(\boldsymbol{q})^{-T}\boldsymbol{g}(\boldsymbol{q}) + \bar{\boldsymbol{\mu}}(\boldsymbol{q},\dot{\boldsymbol{q}}) \begin{pmatrix} \dot{\boldsymbol{x}} \\ \dot{\boldsymbol{n}} \end{pmatrix} - \begin{pmatrix} \boldsymbol{F}_{ext,x} \\ \boldsymbol{F}_{ext,n} \end{pmatrix} \\ &+ \boldsymbol{\Lambda}_e(\boldsymbol{q}) \begin{bmatrix} \boldsymbol{\Lambda}_{e,x}(\boldsymbol{q}) & \boldsymbol{0} \\ \boldsymbol{0} & \boldsymbol{\Lambda}_{e,n}(\boldsymbol{q}) \end{bmatrix}^{-1} \begin{pmatrix} \bar{\boldsymbol{F}}_x + \boldsymbol{F}_{ext,x} \\ \bar{\boldsymbol{F}}_n + \boldsymbol{F}_{ext,n} \end{pmatrix} , \end{aligned} \tag{4.32}$$

where $\bar{\boldsymbol{\mu}}(\boldsymbol{q},\dot{\boldsymbol{q}})$ is given by

$$\bar{\boldsymbol{\mu}}(\boldsymbol{q},\dot{\boldsymbol{q}}) = \boldsymbol{\mu}_e(\boldsymbol{q},\dot{\boldsymbol{q}}) - \boldsymbol{\Lambda}_e(\boldsymbol{q}) \begin{bmatrix} \boldsymbol{\Lambda}_{e,x}(\boldsymbol{q}) & \boldsymbol{0} \\ \boldsymbol{0} & \boldsymbol{\Lambda}_{e,n}(\boldsymbol{q}) \end{bmatrix}^{-1} \begin{bmatrix} \boldsymbol{\mu}_{e,xx}(\boldsymbol{q},\dot{\boldsymbol{q}}) & \boldsymbol{0} \\ \boldsymbol{0} & \boldsymbol{\mu}_{e,nn}(\boldsymbol{q},\dot{\boldsymbol{q}}) \end{bmatrix} ,$$

and $\bar{\boldsymbol{F}}_x$ and $\bar{\boldsymbol{F}}_n$ are intermediate control inputs. Therefore, one obtains the following system equations

$$\boldsymbol{\Lambda}_{e,x}(\boldsymbol{q})\ddot{\boldsymbol{x}} + \boldsymbol{\mu}_{e,xx}(\boldsymbol{q},\dot{\boldsymbol{q}})\dot{\boldsymbol{x}} = \bar{\boldsymbol{F}}_x + \boldsymbol{F}_{ext,x} , \tag{4.33}$$

$$\boldsymbol{\Lambda}_{e,n}(\boldsymbol{q})\ddot{\boldsymbol{n}} + \boldsymbol{\mu}_{e,nn}(\boldsymbol{q},\dot{\boldsymbol{q}})\dot{\boldsymbol{n}} = \bar{\boldsymbol{F}}_n + \boldsymbol{F}_{ext,n} . \tag{4.34}$$

Notice that these equations are still coupled due to the dependence of the inertia matrix and the Coriolis/centrifugal matrix on the joint positions and velocities. One can now design a nullspace controller for the coordinates $\boldsymbol{n}$ via the input $\bar{\boldsymbol{F}}_n$ and a Cartesian impedance controller via the input $\bar{\boldsymbol{F}}_x$ in the same way as it was done in Section 3.2. With the positive definite stiffness and damping matrices $\boldsymbol{K}_d$, $\boldsymbol{D}_d$, $\boldsymbol{K}_n$, and $\boldsymbol{D}_n$ for the Cartesian impedance and the nullspace impedance respectively, the complete control law is then given by (4.28), (4.32), and

$$\begin{pmatrix} \bar{\boldsymbol{F}}_x \\ \bar{\boldsymbol{F}}_n \end{pmatrix} = \begin{pmatrix} \boldsymbol{\Lambda}_{e,x}(\boldsymbol{q})\ddot{\boldsymbol{x}}_d + \boldsymbol{\mu}_{e,xx}(\boldsymbol{q},\dot{\boldsymbol{q}})\dot{\boldsymbol{x}}_d - \boldsymbol{K}_d\tilde{\boldsymbol{x}} - \boldsymbol{D}_d\dot{\tilde{\boldsymbol{x}}} \\ -\boldsymbol{K}_n\boldsymbol{n} - \boldsymbol{D}_n\dot{\boldsymbol{n}} \end{pmatrix} . \tag{4.35}$$

For the stability properties of the resulting closed loop system one can directly refer to Proposition 3.4 and Proposition 3.5, since the closed loop system has the same form as the desired dynamics from Section 3.2. Notice also that this controller contains a feedback of the external torques $\boldsymbol{\tau}_{ext}$ in (4.32), since (4.32) *reshapes* the effective inertia matrix. From a practical point of view, however,

[10] Notice the relation of this controller to the controller from Section 3.1.3.

usually only external forces and torques which act on the end-effector can be measured by a force-torque-sensor mounted at the tip of the robot. But for the implementation of the control law (4.32) also the nullspace forces $\boldsymbol{F}_{ext,n}$ are needed. This problem, which has already been discussed in Section 3.2 in more detail, and the singularities[11] of $\boldsymbol{J}_e(\boldsymbol{q})$, are the main limitations of this approach. In contrast to this in the *Joint Space Decomposition Method* a block-diagonal (hence decoupled) inertia matrix can be achieved by the use of a dynamically consistent nullspace projection matrix without feedback of the external forces.

4.4 Joint Space Decomposition Method

The *Joint Space Decomposition Method* does not augment the task coordinates $\boldsymbol{x}$, but the task velocities $\dot{\boldsymbol{x}}$. This is done by introducing additional nullspace velocities $\boldsymbol{v}_n = \boldsymbol{N}(\boldsymbol{q})\dot{\boldsymbol{q}}$ via an appropriate matrix $\boldsymbol{N}(\boldsymbol{q})$. Together with $\dot{\boldsymbol{x}}$ one has

$$\begin{pmatrix} \dot{\boldsymbol{x}} \\ \boldsymbol{v}_n \end{pmatrix} = \boldsymbol{J}_N(\boldsymbol{q})\dot{\boldsymbol{q}} = \begin{pmatrix} \boldsymbol{J}(\boldsymbol{q}) \\ \boldsymbol{N}(\boldsymbol{q}) \end{pmatrix} \dot{\boldsymbol{q}} \ . \tag{4.36}$$

The nullspace matrix $\boldsymbol{N}(\boldsymbol{q})$ has to be chosen such that the extended Jacobian $\boldsymbol{J}_N(\boldsymbol{q})$ is invertible.

In principle, one can also integrate the nullspace velocity $\boldsymbol{v}_n(t)$ in order to get new nullspace coordinates $\boldsymbol{n}(t) = \int_0^t \boldsymbol{v}_n(\boldsymbol{q}(\tau))\mathrm{d}\tau$. The construction of nullspace coordinates then allows one to use all techniques developed for the non-redundant case also in the redundant case, analogously to the *Task Space Augmentation Method*. This, however, has two important implications: Firstly, the quantity $\boldsymbol{n}(t)$ does not have any useful geometric meaning. Therefore, it is not possible to formulate a meaningful desired trajectory $\boldsymbol{n}_d(t)$. Instead, the desired trajectory has to be formulated as a desired velocity[12] $\boldsymbol{v}_{n,d}(t)$ [HHS89]. Secondly, due to the integration of the nullspace velocity, the resulting controllers do not describe a static state feedback, but rather a dynamic state feedback.

It is important to notice at this point that in general the nullspace base matrix $\boldsymbol{Z}(\boldsymbol{q})$ cannot be represented as the differential of a function $\boldsymbol{n}(\boldsymbol{q})$ because this would necessarily require $\boldsymbol{z}_k(\boldsymbol{q})$ to fulfill the integrability conditions $\frac{\partial \boldsymbol{z}_{k,i}(\boldsymbol{q})}{\partial q_j} = \frac{\partial \boldsymbol{z}_{k,j}(\boldsymbol{q})}{\partial q_i}$. Of course, this is generally not satisfied.

In order to get a dynamically consistent nullspace matrix, one can choose[13]

$$\boldsymbol{N}(\boldsymbol{q}) = \left(\boldsymbol{Z}(\boldsymbol{q})\boldsymbol{M}(\boldsymbol{q})\boldsymbol{Z}(\boldsymbol{q})^T\right)^{-1} \boldsymbol{Z}(\boldsymbol{q})\boldsymbol{M}(\boldsymbol{q}) \ , \tag{4.37}$$

[11] Notice again that there will be additional singularities from the particular choice of $\boldsymbol{n}(\boldsymbol{q})$ besides the singularities of $\boldsymbol{J}(\boldsymbol{q})$.

[12] Some authors, for instance, use a desired nullspace velocity of the form $\boldsymbol{v}_{n,d} = -\boldsymbol{Z}(\boldsymbol{q})\frac{\partial V_n(\boldsymbol{q})}{\partial \boldsymbol{q}}^T$ with a given target function $V_n(\boldsymbol{q})$ [HHS89].

[13] In Park's *inertially weighted kinematically decoupled joint space decomposition* (*KD-JSD*) [PCY99] the choice $\boldsymbol{N}(\boldsymbol{q}) = \boldsymbol{Z}(\boldsymbol{q})\boldsymbol{M}(\boldsymbol{q})$ was made instead. The additional factor $(\boldsymbol{Z}(\boldsymbol{q})\boldsymbol{M}(\boldsymbol{q})\boldsymbol{Z}(\boldsymbol{q})^T)^{-1}$ acts as a normalization. Notice also the similarity to $\boldsymbol{P}_3(\boldsymbol{q})$ in (4.23).

where $\boldsymbol{Z}(\boldsymbol{q})$ is a nullspace base matrix of $\boldsymbol{J}(\boldsymbol{q})$. Notice that for the computation of $\boldsymbol{N}(\boldsymbol{q})$ the matrix $\boldsymbol{Z}(\boldsymbol{q})\boldsymbol{M}(\boldsymbol{q})\boldsymbol{Z}(\boldsymbol{q})^T$ must be inverted which is only an $r \times r$ matrix. The invertibility of this extended Jacobian $\boldsymbol{J}_N(\boldsymbol{q})$ is shown in Section A.5. Therein it is also shown that the inverse of $\boldsymbol{J}_N(\boldsymbol{q})$ is given by

$$\boldsymbol{J}_N(\boldsymbol{q})^{-1} = \left[\boldsymbol{J}^{M+}(\boldsymbol{q}) \; \boldsymbol{Z}(\boldsymbol{q})^T\right] , \tag{4.38}$$

where the *weighted pseudo-inverse* $\boldsymbol{J}^{M+}(\boldsymbol{q})$ is defined as

$$\boldsymbol{J}^{M+}(\boldsymbol{q}) = \boldsymbol{M}(\boldsymbol{q})^{-1}\boldsymbol{J}(\boldsymbol{q})^T(\boldsymbol{J}(\boldsymbol{q})\boldsymbol{M}(\boldsymbol{q})^{-1}\boldsymbol{J}(\boldsymbol{q})^T)^{-1} . \tag{4.39}$$

The joint velocity $\dot{\boldsymbol{q}}$ can thus be computed from the Cartesian velocity $\dot{\boldsymbol{x}}$ and the nullspace velocity $\boldsymbol{v}_n$ via

$$\dot{\boldsymbol{q}} = \boldsymbol{J}^{M+}(\boldsymbol{q})\dot{\boldsymbol{x}} + \boldsymbol{Z}(\boldsymbol{q})^T\boldsymbol{v}_n . \tag{4.40}$$

The dynamical equations (4.3) can be written in the extended velocity coordinates via (3.10) and (3.11). The joint torques $\boldsymbol{\tau}$ are expressed in terms of a Cartesian force $\boldsymbol{F}_x$ and a nullspace force $\boldsymbol{F}_n$ which are related to $\boldsymbol{\tau}$ via

$$\boldsymbol{\tau} = \left[\boldsymbol{J}(\boldsymbol{q})^T \; \boldsymbol{N}(\boldsymbol{q})^T\right] \begin{pmatrix} \boldsymbol{F}_x \\ \boldsymbol{F}_n \end{pmatrix} . \tag{4.41}$$

The generalized forces $\boldsymbol{F}_x$ and $\boldsymbol{F}_n$ are used as new inputs. In the same way the external torques $\boldsymbol{\tau}_{ext}$ are decomposed in terms of generalized external forces $\boldsymbol{F}_{ext,x}$ and $\boldsymbol{F}_{ext,n}$. The dynamical equations then have the form

$$\boldsymbol{\Lambda}_N(\boldsymbol{q}) \begin{pmatrix} \ddot{\boldsymbol{x}} \\ \dot{\boldsymbol{v}}_n \end{pmatrix} + \boldsymbol{\mu}_N(\boldsymbol{q},\dot{\boldsymbol{q}}) \begin{pmatrix} \dot{\boldsymbol{x}} \\ \boldsymbol{v}_n \end{pmatrix} + \boldsymbol{J}_N(\boldsymbol{q})^{-T}\boldsymbol{g}(\boldsymbol{q}) = \begin{pmatrix} \boldsymbol{F}_x + \boldsymbol{F}_{ext,x} \\ \boldsymbol{F}_n + \boldsymbol{F}_{ext,n} \end{pmatrix} \tag{4.42}$$

Since $\boldsymbol{N}(\boldsymbol{q})$ is dynamically consistent[14], the inertia matrix $\boldsymbol{\Lambda}_N(\boldsymbol{q})$ is block-diagonal

$$\boldsymbol{\Lambda}_N(\boldsymbol{q}) = \boldsymbol{J}_N(\boldsymbol{q})^{-T}\boldsymbol{M}(\boldsymbol{q})\boldsymbol{J}_N(\boldsymbol{q})^{-1} \tag{4.43}$$

$$= \begin{bmatrix} (\boldsymbol{J}(\boldsymbol{q})\boldsymbol{M}(\boldsymbol{q})^{-1}\boldsymbol{J}(\boldsymbol{q})^T)^{-1} & \boldsymbol{0} \\ \boldsymbol{0} & \boldsymbol{Z}(\boldsymbol{q})\boldsymbol{M}(\boldsymbol{q})\boldsymbol{Z}(\boldsymbol{q})^T \end{bmatrix} \tag{4.44}$$

$$= \begin{bmatrix} \boldsymbol{\Lambda}_x(\boldsymbol{q}) & \boldsymbol{0} \\ \boldsymbol{0} & \boldsymbol{\Lambda}_n(\boldsymbol{q}) \end{bmatrix} , \tag{4.45}$$

[14] I.e. it fulfills the condition $\boldsymbol{J}(\boldsymbol{q})\boldsymbol{M}(\boldsymbol{q})^{-1}\boldsymbol{N}(\boldsymbol{q})^T = \boldsymbol{0}$.

and $\boldsymbol{\mu}_N(\boldsymbol{q},\dot{\boldsymbol{q}})$ has the form[15]

$$\boldsymbol{\mu}_N = \begin{bmatrix} \boldsymbol{\mu}_x & \boldsymbol{\mu}_{xn} \\ \boldsymbol{\mu}_{nx} & \boldsymbol{\mu}_n \end{bmatrix} \tag{4.46}$$

$$= \begin{bmatrix} \boldsymbol{\Lambda}_x \left(\boldsymbol{J}\boldsymbol{M}^{-1}\boldsymbol{C} - \dot{\boldsymbol{J}}\right)\boldsymbol{J}^{M+} & \boldsymbol{\Lambda}_x \left(\boldsymbol{J}\boldsymbol{M}^{-1}\boldsymbol{C} - \dot{\boldsymbol{J}}\right)\boldsymbol{Z}^T \\ -\left(\dot{\boldsymbol{Z}}\boldsymbol{M} + \boldsymbol{Z}\boldsymbol{C}^T\right)\boldsymbol{J}^{M+} & \boldsymbol{\Lambda}_n\boldsymbol{Z}\left(\boldsymbol{C}\boldsymbol{Z}^T + \boldsymbol{M}\dot{\boldsymbol{Z}}^T\right) \end{bmatrix} . \tag{4.47}$$

Notice that the equality $\boldsymbol{\mu}_{xn}(\boldsymbol{q},\dot{\boldsymbol{q}}) = -\boldsymbol{\mu}_{nx}(\boldsymbol{q},\dot{\boldsymbol{q}})^T$ holds[16], which corresponds to Lemma 3.2 since $\boldsymbol{\Lambda}_N(\boldsymbol{q})$ is block-diagonal.

4.4.1 Controller Design

In contrast to other redundancy resolution methods in the literature, the following control algorithm focuses on a Cartesian impedance controller without shaping of the inertia matrix (for the same reasons as discussed in Section 3.2). Furthermore, the goal of the nullspace impedance controller is to achieve a dynamical relation between the nullspace motion and the external (generalized) forces which can be specified in terms of stiffness and damping coefficients.

The following analysis will be based on the dynamical equations (4.42) with the state variables[17] $\boldsymbol{q}$, $\dot{\boldsymbol{x}}$, and $\boldsymbol{v}_n$.

In the dynamical equations (4.42) the couplings between Cartesian coordinates and nullspace coordinates via off-diagonal terms in the inertia matrix have already been eliminated by the use of a dynamically consistent nullspace matrix $\boldsymbol{N}(\boldsymbol{q})$. Additionally, there are still coupling terms in the Coriolis/centrifugal matrix $\boldsymbol{\mu}_N(\boldsymbol{q},\dot{\boldsymbol{x}},\boldsymbol{v}_n)$ present. These terms can, together with the gravity torques, be eliminated without force feedback by a feedback law of the form

$$\begin{pmatrix} \boldsymbol{F}_x \\ \boldsymbol{F}_n \end{pmatrix} = \boldsymbol{J}_N(\boldsymbol{q})^{-T}\boldsymbol{g}(\boldsymbol{q}) + \begin{pmatrix} \bar{\boldsymbol{F}}_x \\ \bar{\boldsymbol{F}}_n \end{pmatrix} + \begin{bmatrix} \boldsymbol{0} & \boldsymbol{\mu}_{xn}(\boldsymbol{q},\dot{\boldsymbol{x}},\boldsymbol{v}_n) \\ \boldsymbol{\mu}_{nx}(\boldsymbol{q},\dot{\boldsymbol{x}},\boldsymbol{v}_n) & \boldsymbol{0} \end{bmatrix} \begin{pmatrix} \dot{\boldsymbol{x}} \\ \boldsymbol{v}_n \end{pmatrix} , \tag{4.48}$$

with the new input variables $\bar{\boldsymbol{F}}_x$ and $\bar{\boldsymbol{F}}_n$. Notice that, from a passivity point of view, the cancelation of the undesired centrifugal and Coriolis terms in (4.48) can be expected to be not very critical even for uncertainties in the model parameters because of the skew-symmetry property $\boldsymbol{\mu}_{xn}(\boldsymbol{q},\dot{\boldsymbol{x}},\boldsymbol{v}_n) = -\boldsymbol{\mu}_{nx}(\boldsymbol{q},\dot{\boldsymbol{x}},\boldsymbol{v}_n)^T$. The closed loop system so far is given by

[15] In order to simplify the notation, the arguments $\boldsymbol{q}$ and $\dot{\boldsymbol{q}}$ are omitted herein.
[16] The direct derivation of this property from (4.47), however, is quite cumbersome.
[17] The terms $\boldsymbol{\mu}_{ij}(\boldsymbol{q},\dot{\boldsymbol{q}})$ will therefore be written as $\boldsymbol{\mu}_{ij}(\boldsymbol{q},\dot{\boldsymbol{x}},\boldsymbol{v}_n)$ in the following.

$$\boldsymbol{\Lambda}_x(\boldsymbol{q})\ddot{\boldsymbol{x}} + \boldsymbol{\mu}_x(\boldsymbol{q}, \dot{\boldsymbol{x}}, \boldsymbol{v}_n)\dot{\boldsymbol{x}} = \bar{\boldsymbol{F}}_x + \boldsymbol{F}_{ext,x} \ , \tag{4.49}$$

$$\boldsymbol{\Lambda}_n(\boldsymbol{q})\dot{\boldsymbol{v}}_n + \boldsymbol{\mu}_n(\boldsymbol{q}, \dot{\boldsymbol{x}}, \boldsymbol{v}_n)\boldsymbol{v}_n = \bar{\boldsymbol{F}}_n + \boldsymbol{F}_{ext,n} \ , \tag{4.50}$$

$$\dot{\boldsymbol{q}} = \boldsymbol{J}^{+M}(\boldsymbol{q})\dot{\boldsymbol{x}} + \boldsymbol{Z}(\boldsymbol{q})^T\boldsymbol{v}_n \ . \tag{4.51}$$

In the following only the regulation case will be treated in which the virtual Cartesian equilibrium position $\boldsymbol{x}_d$ is fixed. For the Cartesian part of the system dynamics a controller analogously to Section 3.2 with symmetric and positive definite stiffness and damping matrices $\boldsymbol{K}_d$ and $\boldsymbol{D}_d$ is used.

$$\bar{\boldsymbol{F}}_x = -\boldsymbol{K}_d(\boldsymbol{f}(\boldsymbol{q}) - \boldsymbol{x}_d) - \boldsymbol{D}_d\dot{\boldsymbol{x}} \tag{4.52}$$

Furthermore, the input $\bar{\boldsymbol{F}}_n$ is used for controlling the nullspace motion. Two different feedback laws for $\bar{\boldsymbol{F}}_n$ are discussed in the following. The first one corresponds to a nullspace controller common[18] in the literature and will therefore be reviewed only briefly. The idea for this nullspace controller is to implement a velocity controller for the desired velocity $\boldsymbol{v}_{n,d} = -\boldsymbol{Z}(\boldsymbol{q})(\frac{\partial V_n(\boldsymbol{q})}{\partial \boldsymbol{q}})^T$, where $V_n(\boldsymbol{q})$ is a given cost function to be optimized locally via nullspace movements. The considered controller is given by

$$\bar{\boldsymbol{F}}_n = \boldsymbol{\Lambda}_n(\boldsymbol{q})\dot{\boldsymbol{v}}_{n,d} + \boldsymbol{\mu}_n(\boldsymbol{q}, \dot{\boldsymbol{x}}, \boldsymbol{v}_n)\boldsymbol{v}_{n,d} - \boldsymbol{D}_n(\boldsymbol{v}_n - \boldsymbol{v}_{n,d}) \ , \tag{4.53}$$

where $\boldsymbol{D}_n \in \mathbb{R}^{r\times r}$ is a positive definite controller gain matrix. The complete closed loop system (with $\boldsymbol{e}_n = \boldsymbol{v}_n - \boldsymbol{v}_{n,d}$) is given by (4.51) and

$$\boldsymbol{\Lambda}_x(\boldsymbol{q})\ddot{\boldsymbol{x}} + (\boldsymbol{\mu}_x(\boldsymbol{q}, \dot{\boldsymbol{x}}, \boldsymbol{v}_n) + \boldsymbol{D}_d)\dot{\boldsymbol{x}} + \boldsymbol{K}_d(\boldsymbol{f}(\boldsymbol{q}) - \boldsymbol{x}_d) = \boldsymbol{F}_{ext,x} \ , \tag{4.54}$$

$$\boldsymbol{\Lambda}_n(\boldsymbol{q})\dot{\boldsymbol{e}}_n + (\boldsymbol{\mu}_n(\boldsymbol{q}, \dot{\boldsymbol{x}}, \boldsymbol{v}_n) + \boldsymbol{D}_n)\boldsymbol{e}_n = \boldsymbol{F}_{ext,n} \ . \tag{4.55}$$

For this system one can show that in case of free motion (i.e. $\boldsymbol{F}_{ext,x} = \boldsymbol{0}$ and $\boldsymbol{F}_{ext,n} = \boldsymbol{0}$) the variables $\dot{\boldsymbol{x}}$, $\boldsymbol{f}(\boldsymbol{q}) - \boldsymbol{x}_d$, and $\boldsymbol{e}_n$ converge to zero. Basically, this was shown in the works [HHS89, OCY98, NSV99] for similar cases. Furthermore, one can also show that, when the Cartesian error is zero (i.e. $\boldsymbol{f}(\boldsymbol{q}) = \boldsymbol{x}_d$), the joint angles $\boldsymbol{q}$ will locally converge to a constraint minimum of $V_n(\boldsymbol{q})$ [HHS89].

One problem of the controller in (4.53) is that it does not allow to specify the nullspace behavior explicitly in terms of stiffness and damping coefficients, but in terms of optimizing a cost function. Therefore, this approach is not really appropriate to implement a nullspace impedance. Furthermore, only convergence (and not stability) can be shown for $\boldsymbol{e}_n$ unless a complete mass decoupling as in [OCY98] is used[19]. Therefore, a different choice for $\bar{\boldsymbol{F}}_n$ will be discussed in the following, which is particularly designed for the problem of implementing a nullspace impedance without feedback of the external forces. Consider therefore the feedback law[20]

$$\bar{\boldsymbol{F}}_n = -\boldsymbol{D}_n\boldsymbol{v}_n + \boldsymbol{u}_n(\boldsymbol{q}) \ , \tag{4.56}$$

---

[18] In the literature this control law usually is used in combination with a complete decoupling of the inertia matrix.

[19] Note that this requires force feedback in case of an impedance controller.

[20] Notice that by replacing $\boldsymbol{u}(\boldsymbol{q})$ with $\boldsymbol{u} = \boldsymbol{\Lambda}_n(\boldsymbol{q})\dot{\boldsymbol{v}}_{n,d} + (\boldsymbol{\mu}_n(\boldsymbol{q}, \dot{\boldsymbol{x}}, \boldsymbol{v}_n) + \boldsymbol{D}_n)\boldsymbol{v}_{n,d}$ one would obtain the same feedback law as in (4.53).

where $\boldsymbol{D}_n$ is a positive definite matrix corresponding to the desired nullspace damping. The term $\boldsymbol{u}_n(\boldsymbol{q})$ will be designed later according to a particular nullspace stiffness term which allows to prove, at least locally, the asymptotic stability of the closed loop system for the case of free motion. The closed loop system for the controller (4.56) is given by

$$\boldsymbol{\Lambda}_x(\boldsymbol{q})\ddot{\boldsymbol{x}} + (\boldsymbol{\mu}_x(\boldsymbol{q},\dot{\boldsymbol{x}},\boldsymbol{v}_n) + \boldsymbol{D}_d)\dot{\boldsymbol{x}} + \boldsymbol{K}_d(\boldsymbol{f}(\boldsymbol{q}) - \boldsymbol{x}_d) = \boldsymbol{F}_{ext,x} \ , \tag{4.57}$$

$$\boldsymbol{\Lambda}_n(\boldsymbol{q})\dot{\boldsymbol{v}}_n + (\boldsymbol{\mu}_n(\boldsymbol{q},\dot{\boldsymbol{x}},\boldsymbol{v}_n) + \boldsymbol{D}_n)\boldsymbol{v}_n - \boldsymbol{u}_n(\boldsymbol{q}) = \boldsymbol{F}_{ext,n} \ , \tag{4.58}$$

together with

$$\dot{\boldsymbol{q}} = \boldsymbol{J}^{+M}(\boldsymbol{q})\dot{\boldsymbol{x}} + \boldsymbol{Z}(\boldsymbol{q})^T\boldsymbol{v}_n \ . \tag{4.59}$$

In the next section the term $\boldsymbol{u}_n(\boldsymbol{q})$ will be designed such that a (local) stability proof for the closed loop system can be given.

4.4.2 Stability Analysis

The stability analysis of (4.57)-(4.59) will be based on the following two theorems concerning the stability analysis with semi-definite Lyapunov functions. The notation of *conditional stability* being used is defined in Appendix A.2.

Theorem 4.2 *Consider a system of the form* $\dot{\boldsymbol{z}} = \boldsymbol{f}(\boldsymbol{z})$, $\boldsymbol{z} \in \mathbb{R}^n$, *with equilibrium point* $\boldsymbol{z}^*$. *Let* $V(\boldsymbol{z})$ *be a* C^1 *positive semi-definite function which has a negative semi-definite time derivative along the solutions of the system, i.e.*

$$\dot{V}(\boldsymbol{z}) = \frac{\partial V(\boldsymbol{z})}{\partial \boldsymbol{z}}\boldsymbol{f}(\boldsymbol{z}) \leq 0 \ . \tag{4.60}$$

Let $\mathcal{A}$ *be the the largest positively invariant set contained in* $\{\boldsymbol{z} \in \mathbb{R}^n | V(\boldsymbol{z}) = 0\}$. *If* $\boldsymbol{z}^*$ *is asymptotically stable conditionally to* $\mathcal{A}$, *then it is a stable equilibrium of* $\dot{\boldsymbol{z}} = \boldsymbol{f}(\boldsymbol{z})$.

Proof. This theorem is due to Iggidr et al. [IKO96]. A proof can also be found in [SJK97].

In [vdS00] a further extension of this theorem for passive systems is given. The particular case of strict output passivity[21], which will be of interest for the stability analysis of (4.57)-(4.59), is stated in the following theorem. The proof of this theorem can be found in [vdS00].

Theorem 4.3 *Let the system*

$$\dot{\boldsymbol{z}} = \boldsymbol{g}_1(\boldsymbol{z}) + \boldsymbol{g}_2(\boldsymbol{z})\boldsymbol{u} \tag{4.61}$$

$$\boldsymbol{y} = \boldsymbol{h}(\boldsymbol{z}) \tag{4.62}$$

with state $\boldsymbol{z} \in \mathbb{R}^n$, *input* $\boldsymbol{u} \in \mathbb{R}^m$, *and output* $\boldsymbol{y} \in \mathbb{R}^m$ *be strict output passive for the output* $\boldsymbol{y} = \boldsymbol{h}(\boldsymbol{z})$. *Let further be* $\mathcal{A}$ *the largest positively invariant set contained in* $\{\boldsymbol{z} \in \mathbb{R}^n | \boldsymbol{h}(\boldsymbol{z}) = \boldsymbol{0}\}$. *If the equilibrium* $\boldsymbol{z}^*$ *is asymptotically stable conditionally to* $\mathcal{A}$, *then it is asymptotically stable for* $\boldsymbol{u} = \boldsymbol{0}$.

[21] The definition of strict output passivity is given in Appendix A.3.

Theorem 4.3 is used in the following stability analysis of (4.57)-(4.59). The nullspace impedance term $\boldsymbol{u}(\boldsymbol{q})$ will be designed such that for a given joint configuration $\boldsymbol{q}_d$, which must satisfy the condition $\boldsymbol{f}(\boldsymbol{q}_d) = \boldsymbol{x}_d$, the point $(\boldsymbol{q}, \dot{\boldsymbol{x}}, \boldsymbol{v}_n) = (\boldsymbol{q}_d, \dot{\boldsymbol{x}}_d, \dot{\boldsymbol{v}}_{n,d})$, with $\dot{\boldsymbol{x}}_d = \boldsymbol{0}$ and $\dot{\boldsymbol{v}}_{n,d} = \boldsymbol{0}$, is an asymptotically stable equilibrium point of (4.57)-(4.59) when no external forces $\boldsymbol{F}_{ext,x}$ and $\boldsymbol{F}_{ext,n}$ act on the system. In the following derivation, the external force $\boldsymbol{F}_{ext,x}$ will correspond to the input $\boldsymbol{u}$ from Theorem 4.3.

Consider the following *positive semi-definite Lyapunov function*

$$S(\boldsymbol{q}, \dot{\boldsymbol{x}}, \boldsymbol{v}_n) = \frac{1}{2}\dot{\boldsymbol{x}}^T \boldsymbol{\Lambda}_x(\boldsymbol{q})\dot{\boldsymbol{x}} + \frac{1}{2}(\boldsymbol{f}(\boldsymbol{q}) - \boldsymbol{x}_d)^T \boldsymbol{K}_d(\boldsymbol{f}(\boldsymbol{q}) - \boldsymbol{x}_d) \ . \tag{4.63}$$

The time derivative of $S(\boldsymbol{q}, \dot{\boldsymbol{x}}, \boldsymbol{v}_n)$ along the solutions of (4.57)-(4.59) is given by

$$\dot{S}(\boldsymbol{q}, \dot{\boldsymbol{x}}, \boldsymbol{v}_n) = -\dot{\boldsymbol{x}}^T \boldsymbol{D}_d \dot{\boldsymbol{x}} - \dot{\boldsymbol{x}}^T \boldsymbol{F}_{ext,x} \ . \tag{4.64}$$

From this one can follow that the condition $\dot{S}(\boldsymbol{q}, \dot{\boldsymbol{x}}, \boldsymbol{v}_n) \leq 0$ from Theorem 4.2 is fulfilled for the case $\boldsymbol{F}_{ext,x} = \boldsymbol{0}$. Furthermore, by considering $S(\boldsymbol{q}, \dot{\boldsymbol{x}}, \boldsymbol{v}_n)$ as a storage function, one can follow that the system is strict output passive with[22] $\beta = \lambda_{min}(\boldsymbol{D}_d)$ for the output $\dot{\boldsymbol{x}}$ with the corresponding input $\boldsymbol{F}_{ext,x}$. Following Theorem 4.3 one must then show that the system is asymptotically stable conditionally to the largest positively invariant set in $\{(\boldsymbol{q}, \dot{\boldsymbol{x}}, \boldsymbol{v}_n) \in \mathbb{R}^{2n} | \dot{\boldsymbol{x}} = \boldsymbol{0}\}$. Consider further the case of free motion, i.e. $\boldsymbol{F}_{ext,x} = \boldsymbol{0}$ and $\boldsymbol{F}_{ext,n} = \boldsymbol{0}$. The largest positively invariant set for which $\dot{\boldsymbol{x}} = \boldsymbol{0}$ holds can be derived from (4.57) and is given by

$$\mathcal{A} = \{(\boldsymbol{q}, \dot{\boldsymbol{x}}, \boldsymbol{v}_n) | \boldsymbol{f}(\boldsymbol{q}) = \boldsymbol{x}_d, \dot{\boldsymbol{x}} = \boldsymbol{0}\} \ . \tag{4.65}$$

Notice also that, since $\mathcal{A}$ is a positively invariant set, all trajectories which start in $\mathcal{A}$ will remain in $\mathcal{A}$. Consider then the Lyapunov function candidate

$$V_{\mathcal{A}}(\boldsymbol{q}, \boldsymbol{v}_n) = \frac{1}{2}\boldsymbol{v}_n^T \boldsymbol{\Lambda}_n(\boldsymbol{q})\boldsymbol{v}_n + \frac{1}{2}(\boldsymbol{q} - \boldsymbol{q}_d)^T \boldsymbol{K}_n(\boldsymbol{q} - \boldsymbol{q}_d) \tag{4.66}$$

with a symmetric and positive definite joint stiffness matrix $\boldsymbol{K}_n$. In the set $\mathcal{A}$ the function $V_{\mathcal{A}}(\boldsymbol{q}, \boldsymbol{v}_n)$ clearly is a positive definite function[23] around $(\boldsymbol{q}_d, \boldsymbol{0}, \boldsymbol{0})$. The time derivative of $V_{\mathcal{A}}(\boldsymbol{q}, \boldsymbol{v}_n)$ in $\mathcal{A}$ along the solutions of (4.57)-(4.59) is given by[24]

$$\begin{aligned} \dot{V}_{\mathcal{A}}(\boldsymbol{q}, \boldsymbol{v}_n) &= -\boldsymbol{v}_n^T \boldsymbol{D}_n \boldsymbol{v}_n + \boldsymbol{v}_n^T \boldsymbol{u}_n(\boldsymbol{q}) + (\boldsymbol{q} - \boldsymbol{q}_d)^T \boldsymbol{K}_n \dot{\boldsymbol{q}} \\ &= -\boldsymbol{v}_n^T \boldsymbol{D}_n \boldsymbol{v}_n + \boldsymbol{v}_n^T \boldsymbol{u}_n(\boldsymbol{q}) + (\boldsymbol{q} - \boldsymbol{q}_d)^T \boldsymbol{K}_n \boldsymbol{Z}(\boldsymbol{q})^T \boldsymbol{v}_n \\ &= -\boldsymbol{v}_n^T \boldsymbol{D}_n \boldsymbol{v}_n + \boldsymbol{v}_n^T \left(\boldsymbol{u}_n(\boldsymbol{q}) + \boldsymbol{Z}(\boldsymbol{q})\boldsymbol{K}_n(\boldsymbol{q} - \boldsymbol{q}_d)\right) \end{aligned} \tag{4.67}$$

With the choice

$$\boldsymbol{u}_n(\boldsymbol{q}) = -\boldsymbol{Z}(\boldsymbol{q})\boldsymbol{K}_n(\boldsymbol{q} - \boldsymbol{q}_d) \tag{4.68}$$

[22] See Appendix A.3 for the definition of strict output passivity.

[23] Notice that it is not a positive definite function in the complete state space $\mathbb{R}^{2n}$, since it is independent of $\dot{\boldsymbol{x}}$.

[24] Remember that $\boldsymbol{q}_d$ must satisfy the condition $\boldsymbol{f}(\boldsymbol{q}_d) = \boldsymbol{x}_d$.

stability conditionally to $\mathcal{A}$ follows from $\dot{V}_A(\boldsymbol{q}, \boldsymbol{v}_n) = -\boldsymbol{v}_n^T \boldsymbol{D}_n \boldsymbol{v}_n \leq 0$. In order to show also asymptotic stability conditionally to $\mathcal{A}$ one can refer to LaSalle's invariance principle. It can be shown that all trajectories in $\mathcal{A}$ will converge to the largest positively invariant set contained in

$$\{(\boldsymbol{q}, \dot{\boldsymbol{x}}, \boldsymbol{v}_n) | \boldsymbol{f}(\boldsymbol{q}) = \boldsymbol{x}_d, \dot{\boldsymbol{x}} = \boldsymbol{0}, \boldsymbol{v}_n = \boldsymbol{0}\} \ . \tag{4.69}$$

Therefore, the joint variables $\boldsymbol{q}$ will converge to a set where the conditions $\boldsymbol{f}(\boldsymbol{q}) = \boldsymbol{x}_d$ and $\boldsymbol{u}_n(\boldsymbol{q}) = -\boldsymbol{Z}(\boldsymbol{q})\boldsymbol{K}_n(\boldsymbol{q} - \boldsymbol{q}_d) = \boldsymbol{0}$ are fulfilled. This holds not only for $\boldsymbol{q} = \boldsymbol{q}_d$, but for all positions $\boldsymbol{q}$ which satisfy $\boldsymbol{Z}(\boldsymbol{q})\boldsymbol{K}_n(\boldsymbol{q} - \boldsymbol{q}_d) = \boldsymbol{0}$. The former corresponds to the desired equilibrium point while the latter corresponds to a point in which the nullspace base vectors in $\boldsymbol{Z}(\boldsymbol{q})$ are orthogonal[25] to $\boldsymbol{K}_n(\boldsymbol{q} - \boldsymbol{q}_d)$. Notice that the point $\boldsymbol{q} = \boldsymbol{q}_d$ corresponds to the global minimum of the cost function $V_q(\boldsymbol{q}) = (\boldsymbol{q} - \boldsymbol{q}_d)^T \boldsymbol{K}_n(\boldsymbol{q} - \boldsymbol{q}_d)$, while the points which satisfy $\boldsymbol{Z}(\boldsymbol{q})\boldsymbol{K}_n(\boldsymbol{q} - \boldsymbol{q}_d) = \boldsymbol{0}$ correspond to the constraint local minima $\arg_{\boldsymbol{q}} \min_{\boldsymbol{f}(\boldsymbol{q})=\boldsymbol{x}_d} V_q(\boldsymbol{q})$. With Theorem 4.3 one can thus conclude that locally the solution curves of the system (4.57)-(4.59) asymptotically approach the set of constraint local minima of $V_q(\boldsymbol{q})$. The results of the above analysis can be summarized as follows.

Proposition 4.4. *Consider the system (4.57)-(4.59) with* $\boldsymbol{u}(\boldsymbol{q}) = -\boldsymbol{Z}(\boldsymbol{q})\boldsymbol{K}_n(\boldsymbol{q} - \boldsymbol{q}_d)$. *The matrices* $\boldsymbol{K}_d, \boldsymbol{D}_d \in \mathbb{R}^{m \times m}$ *and* $\boldsymbol{K}_n, \boldsymbol{D}_n \in \mathbb{R}^{r \times r}$ *are all assumed to be symmetric and positive definite. Suppose that the Jacobian matrix* $\boldsymbol{J}(\boldsymbol{q})$ *is non-singular in the considered workspace and the joint configuration* $\boldsymbol{q}_d$ *satisfies* $\boldsymbol{f}(\boldsymbol{q}_d) = \boldsymbol{x}_d$. *Then the system is strict output passive for the input* $\boldsymbol{F}_{ext,x}$ *and the output* $\dot{\boldsymbol{x}}$. *Furthermore, locally the solution curves asymptotically approach the set of constraint local minima of* $V_q(\boldsymbol{q}) = (\boldsymbol{q} - \boldsymbol{q}_d)^T \boldsymbol{K}_n(\boldsymbol{q} - \boldsymbol{q}_d)$. *In particular, the point* $\boldsymbol{q}_d$ *which is an isolated constraint minimum of* $V_q(\boldsymbol{q})$ *is asymptotically stable.*

Discussion of the Derived Controller

By summing up all the previous parts, one can see that the complete controller equations are given by

$$\boldsymbol{\tau} = \boldsymbol{g}(\boldsymbol{q}) + \boldsymbol{J}(\boldsymbol{q})^T \hat{\boldsymbol{F}}_x + \boldsymbol{N}(\boldsymbol{q})^T \hat{\boldsymbol{F}}_n \tag{4.70}$$

$$\hat{\boldsymbol{F}}_x = \boldsymbol{\mu}_{xn}(\boldsymbol{q}, \dot{\boldsymbol{x}}, \boldsymbol{v}_n)\boldsymbol{v}_n - \boldsymbol{D}_d \dot{\boldsymbol{x}} - \boldsymbol{K}_d(\boldsymbol{f}(\boldsymbol{q}) - \boldsymbol{x}_d) \tag{4.71}$$

$$\hat{\boldsymbol{F}}_n = \boldsymbol{\mu}_{nx}(\boldsymbol{q}, \dot{\boldsymbol{x}}, \boldsymbol{v}_n)\dot{\boldsymbol{x}} - \boldsymbol{D}_n \boldsymbol{v}_n - \boldsymbol{Z}(\boldsymbol{q})\boldsymbol{K}_n(\boldsymbol{q} - \boldsymbol{q}_d) \ . \tag{4.72}$$

It is worth remarking that the last term $\boldsymbol{N}(\boldsymbol{q})^T \boldsymbol{Z}(\boldsymbol{q})\boldsymbol{K}_n(\boldsymbol{q} - \boldsymbol{q}_d)$, which actually represents the implementation of the nullspace stiffness, corresponds exactly to the relevant term of (4.16) when the projection matrix from (4.23) is used.

The objective of this section was to derive a controller which implements a nullspace impedance (in terms of stiffness and damping) decoupled from the Cartesian dynamics while avoiding the feedback of external forces. It should be mentioned that if either this decoupling is not required or the use of feedback of

[25] With respect to the Euclidean metric.

the external forces is possible[26], then one can instead use the controller which was discussed in Section 4.3. The most critical point in the stability analysis of this section is that it allows only a local stability statement and it is not clear how to get an estimate for the range of attraction. Notice therefore also that the function $S_A(\boldsymbol{q}, \dot{\boldsymbol{x}}, \boldsymbol{v}_n) = S(\boldsymbol{q}, \dot{\boldsymbol{x}}, \boldsymbol{v}_n) + V_A(\boldsymbol{q}, \boldsymbol{v}_n)$ cannot be used as a Lyapunov function for the closed loop system. The time derivative of $S_A(\boldsymbol{q}, \dot{\boldsymbol{x}}, \boldsymbol{v}_n)$ along the solutions of (4.57)-(4.59) is given by

$$\dot{S}_A = -\begin{pmatrix} \dot{\boldsymbol{x}} \\ \boldsymbol{v}_n \end{pmatrix}^T \begin{bmatrix} \boldsymbol{D}_d & \boldsymbol{0} \\ \boldsymbol{0} & \boldsymbol{D}_n \end{bmatrix} \begin{pmatrix} \dot{\boldsymbol{x}} \\ \boldsymbol{v}_n \end{pmatrix} + (\boldsymbol{q} - \boldsymbol{q}_d)^T \boldsymbol{K}_n \boldsymbol{J}^{\boldsymbol{M}+}(\boldsymbol{q})\dot{\boldsymbol{x}} \ .$$

The additional term $(\boldsymbol{q} - \boldsymbol{q}_d)^T \boldsymbol{K}_n \boldsymbol{J}^{\boldsymbol{M}+}(\boldsymbol{q})\dot{\boldsymbol{x}}$, which prevents the derivative $\dot{S}_A$ from being negative semi-definite, vanishes when the system is constrained to the set $\mathcal{A}$ and therefore did not disturb the above stability analysis based on Theorem 4.3. Furthermore, the analysis was restricted to the regulation case with a constant virtual equilibrium point $\boldsymbol{x}_d$. Notice also that these restrictions are not given for the controller from Section 4.3.

Nevertheless, the analysis of this section allows to give a stability proof for a nullspace impedance controller, which is, apart from the cancellation of some coupling terms in (4.48), identical to the simple dynamically consistent projection based controller. Moreover, also strict output passivity was shown.

4.5 Summary

In this chapter the impedance control problem for a redundant robot was discussed. Based on the discussion of the last chapter a desired Cartesian impedance without inertia shaping is considered as the main task.

As a preliminary step the construction of a nullspace base matrix was discussed. Based on the nullspace base matrix different nullspace projection matrices were investigated with respect to the implementation of nullspace stiffness and damping. The notation of *dynamical consistency* of such a projection matrix was discussed and incorporated in the proposed projection matrices by the choice of an appropriate metric. The resulting dynamically consistent projection matrix is equivalent to the projection matrix from the operational space formulation, but has the advantage that for its implementation only the inversion of an $r \times r$ matrix and no inversion of the inertia matrix is needed. The disadvantage of using such a projection matrix for the implementation of a nullspace stiffness and damping lies in the fact that it does not enable a consistent stability analysis.

As an alternative approach for the implementation of a nullspace stiffness and damping, the *Task Space Augmentation Method* was discussed. This method allows to extend the Cartesian impedance control techniques from the last chapter to the redundant case by choosing additional nullspace coordinates. Although

[26] These requirements depend of course on the considered application.

theoretically well funded, the main problem in applying this method is the choice of appropriate coordinates. Moreover, the resulting Cartesian dynamics and the nullspace dynamics usually are by no means decoupled from each other. A decoupling of these dynamics is only possible in case that all external forces, i.e. Cartesian forces and nullspace forces, can be measured.

Finally, the *Joint Space Decomposition Method* was discussed as a third approach. Based on this method a nullspace stiffness controller was proposed for which a local stability proof was presented. The proposed controller is quite similar to the use of the dynamically consistent nullspace projection, but herein an additional term is included which decouples the Cartesian dynamics and the nullspace dynamics by eliminating the cross diagonal parts in the Coriolis/centrifugal matrix. The use of this term, and the resulting decoupling, allowed to apply some results from the stability theory with semi-definite Lyapunov functions and thus facilitated the stability proof.

5 The Singular Perturbation Approach

In the previous chapters the impedance control problem was analyzed for a rigid body robot model, in which the joint elasticity was neglected. In contrast to this, the flexible joint robot model from Section 2.2.3 will be considered now.

This chapter deals with the singular perturbation approach. This is an approximate method designed for systems which possess a so-called *two-time-scale-property*. This means that the system can be virtually split up into two coupled subsystems which describe a faster and a slower part. The singular perturbation theory then allows to make stability conclusions out of the properties of the separated subsystems, similarly to the procedure of the analysis of cascaded control systems in the linear case. It also provides the theoretical justification for neglecting un-modeled "high frequency" dynamics.

In the case of a robot with flexible joints, the faster part of the model is formed by an inner torque control loop, while the slower part is formed by the rigid body robot dynamics. The singular perturbation method allows to extend all the control methods for rigid body robots, which were discussed in the previous chapters, to the flexible joint case.

In Section 5.1 the general singular perturbation theory will be shortly reviewed. Its application to the flexible joint model is then described in Section 5.2. Based on that, the commonly used *composite controller design* and a modified controller formulation from [OASH02], which fits better the flexible joint robot model, are described in Section 5.3 and 5.4.

5.1 Singular Perturbation Theory

Herein only the basic idea of the singular perturbation theory shall be given. A more comprehensive treatment of this theory can be found in [Kha02] and [KKO86].

The singular perturbation theory was designed in order to analyze models, which depend on a small scalar parameter $\epsilon \in \mathbb{R}$, and can be written in the form

$$\dot{\boldsymbol{z}}_1 = \boldsymbol{f}(\boldsymbol{z}_1, \boldsymbol{z}_2, \epsilon, t) \ , \tag{5.1}$$

$$\epsilon\dot{\boldsymbol{z}}_2 = \boldsymbol{g}(\boldsymbol{z}_1, \boldsymbol{z}_2, \epsilon, t) \ , \tag{5.2}$$

C. Ott: Cartesian Imped. Cntrl. of Redund. & Flexible-Joint Robots, STAR 49, pp. 65–75, 2008.
springerlink.com

where $\boldsymbol{z}_1 \in \mathbb{R}^{n_1}$ and $\boldsymbol{z}_2 \in \mathbb{R}^{n_2}$ are the state variables of the system. For a small value of ϵ, the system has the above mentioned two-time-scale-property and the two equations describe the slower part (5.1) and the faster part (5.2) of the system.

If ϵ is set to zero (i.e. assuming the fast part of the model to be in steady state), (5.2) gets an algebraic equation and therefore the order of the system changes. In the following it is assumed that one can solve this resulting algebraic equation uniquely[1] for the state variables $\boldsymbol{z}_2$. The solution of the system for $\epsilon = 0$ will be denoted by $\bar{\boldsymbol{z}}_1$ and $\bar{\boldsymbol{z}}_2$. The relationship between $\bar{\boldsymbol{z}}_1$ and $\bar{\boldsymbol{z}}_2$ is given by the function $\bar{\boldsymbol{z}}_2 = \boldsymbol{h}(\bar{\boldsymbol{z}}_1, t)$ which is the solution of the equation

$$\mathbf{0} = \boldsymbol{g}(\bar{\boldsymbol{z}}_1, \bar{\boldsymbol{z}}_2, \mathbf{0}, t) \quad \Rightarrow \quad \bar{\boldsymbol{z}}_2 = \boldsymbol{h}(\bar{\boldsymbol{z}}_1, t) \tag{5.3}$$

for $\bar{\boldsymbol{z}}_2$. By setting the variable ϵ to zero and combining (5.1) with (5.3) one therefore gets the so-called *quasi-steady state model*[2]

$$\dot{\bar{\boldsymbol{z}}}_1 = \boldsymbol{f}(\bar{\boldsymbol{z}}_1, \boldsymbol{h}(\bar{\boldsymbol{z}}_1, t), 0, t) \ . \tag{5.4}$$

The model (5.4) basically describes the system behavior at a *slow* time scale.

For the further analysis a change of coordinates for the fast part of the system is introduced. The new coordinates $\boldsymbol{y} \in \mathbb{R}^{n_2}$ represent the deviation of $\boldsymbol{z}_2$ from its steady-state value $\boldsymbol{h}(\boldsymbol{z}_1, t)$, i.e.

$$\boldsymbol{y} = \boldsymbol{z}_2 - \boldsymbol{h}(\boldsymbol{z}_1, t) \ . \tag{5.5}$$

The complete model can then be rewritten in the new state variables $\boldsymbol{z}_1$ and $\boldsymbol{y}$ as follows

$$\dot{\boldsymbol{z}}_1 = \boldsymbol{f}(\boldsymbol{z}_1, \boldsymbol{y} + \boldsymbol{h}(\boldsymbol{z}_1, t), \epsilon, t) \ , \tag{5.6}$$

$$\begin{aligned} \epsilon\dot{\boldsymbol{y}} &= \boldsymbol{g}(\boldsymbol{z}_1, \boldsymbol{y} + \boldsymbol{h}(\boldsymbol{z}_1, t), \epsilon, t) - \epsilon\frac{\partial \boldsymbol{h}(\boldsymbol{z}_1, t)}{\partial t} \\ &\quad -\epsilon\frac{\partial \boldsymbol{h}(\boldsymbol{z}_1, t)}{\partial \boldsymbol{z}_1}\boldsymbol{f}(\boldsymbol{z}_1, \boldsymbol{y} + \boldsymbol{h}(\boldsymbol{z}_1, t), \epsilon, t) \ . \end{aligned} \tag{5.7}$$

Obviously, the quasi-steady state model can be obtained from (5.6) by setting $\boldsymbol{y}$ to zero. In order to get also a model for the system behavior at a *fast* time scale, a change of the time base is performed by introducing the new time variable $\nu = (t - t_0)/\epsilon$, where t_0 is an (arbitrary) initial point in time. The parameter ϵ then describes the scaling between the time bases t and ν. With

$$\mathrm{d}t = \epsilon \mathrm{d}\nu \ , \tag{5.8}$$

equation (5.7) can be written in the new time base as

$$\begin{aligned} \frac{\mathrm{d}\boldsymbol{y}}{\mathrm{d}\nu} &= \boldsymbol{g}(\boldsymbol{z}_1, \boldsymbol{y} + \boldsymbol{h}(\boldsymbol{z}_1, t), \epsilon, t) - \epsilon\frac{\partial \boldsymbol{h}(\boldsymbol{z}_1, t)}{\partial t} \\ &\quad -\epsilon\frac{\partial \boldsymbol{h}(\boldsymbol{z}_1, t)}{\partial \boldsymbol{z}_1}\boldsymbol{f}(\boldsymbol{z}_1, \boldsymbol{y} + \boldsymbol{h}(\boldsymbol{z}_1, t), \epsilon, t) \ . \end{aligned} \tag{5.9}$$

[1] The system is then said to be in *standard* singular perturbation form.
[2] Also denoted as *slow model* or *reduced model* in the literature.

The variables t and $\boldsymbol{z}_1$ are slowly varying, since in the ν time scale they are given by $t = t_0 + \epsilon\nu$ and $\boldsymbol{z}_1 = \boldsymbol{z}_1(t_0 + \epsilon\nu)$. Setting $\epsilon = 0$ freezes these variables at $t = t_0$ and $\boldsymbol{z}_1 = \boldsymbol{z}_1(t_0)$, and reduces (5.9) to

$$\frac{\mathrm{d}\boldsymbol{y}}{\mathrm{d}\nu} = \boldsymbol{g}(\boldsymbol{z}_1(t_0), \boldsymbol{y} + \boldsymbol{h}(\boldsymbol{z}_1(t_0), t_0), 0, t_0) \ . \tag{5.10}$$

In order to analyze the situation when the slowly varying variables $(t, \boldsymbol{z}_1)$ move away from their initial values $(t_0, \boldsymbol{z}_1(t_0))$, it is assumed that the solution $\bar{\boldsymbol{z}}_1(t)$ of (5.4) is defined for $t \in [0, t_1]$ and $\bar{\boldsymbol{z}}_1(t) \in D_1 \subset \mathbb{R}^{n_1}$ for some domain D_1. Equation (5.10) is then written in the form

$$\frac{\mathrm{d}\boldsymbol{y}}{\mathrm{d}\nu} = \boldsymbol{g}(\boldsymbol{z}_1, \boldsymbol{y} + \boldsymbol{h}(\boldsymbol{z}_1, t), 0, t) \ , \tag{5.11}$$

which is the so-called *boundary layer model.* Note that t and $\boldsymbol{z}_1$ are treated as fixed parameters in (5.11). In the following an important theorem due to Tychonov is reported[3]. This theorem is based on equation (5.4) and (5.11) and will be useful for the justification of the controller design. The theorem requires the boundary layer model (5.11) to be exponentially stable uniformly in the frozen parameters, as stated in the following definition.

Definition 5.1. *The equilibrium point $\boldsymbol{y} = \boldsymbol{0}$ of the boundary layer system (5.11) is exponentially stable, uniformly in $(t, \boldsymbol{z}_1) \in [0, t_1] \times D_1$, if there exist positive constants k, γ, and ρ_0 such that the solutions of (5.11) satisfy*

$$||\boldsymbol{y}(\nu)|| \leq k||\boldsymbol{y}(0)|| \exp(-\gamma\nu) \quad \forall||\boldsymbol{y}(0)|| < \rho_0, \ \forall(t, \boldsymbol{z}_1) \in [0, t_1] \times D_1, \ \forall\nu \geq 0 \ .$$

Theorem 5.2. *(Tychonov's Theorem [Kha02])*

Given the system (5.1) and (5.2) with initial state $\boldsymbol{z}_1(t_0, \epsilon) = \boldsymbol{z}_{1,0}(\epsilon)$ and $\boldsymbol{z}_2(t_0, \epsilon) = \boldsymbol{z}_{2,0}(\epsilon)$. Let $\boldsymbol{z}_2 = \boldsymbol{h}(\boldsymbol{z}_1, t)$ be an isolated root of $\boldsymbol{0} = \boldsymbol{g}(\boldsymbol{z}_1, \boldsymbol{z}_2, 0, t)$. Assume that for all

$$(t, \boldsymbol{z}_1, \boldsymbol{z}_2 - \boldsymbol{h}(\boldsymbol{z}_1, t), \epsilon) \in [0, t_1[\times D_1 \times D_y \times [0, \epsilon_0]$$

in some domains $D_1 \subset \mathbb{R}^{n_1}$ and $D_y \subset \mathbb{R}^{n_2}$, in which D_1 is convex and D_y contains the origin, the following conditions hold:

- *The functions $\boldsymbol{f}$, $\boldsymbol{g}$, their first partial derivatives with respect to $(\boldsymbol{z}_1, \boldsymbol{z}_2, \epsilon)$, and the first partial derivative of $\boldsymbol{g}$ with respect to t are continuous. The function $\boldsymbol{h}(\boldsymbol{z}_1, t)$ and the Jacobian $\partial\boldsymbol{g}(\boldsymbol{z}_1, \boldsymbol{z}_2, 0, t)/\partial\boldsymbol{z}_2$ have continuous first partial derivatives with respect to their arguments. The initial data $\boldsymbol{z}_{1,0}(\epsilon)$ and $\boldsymbol{z}_{2,0}(\epsilon)$ are smooth functions of ϵ.*
- *The reduced problem (5.4) has a unique solution $\bar{\boldsymbol{z}}_1(t) \in S$ for $t \in [t_0, t_1]$, where S is a compact subset of D_1.*
- *The origin of the boundary layer model (5.11) is exponentially stable, uniformly in $(t, \boldsymbol{z}_1)$. Let $R_y \subset D_y$ be the region of attraction of (5.10) and Ω_y be a compact subset of R_y.*

[3] Notice that there are different versions of Tychonov's theorem in the literature. The one presented here is taken from [Kha02].

Then, there exists a positive constant ϵ^* *such that for all initial conditions of the form* $\boldsymbol{z}_{2,0}(0) - \boldsymbol{h}(\boldsymbol{z}_{1,0}(0), t_0) \in \Omega_y$ *and* $0 < \epsilon < \epsilon^*$*, the singular perturbation problem (5.1)-(5.2) has a unique solution* $\boldsymbol{z}_1(t,\epsilon)$*,* $\boldsymbol{z}_2(t,\epsilon)$ *on* $[t_0, t_1]$*, and*

$$\boldsymbol{z}_1(t,\epsilon) - \bar{\boldsymbol{z}}_1(t) = O(\epsilon)$$

$$\boldsymbol{z}_2(t,\epsilon) - \boldsymbol{h}(\bar{\boldsymbol{z}}_1(t), t) - \hat{\boldsymbol{y}}(t/\epsilon) = O(\epsilon)$$

hold uniformly for $t \in [t_0, t_1]$*, where* $\hat{\boldsymbol{y}}(\nu)$ *is the solution of the boundary-layer model (5.11). Moreover, given any* $t_b > t_0$*, there is* $\epsilon^{**} \le \epsilon^*$ *such that*

$$\boldsymbol{z}_2(t,\epsilon) - \boldsymbol{h}(\bar{\boldsymbol{z}}_1(t), t) = O(\epsilon)$$

holds uniformly for $t \in [t_b, t_1]$ *whenever* $\epsilon < \epsilon^{**}$.

For a proof of the above theorem, see[4] [Kha02].

Notice that Tychonov's theorem does only give a statement on a finite time interval. There are extensions of this theorem to the infinite time interval, but they require exponential stability of the quasi-steady state model[5]. In case of the impedance control problem from Section 3.2 one cannot show exponential stability of the desired dynamics even in the rigid body case. Therefore, the argumentation for the controller design in the next section will be based on Tychonov's theorem only. Notice that, although this theorem is very important from a conceptual point of view, since it allows to give a theoretical foundation of many physically argued modeling reduction techniques, its value for stability conclusions, however, is limited.

In the next section it will be shown, how the singular perturbation theory can be applied to the model of a flexible joint robot.

5.2 Singular Perturbation Model of a Flexible Joint Robot

The singular perturbation theory, as outlined in the last section, shall now be applied to the flexible joint robot model from Section 2.2.3. Using the notation of Section 2.2.3 the reduced model of a flexible joint robot without damping is given by

$$\boldsymbol{M}(\boldsymbol{q})\ddot{\boldsymbol{q}} + \boldsymbol{C}(\boldsymbol{q},\dot{\boldsymbol{q}})\dot{\boldsymbol{q}} + \boldsymbol{g}(\boldsymbol{q}) = \boldsymbol{K}(\boldsymbol{\theta} - \boldsymbol{q}) + \boldsymbol{\tau}_{ext} \;, \tag{5.12}$$

$$\boldsymbol{B}\ddot{\boldsymbol{\theta}} + \boldsymbol{K}(\boldsymbol{\theta} - \boldsymbol{q}) = \boldsymbol{\tau}_m \;. \tag{5.13}$$

---

4 This proof implicitly gives also a guideline of how to find ϵ^*, if proper Lyapunov functions of the quasi-steady state model and the boundary layer system can be found.

5 Notice also that there also exist results within the singular perturbation theory, which do not require exponential stability of the quasi-steady state model but, instead, require some special *interconnection conditions* of Lyapunov functions for the steady state system and the boundary layer system, see, e.g., [Kha02] for more details on this. For an application to the flexible joint robot model these results, however, have not yet been utilized so far and will therefore be not reported here.

In order to bring this system in standard singular perturbation form, one can perform the following change of coordinates. As coordinates $\boldsymbol{z}_1$ and $\boldsymbol{z}_2$ the joint angles $\boldsymbol{q}$ and the joint torques $\boldsymbol{\tau} = \boldsymbol{K}(\boldsymbol{\theta} - \boldsymbol{q})$ as well as their first derivatives with respect to time are chosen

$$\boldsymbol{z}_1 = \begin{pmatrix} \boldsymbol{q} \\ \dot{\boldsymbol{q}} \end{pmatrix} , \qquad \boldsymbol{z}_2 = \begin{pmatrix} \boldsymbol{\tau} \\ \dot{\boldsymbol{\tau}} \end{pmatrix} . \tag{5.14}$$

Therefore, one may substitute $\boldsymbol{K}(\boldsymbol{\theta} - \boldsymbol{q}) = \boldsymbol{\tau}$ and $\ddot{\boldsymbol{\theta}} = \boldsymbol{K}^{-1}\ddot{\boldsymbol{\tau}} + \ddot{\boldsymbol{q}}$ in equation (5.12) and (5.13) in order to obtain

$$\boldsymbol{M}(\boldsymbol{q})\ddot{\boldsymbol{q}} + \boldsymbol{C}(\boldsymbol{q}, \dot{\boldsymbol{q}})\dot{\boldsymbol{q}} + \boldsymbol{g}(\boldsymbol{q}) = \boldsymbol{\tau} + \boldsymbol{\tau}_{ext} , \tag{5.15}$$

$$\boldsymbol{B}\boldsymbol{K}^{-1}\ddot{\boldsymbol{\tau}} + \boldsymbol{\tau} = \boldsymbol{\tau}_m - \boldsymbol{B}\ddot{\boldsymbol{q}} . \tag{5.16}$$

The second derivative of the link angles $\boldsymbol{q}$ in (5.16) can be eliminated by using (5.15), because the inertia matrix $\boldsymbol{M}(\boldsymbol{q})$ is positive definite and therefore always invertible. In order to bring the system in singular perturbation form, the parameter ϵ must be introduced. The stiffness values K_i of the joints usually are very high. This is incorporated into the model by formally replacing the matrix $\boldsymbol{K}$ by $\boldsymbol{K} = \frac{\boldsymbol{K}_\epsilon}{\epsilon^2}$ with a positive definite diagonal matrix $\boldsymbol{K}_\epsilon$. A small value of ϵ represents a high joint stiffness value.

Notice also that the general singular perturbation approach was described for a free system in the last section, in which no control inputs were present. In the case of the flexible joint robot model the motor torques $\boldsymbol{\tau}_m$, which are the control inputs, are considered to be a (so far unknown) function of the state variables $\boldsymbol{z}_1$ and $\boldsymbol{z}_2$. Furthermore, the external torques $\boldsymbol{\tau}_{ext}$ are treated as disturbance inputs.

With the above mentioned considerations one can rewrite the model in (standard) singular perturbation form as

$$\boldsymbol{M}(\boldsymbol{q})\ddot{\boldsymbol{q}} + \boldsymbol{C}(\boldsymbol{q}, \dot{\boldsymbol{q}})\dot{\boldsymbol{q}} + \boldsymbol{g}(\boldsymbol{q}) = \boldsymbol{\tau} + \boldsymbol{\tau}_{ext} , \tag{5.17}$$

$$\begin{aligned} \epsilon^2\ddot{\boldsymbol{\tau}} + \boldsymbol{K}_\epsilon(\boldsymbol{B}^{-1} + \boldsymbol{M}(\boldsymbol{q})^{-1})\boldsymbol{\tau} = {} & \boldsymbol{K}_\epsilon\boldsymbol{B}^{-1}\boldsymbol{\tau}_m + \\ & \boldsymbol{K}_\epsilon\boldsymbol{M}(\boldsymbol{q})^{-1}\left(\boldsymbol{C}(\boldsymbol{q}, \dot{\boldsymbol{q}})\dot{\boldsymbol{q}} + \boldsymbol{g}(\boldsymbol{q}) - \boldsymbol{\tau}_{ext}\right) . \end{aligned} \tag{5.18}$$

By rewriting this equation as a set of first order differential equations, one can easily see that, due to the introduction of the parameter ϵ, the model indeed is now written in the same form as equation (5.1) and (5.2).

By following the procedure of the last section the *quasi-steady state* of the torque $\bar{\boldsymbol{\tau}} = \boldsymbol{h}(\bar{\boldsymbol{q}}, \dot{\bar{\boldsymbol{q}}}, t)$ is computed by setting ϵ to zero (as it was done in (5.3)):

$$\begin{aligned} \boldsymbol{h}(\bar{\boldsymbol{q}}, \dot{\bar{\boldsymbol{q}}}, t) &= (\boldsymbol{I} + \boldsymbol{B}\boldsymbol{M}(\bar{\boldsymbol{q}})^{-1})^{-1}\left(\bar{\boldsymbol{\tau}}_m + \boldsymbol{B}\boldsymbol{M}(\bar{\boldsymbol{q}})^{-1}\left(\boldsymbol{C}(\bar{\boldsymbol{q}}, \dot{\bar{\boldsymbol{q}}})\dot{\bar{\boldsymbol{q}}} + \boldsymbol{g}(\bar{\boldsymbol{q}}) - \bar{\boldsymbol{\tau}}_{ext}\right)\right) \\ &= (\boldsymbol{I} + \boldsymbol{B}\boldsymbol{M}(\bar{\boldsymbol{q}})^{-1})^{-1}\bar{\boldsymbol{\tau}}_m + \\ &\quad\; (\boldsymbol{I} + \boldsymbol{M}(\bar{\boldsymbol{q}})\boldsymbol{B}^{-1})^{-1}\left(\boldsymbol{C}(\bar{\boldsymbol{q}}, \dot{\bar{\boldsymbol{q}}})\dot{\bar{\boldsymbol{q}}} + \boldsymbol{g}(\bar{\boldsymbol{q}}) - \bar{\boldsymbol{\tau}}_{ext}\right) . \end{aligned} \tag{5.19}$$

The variables $\bar{\boldsymbol{\tau}}_m$ and $\bar{\boldsymbol{\tau}}_{ext}$ correspond to the input variables $\boldsymbol{\tau}_m$ and $\boldsymbol{\tau}_{ext}$, when the parameter ϵ is set to zero.

5.2.1 The Quasi-steady State Model

The quasi-steady state model follows from substituting $\bar{\boldsymbol{\tau}}$ in equation (5.17). The resulting equation can be simplified by pre-multiplying it by the factor $(\boldsymbol{I}+\boldsymbol{B}\boldsymbol{M}(\bar{\boldsymbol{q}})^{-1})$. Therefore, one obtains the system equation

$$(\boldsymbol{M}(\bar{\boldsymbol{q}})+\boldsymbol{B})\,\ddot{\bar{\boldsymbol{q}}}+\boldsymbol{C}(\bar{\boldsymbol{q}},\dot{\bar{\boldsymbol{q}}})\dot{\bar{\boldsymbol{q}}}+\boldsymbol{g}(\bar{\boldsymbol{q}})=\bar{\boldsymbol{\tau}}_m+\bar{\boldsymbol{\tau}}_{ext}\ . \tag{5.20}$$

Notice that this model has the form of the rigid body model with the inertia matrix $\boldsymbol{M}(\bar{\boldsymbol{q}})+\boldsymbol{B}$.

5.2.2 The Boundary Layer System

For the derivation of the boundary layer model one introduces the new coordinates $\boldsymbol{y}\in\mathbb{R}^n$ (see equation (5.5)) via

$$\boldsymbol{y}=\boldsymbol{\tau}-\boldsymbol{h}(\boldsymbol{q},\dot{\boldsymbol{q}},t) \tag{5.21}$$

which correspond to the deviation of the actual torque from its *quasi-steady state value*. As it was done in the general case one substitutes these new coordinates into equation (5.18), introduces the fast time scale $\nu=(t-t_0)/\epsilon$, and sets the parameter ϵ to zero. This leads to

$$\frac{\mathrm{d}^2\boldsymbol{y}}{\mathrm{d}\nu^2}+\boldsymbol{K}_\epsilon(\boldsymbol{B}^{-1}+\boldsymbol{M}(\boldsymbol{q})^{-1})(\boldsymbol{y}+\boldsymbol{h}(\boldsymbol{q},\dot{\boldsymbol{q}},t))=\boldsymbol{K}_\epsilon\boldsymbol{B}^{-1}\boldsymbol{\tau}_m+ \\ \boldsymbol{K}_\epsilon\boldsymbol{M}(\boldsymbol{q})^{-1}(\boldsymbol{C}(\boldsymbol{q},\dot{\boldsymbol{q}})\dot{\boldsymbol{q}}+\boldsymbol{g}(\boldsymbol{q})-\boldsymbol{\tau}_{ext})\ .$$

Herein the state variables and the disturbance inputs $\boldsymbol{\tau}_{ext}$ from the slow part of the system are treated as constant with respect to the time variable ν. Using the identity

$$(\boldsymbol{B}^{-1}+\boldsymbol{M}(\boldsymbol{q})^{-1})\boldsymbol{h}(\boldsymbol{q},\dot{\boldsymbol{q}},t)=\boldsymbol{B}^{-1}\bar{\boldsymbol{\tau}}_m+\boldsymbol{M}(\boldsymbol{q})^{-1}(\boldsymbol{C}(\boldsymbol{q},\dot{\boldsymbol{q}})\dot{\boldsymbol{q}}+\boldsymbol{g}(\boldsymbol{q})-\bar{\boldsymbol{\tau}}_{ext})\ ,$$

which follows directly from the definition of $\boldsymbol{h}(\boldsymbol{q},\dot{\boldsymbol{q}},t)$ in (5.19), one finally gets

$$\frac{\mathrm{d}^2\boldsymbol{y}}{\mathrm{d}\nu^2}+\boldsymbol{K}_\epsilon(\boldsymbol{B}^{-1}+\boldsymbol{M}(\boldsymbol{q})^{-1})\boldsymbol{y}=\boldsymbol{K}_\epsilon\boldsymbol{B}^{-1}\,(\boldsymbol{\tau}_m-\bar{\boldsymbol{\tau}}_m)\ . \tag{5.22}$$

Notice that, since the variable $\boldsymbol{q}$ is considered as constant in this equation, the boundary layer system of the flexible joint robot model is linear and time invariant.

5.3 Controller Design

In the following the design idea of the *composite feedback design method* is described. This design methodology has been used by many authors considering the controller design for flexible joint robots based on the singular perturbation approach. Finally, an alternative controller structure is proposed, which has some advantages compared to the common composite design.

The idea of the composite controller design method (Figure 5.1) is to split up the control input $\boldsymbol{\tau}_m$ into a slow and a fast part $\boldsymbol{\tau}_{m,slow}$ and $\boldsymbol{\tau}_{m,fast}$, i.e. $\boldsymbol{\tau}_m = \boldsymbol{\tau}_{m,slow} + \boldsymbol{\tau}_{m,fast}$. These two control inputs should be designed such that the slow part affects the quasi-steady state model only, and the fast part affects the boundary layer system only. Then the two control inputs can be used in order to control the two subsystems separately. Formally, this can be done by choosing $\boldsymbol{\tau}_{m,slow} = \boldsymbol{\tau}_m|_{\epsilon\to 0} = \bar{\boldsymbol{\tau}}$ and $\boldsymbol{\tau}_{m,fast} = \boldsymbol{\tau}_m - \bar{\boldsymbol{\tau}}$.

This separation of the control input into two parts has two important implications for the design. On the one hand the control input $\boldsymbol{\tau}_{m,slow}$ must be chosen such that the system is in standard form, i.e. the *quasi-steady state* $\bar{\boldsymbol{\tau}}$ must have a unique solution. This can be achieved here for instance if $\boldsymbol{\tau}_{m,slow}$ is a function of the state $(\boldsymbol{q}, \dot{\boldsymbol{q}})$ of the quasi-steady state model only and does not contain a feedback of the joint torque $\boldsymbol{\tau}$. On the other hand the control input $\boldsymbol{\tau}_{m,fast}$ should be designed such that it fulfills the condition $\boldsymbol{\tau}_{m,fast}|_{\epsilon\to 0} = \boldsymbol{0}$, i.e. it should not affect the quasi-steady state model.

While the boundary layer system, which describes the torque dynamics in the new variables $\boldsymbol{y}$, is linear and time invariant, the quasi-steady state model has the form of an equivalent rigid body model. Therefore, classical controller methods from the rigid body robotics literature may be applied to design $\boldsymbol{\tau}_{m,slow}$.

According to Tychonov's theorem, the goal for the controller design is to get an exponentially stable boundary layer system on the one hand. On the other hand, convergence of the tracking error is desired for the quasi-steady state system. Notice that Tychonov's theorem ensures only that the tracking error will not deviate more than of order $O(\epsilon)$ from its quasi-steady state on a finite time interval. Also this conclusion is only valid for a sufficiently small value of ϵ, which means that the boundary layer system has to be considerably *faster* than the quasi-steady state system.

In most of the previous works concerning the singular perturbation approach for flexible joint robots, the fast part of the motor torque was chosen as a term of the form $\boldsymbol{\tau}_{m,fast} = -\epsilon \boldsymbol{D}_\tau \dot{\boldsymbol{y}}$ with $\boldsymbol{D}_\tau$ as a positive definite damping matrix. Additionally, the slow part $\boldsymbol{\tau}_{m,slow}$ was designed according to a control law $\boldsymbol{\tau}_{rigid}(\boldsymbol{q}, \dot{\boldsymbol{q}})$ for the rigid robot model (5.20). Notice that the term $-\epsilon \boldsymbol{D}_\tau \dot{\boldsymbol{y}}$ may be modified to $-\epsilon \boldsymbol{D}_\tau \dot{\boldsymbol{\tau}}$ because the time t and the state variables of the quasi-steady state system, $\boldsymbol{q}$ and $\dot{\boldsymbol{q}}$, are treated as fixed parameters in the boundary layer system. In both cases it is ensured that the condition $\boldsymbol{\tau}_{m,fast}|_{\epsilon\to 0} = \boldsymbol{0}$ holds. A composite feedback controller for the flexible joint robot model is then given by

$$\boldsymbol{\tau}_m = \boldsymbol{\tau}_{rigid}(\boldsymbol{q}, \dot{\boldsymbol{q}}) - \epsilon \boldsymbol{D}_\tau \dot{\boldsymbol{\tau}} \ . \tag{5.23}$$

The problem with this approach is that the control law (5.23) does not really *improve* the dynamics of the boundary layer system but only adds damping to it. An alternative would be to use also the torque error $\boldsymbol{K}_\tau \boldsymbol{y} = \boldsymbol{K}_\tau(\boldsymbol{\tau} - \boldsymbol{h}(\boldsymbol{q}, \dot{\boldsymbol{q}}, t))$

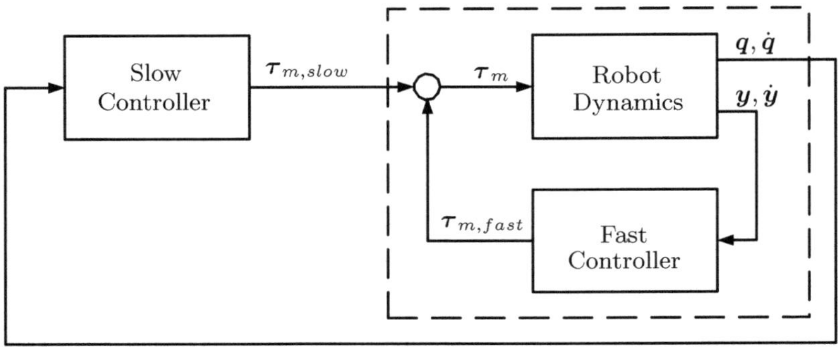

Fig. 5.1. Composite control

in the feedback, with a positive definite controller gain matrix $\boldsymbol{K}_\tau$, in order to enhance the performance of the boundary layer dynamics. However, for the implementation of such a feedback one has to compute the steady state $\boldsymbol{h}(\boldsymbol{q}, \dot{\boldsymbol{q}}, t)$ explicitly. This is in many cases not desired because it requires to measure the external forces $\boldsymbol{\tau}_{ext}$ and to invert[6] the inertia matrix $\boldsymbol{M}(\boldsymbol{q})$.

5.4 Modified Controller Design

In the following, a modification of the controller is proposed according to [OASH02] such that an explicit computation of $\boldsymbol{h}(\boldsymbol{q}, \dot{\boldsymbol{q}}, t)$ can be avoided[7].

Let the control input be decomposed as

$$\boldsymbol{\tau}_m = \boldsymbol{\tau}_{ss}(\boldsymbol{q}, \dot{\boldsymbol{q}}) - \boldsymbol{K}_\tau \boldsymbol{\tau} - \epsilon \boldsymbol{D}_\tau \dot{\boldsymbol{\tau}} \ , \tag{5.24}$$

where $\boldsymbol{K}_\tau$ and $\boldsymbol{D}_\tau$ are positive definite controller gain matrices and $\boldsymbol{\tau}_{ss}(\boldsymbol{q}, \dot{\boldsymbol{q}})$ is an intermediate control input for the quasi-steady state system which will be defined later. Formally, in (5.24) the term $\boldsymbol{K}_\tau \boldsymbol{\tau}$ was used instead of the term $\boldsymbol{K}_\tau \boldsymbol{y}$. This term clearly affects both, the fast part of the control input $\boldsymbol{\tau}_{m,fast}$ and the slow part $\boldsymbol{\tau}_{m,slow}$. In the following the closed loop quasi-steady state system and the boundary layer system are given for this control law and it is shown, how the term $\boldsymbol{\tau}_{ss}(\boldsymbol{q}, \dot{\boldsymbol{q}})$ should be chosen for the Cartesian impedance control problem from Section 3.2.

[6] At this point it should again be mentioned that this inversion may be numerically ill-posed. While the invertibility is ensured by the fact that the inertia matrix is positive definite, its eigenvalues usually can have different orders of magnitude and $\boldsymbol{M}(\boldsymbol{q})$ may contain very small eigenvalues in certain joint configurations.

[7] The following modification of the control law was used in [OASH02]. Therein a control problem was analyzed, in which the explicit computation of $\boldsymbol{h}(\boldsymbol{q}, \dot{\boldsymbol{q}}, t)$ was not possible because the dynamical parameters of the rigid body part were assumed to be unknown.

5.4.1 The Modified Quasi-steady State System

The slow part of the control input $\boldsymbol{\tau}_{m,slow}$ follows by setting ϵ to zero[8] in the expression (5.24) for $\boldsymbol{\tau}_m$. This leads to

$$\begin{aligned}\boldsymbol{\tau}_{m,slow} &= \boldsymbol{\tau}_{ss}(\bar{\boldsymbol{q}}, \dot{\bar{\boldsymbol{q}}}) - \boldsymbol{K}_\tau \boldsymbol{h}(\bar{\boldsymbol{q}}, \dot{\bar{\boldsymbol{q}}}, t) \\ &= \boldsymbol{\tau}_{ss}(\bar{\boldsymbol{q}}, \dot{\bar{\boldsymbol{q}}}) - \boldsymbol{K}_\tau \bar{\boldsymbol{\tau}} \ . \end{aligned} \tag{5.25}$$

In (5.25) one can see that $\boldsymbol{\tau}_{m,slow}$ now also contains the quasi-steady state torque $\bar{\boldsymbol{\tau}}$ and the expression for $\bar{\boldsymbol{\tau}}$ from[9] (5.19) becomes an implicit equation. However, one can solve this equation easily for the *steady state* torque $\bar{\boldsymbol{\tau}}$ and obtains

$$\begin{aligned}\bar{\boldsymbol{\tau}} = {} & (\boldsymbol{I} + \boldsymbol{K}_\tau + \boldsymbol{B}\boldsymbol{M}(\bar{\boldsymbol{q}})^{-1})^{-1} \\ & \left(\bar{\boldsymbol{\tau}}_{ss}(\bar{\boldsymbol{q}}, \dot{\bar{\boldsymbol{q}}}) + \boldsymbol{B}\boldsymbol{M}(\bar{\boldsymbol{q}})^{-1}\left(\boldsymbol{C}(\bar{\boldsymbol{q}}, \dot{\bar{\boldsymbol{q}}})\dot{\bar{\boldsymbol{q}}} + \boldsymbol{g}(\bar{\boldsymbol{q}}) - \bar{\boldsymbol{\tau}}_{ext}\right)\right)\end{aligned}$$

Using this relation together with (5.20), one finally gets the following steady state system

$$(\boldsymbol{M}(\bar{\boldsymbol{q}}) + (\boldsymbol{I} + \boldsymbol{K}_\tau)^{-1}\boldsymbol{B})\ddot{\bar{\boldsymbol{q}}} + \boldsymbol{C}(\bar{\boldsymbol{q}}, \dot{\bar{\boldsymbol{q}}})\dot{\bar{\boldsymbol{q}}} + \boldsymbol{g}(\bar{\boldsymbol{q}}) = (\boldsymbol{I} + \boldsymbol{K}_\tau)^{-1}\boldsymbol{\tau}_{ss}(\bar{\boldsymbol{q}}, \dot{\bar{\boldsymbol{q}}}) \ . \tag{5.26}$$

The function $\boldsymbol{\tau}_{ss}(\bar{\boldsymbol{q}}, \dot{\bar{\boldsymbol{q}}})$ can now be chosen as $\boldsymbol{\tau}_{ss}(\bar{\boldsymbol{q}}, \dot{\bar{\boldsymbol{q}}}) = (\boldsymbol{I} + \boldsymbol{K}_\tau)\boldsymbol{\tau}_d$ with $\boldsymbol{\tau}_d$ as a new control input which can be designed according to a control law for rigid robots, like for instance the controller from Section 3.2. With this substitution the complete control law (5.24) can be written in the form (see also Figure 5.2)

$$\boldsymbol{\tau}_m = \boldsymbol{\tau}_d - \boldsymbol{K}_\tau(\boldsymbol{\tau} - \boldsymbol{\tau}_d) - \epsilon \boldsymbol{D}_\tau \dot{\boldsymbol{\tau}} \ , \tag{5.27}$$

and the closed loop system is given by

$$(\boldsymbol{M}(\bar{\boldsymbol{q}}) + (\boldsymbol{I} + \boldsymbol{K}_\tau)^{-1}\boldsymbol{B})\ddot{\bar{\boldsymbol{q}}} + \boldsymbol{C}(\bar{\boldsymbol{q}}, \dot{\bar{\boldsymbol{q}}})\dot{\bar{\boldsymbol{q}}} + \boldsymbol{g}(\bar{\boldsymbol{q}}) = \boldsymbol{\tau}_d \ . \tag{5.28}$$

In contrast to (5.20) the inertia matrix of the new quasi-steady state system depends now also on the controller gain matrix $\boldsymbol{K}_\tau$, and thus, by $\boldsymbol{K}_\tau$ the apparent motor inertia is scaled. This physical interpretation of torque feedback will be utilized later in Chapter 7 for the design of a passivity based controller. Notice that $\boldsymbol{\tau}_d$ does not correspond to the steady state value $\bar{\boldsymbol{\tau}}$ of the boundary layer system as one would expect from (5.27).

5.4.2 The Modified Boundary Layer System

The new boundary layer system follows directly from (5.22) and (5.24). By considering that the fast part of the control input has the form

$$\boldsymbol{\tau}_{m,fast} = -\boldsymbol{K}_\tau(\boldsymbol{\tau} - \bar{\boldsymbol{\tau}}) - \epsilon \boldsymbol{D}_\tau \dot{\boldsymbol{\tau}} = -\boldsymbol{K}_\tau \boldsymbol{y} - \epsilon \boldsymbol{D}_\tau \dot{\boldsymbol{\tau}} \ , \tag{5.29}$$

[8] This also means $\boldsymbol{y} \to \boldsymbol{0}$.

[9] Notice that $\boldsymbol{\tau}_{m,slow} = \bar{\boldsymbol{\tau}}_m$.

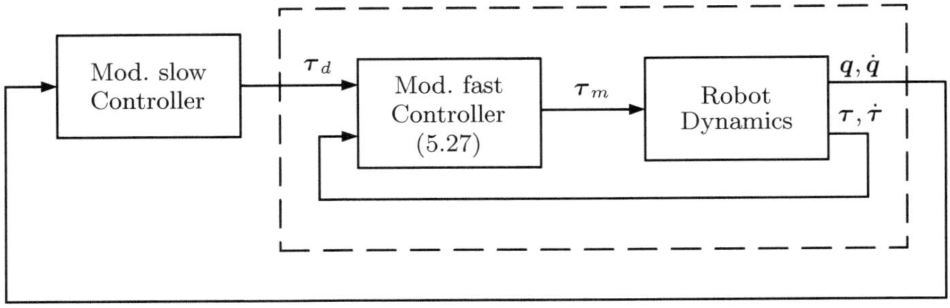

Fig. 5.2. Modified controller

one gets the following boundary layer system

$$\frac{\mathrm{d}^2\boldsymbol{y}}{\mathrm{d}\nu^2} + \boldsymbol{K}_\epsilon \boldsymbol{B}^{-1}\boldsymbol{D}_\tau \frac{\mathrm{d}\boldsymbol{y}}{\mathrm{d}\nu} + \boldsymbol{K}_\epsilon \boldsymbol{B}^{-1}(\boldsymbol{I} + \boldsymbol{B}\boldsymbol{M}(\boldsymbol{q})^{-1} + \boldsymbol{K}_\tau)\boldsymbol{y} = \boldsymbol{0} \ . \tag{5.30}$$

Now the positive definite matrices $\boldsymbol{K}_\tau$ and $\boldsymbol{D}_\tau$ may be used for shaping the dynamics of the linear and time-invariant boundary layer system in order to obtain exponential stability according to Tychonov's theorem.

5.4.3 Design of the Cartesian Impedance Controller

The impedance control problem from Section 3.2 shall now be treated for the flexible joint case, based on the singular perturbation approach.

From the analysis of this chapter it follows that the controller of the rigid body case (3.18) can now be applied to the quasi-steady state system via the intermediate control input $\boldsymbol{\tau}_d$ together with the inner loop controller from (5.27). In the design it is now important to compute all the components from (3.18), like the Cartesian inertia matrix $\boldsymbol{\Lambda}(\boldsymbol{x})$ and the corresponding Coriolis/centrifugal matrix $\boldsymbol{\mu}(\boldsymbol{x}, \dot{\boldsymbol{x}})$, based on the modified quasi-steady state system in equation (5.28). All the stability considerations of Chapter 3 can then be readily applied to the quasi-steady state system. In the non-redundant and non-singular case the complete controller finally is given by

$$\boldsymbol{\tau}_m = \boldsymbol{\tau}_d - \boldsymbol{K}_\tau(\boldsymbol{\tau} - \boldsymbol{\tau}_d) - \epsilon \boldsymbol{D}_\tau \dot{\boldsymbol{\tau}} \tag{5.31}$$

$$\boldsymbol{\tau}_d = \boldsymbol{g}(\boldsymbol{q}) + \boldsymbol{J}(\boldsymbol{q})^T \left(\boldsymbol{\Lambda}(\boldsymbol{x})\ddot{\boldsymbol{x}}_d + \boldsymbol{\mu}(\boldsymbol{x}, \dot{\boldsymbol{x}})\dot{\boldsymbol{x}}_d - \boldsymbol{K}_d\tilde{\boldsymbol{x}} - \boldsymbol{D}_d\dot{\tilde{\boldsymbol{x}}}\right) \ , \tag{5.32}$$

where the Cartesian error $\tilde{\boldsymbol{x}}$, the desired position $\boldsymbol{x}_d$, and the Jacobian $\boldsymbol{J}(\boldsymbol{q})$ are the same as in Chapter 3. The matrices $\boldsymbol{K}_d$ and $\boldsymbol{D}_d$ are again the symmetric and positive definite matrices of the desired stiffness and damping.

Notice also that all the controllers from Chapter 3 and Chapter 4, which were designed for the rigid body case, can be applied also to the singular perturbation model of the flexible joint robot via the control input $\boldsymbol{\tau}_d$. While the singular perturbation approach is quite flexible from a practical point of view, its main limitation clearly lies in the fact that all the argumentation is based only on Tychonov's theorem which does not give a rigorous stability statement.

Also it seems that none of the more comprehensive theorems in the singular perturbation theory can be applied to the considered impedance control problem. The reason is that they require exponential stability not only of the boundary layer system but also of the quasi-steady state system. For the considered desired dynamical impedance behavior of Section 3.2 exponential stability (in case of free motion), however, cannot be shown.

5.5 Summary

In this chapter the singular perturbation approach and its application to the reduced flexible joint robot model was discussed. This method allows to split up the model into two separate subsystems for the fast and the slow part of the dynamics. The fast part of the flexible joint robot model consists of the torque dynamics, while the slow part has the structure of a rigid body robot model. After discussing the commonly used composite controller, a modified control law is proposed which allows to improve the torque dynamics but avoids the computation of the *quasi-steady state* $\boldsymbol{h}(\boldsymbol{q}, \dot{\boldsymbol{q}}, t)$ of the joint torques. The slow part of the system dynamics with the proposed controller also has the structure of a (modified) rigid body system. The controller can thus be seen as a cascade of an inner loop torque controller and an outer loop impedance controller designed for a rigid body robot model. The main disadvantage of the singular perturbation approach is the weak theoretical justification in terms of stability, which is due to the limitations of Tychonov's theorem. Therefore, some more sophisticated approaches are discussed in the next chapter which will bring along stronger stability statements but will also lead to more complex controller structures.

6 Controller Design Based on the Cascaded Structure

In the previous chapter a quite general approach to the control of flexible joint robots was treated, which was not at all restricted to the special problem of Cartesian impedance control. The main disadvantage of the singular perturbation approach is that only very weak stability statements can be made. The reason for this is that in the singular perturbation analysis some coupling terms between the slow and the fast dynamics are neglected. In this chapter it will be investigated, in which way the controller must be modified in order to allow *stronger* stability statements.

Two different approaches will be analyzed. The first one, which is discussed in Section 6.1, is based on the control theory for cascaded systems. While the flexible joint robot model from Section 2.2.3 is not in cascaded form, it can be brought into this form by an inner feedback loop. The second method is based on the *backstepping* approach and will be discussed in Section 6.2.

6.1 Decoupling Based Approach

First, an overview of some relevant results concerning the control of cascaded systems will be given in Section 6.1.1. Then the application to the flexible joint robot model will be presented in Section 6.1.2.

6.1.1 Control Theory for Cascaded Systems

Consider an autonomous triangular system (see Figure 6.1) of the form

$$\dot{\boldsymbol{z}}_1 = \boldsymbol{f}_1(\boldsymbol{z}_1, \boldsymbol{z}_2) \ , \tag{6.1}$$

$$\dot{\boldsymbol{z}}_2 = \boldsymbol{f}_2(\boldsymbol{z}_2) \ , \tag{6.2}$$

where $\boldsymbol{z}_1 \in \mathbb{R}^{n_1}$ and $\boldsymbol{z}_2 \in \mathbb{R}^{n_2}$ are the state variables. It is assumed that the functions $\boldsymbol{f}_1(\boldsymbol{z}_1, \boldsymbol{z}_2)$ and $\boldsymbol{f}_2(\boldsymbol{z}_2)$ are locally Lipschitz and that all solutions exist for $t \geq 0$. Furthermore, it is assumed that the origin is an equilibrium point of

C. Ott: Cartesian Imped. Cntrl. of Redund. & Flexible-Joint Robots, STAR 49, pp. 77–92, 2008.
springerlink.com

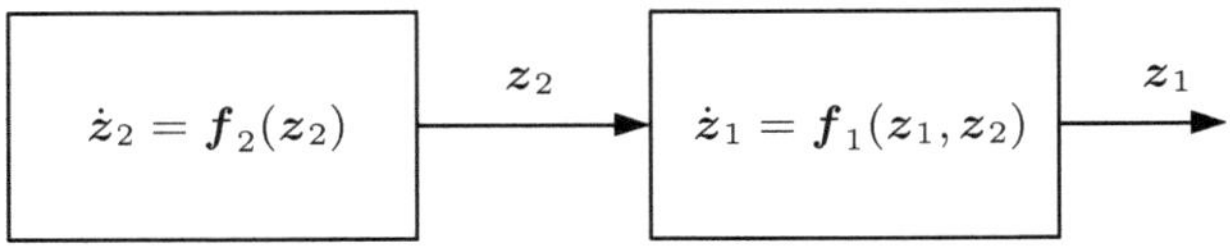

Fig. 6.1. Cascaded system structure

(6.1)-(6.2), i.e. $\boldsymbol{f}_1(\boldsymbol{0},\boldsymbol{0}) = \boldsymbol{0}$ and $\boldsymbol{f}_2(\boldsymbol{0}) = \boldsymbol{0}$. In the following the situation is analyzed when the uncoupled systems (6.2) and

$$\dot{\boldsymbol{z}}_1 = \boldsymbol{f}_1(\boldsymbol{z}_1, \boldsymbol{0}) \tag{6.3}$$

are globally asymptotically stable. Then the question arises under which conditions also the coupled system (6.1)-(6.2) will be asymptotically stable. Locally this is always true (see the references in [SS90]). In order to ensure that this holds also globally it was proven in [SS90] that it is sufficient to show that all solutions of the coupled system remain bounded. This is formulated in a more general form[1] in the following theorem, taken from [SS90].

Theorem 6.1 *If the system system (6.3) is globally asymptotically stable, and if (6.2) is asymptotically stable with region of attraction $\mathcal{A} \subseteq \mathbb{R}^{n_2}$, and if every solution curve of (6.1)-(6.2) with initial point in $\mathbb{R}^{n_1} \times \mathcal{A}$ is bounded for $t > 0$, then the system (6.1)-(6.2) is asymptotically stable with region of attraction $\mathbb{R}^{n_1} \times \mathcal{A}$.*

The proof of this theorem can be found in [SS90]. Notice that Theorem 6.1 also handles the case of global asymptotic stability, if the region of attraction of (6.2) is the whole state space $\mathbb{R}^{n_2}$.

Theorem 6.1 considers the autonomous case. The impedance control problem in Section 3.2, on the other hand, leads to a time-varying control problem even in the case of a rigid body robot. Therefore, also the time-varying case shall be considered. The relevant definitions and lemmata for the stability analysis of time-varying systems can be found in Appendix A.1.

An extension of Theorem 6.1 to a special class of time-varying systems was presented in [Lor01]. This contribution from [Lor01] is presented here in form of the following theorem.

Theorem 6.2 *Consider the system*

$$\dot{\boldsymbol{z}}_1 = \boldsymbol{f}_1(t, \boldsymbol{z}_1) + \boldsymbol{h}(t, \boldsymbol{z})\boldsymbol{z}_2 \tag{6.4}$$

$$\dot{\boldsymbol{z}}_2 = \boldsymbol{f}_2(t, \boldsymbol{z}_2) \tag{6.5}$$

with state $\boldsymbol{z} = (\boldsymbol{z}_1, \boldsymbol{z}_2) \in \mathbb{R}^{n_1} \times \mathbb{R}^{n_2}$. The functions $\boldsymbol{f}_1(t, \boldsymbol{z}_1)$, $\boldsymbol{f}_2(t, \boldsymbol{z}_2)$ and $\boldsymbol{h}(t, \boldsymbol{z})$ are continuous in their arguments, locally Lipschitz in $\boldsymbol{z}$, uniformly in

[1] Considering also the semi-global case in which the region of attraction is not the whole state space.

t, and $\boldsymbol{f}_1(t, \boldsymbol{z}_1)$ *is continuously differentiable in both arguments. This system is uniformly globally asymptotically stable if and only if the following holds:*

- *There exists a nondecreasing function* $H(\cdot)$ *such that*

$$||\boldsymbol{h}(\boldsymbol{z}, t)|| \leq H(||\boldsymbol{z}||) \ . \tag{6.6}$$

- *The systems*

$$\dot{\boldsymbol{z}}_1 = \boldsymbol{f}_1(t, \boldsymbol{z}_1)$$
$$\dot{\boldsymbol{z}}_2 = \boldsymbol{f}_2(t, \boldsymbol{z}_2)$$

  are uniformly globally asymptotically stable.
- *The solutions of (6.4)-(6.5) are uniformly globally bounded.*

Theorem 6.2 gives necessary and sufficient conditions from which one can conclude uniform global asymptotic stability of the triangular system (6.4)-(6.5). Notice that, if the uncoupled systems can even be shown to be exponentially stable, then there exist also other results in the literature which can be applied to such a triangular system (see [Vid93]). The impedance control problem from Section 3.2 though led to a closed loop system for which only (uniform global) asymptotic stability, instead of exponential stability, can be shown (see Proposition 3.4). In the following it will be shown how Theorem 6.2 can be used for the design of a Cartesian impedance controller for the flexible joint robot model.

6.1.2 Decoupling of the Torque Dynamics

The reduced model of a flexible joint robot (without damping), as derived in Section 2.2.3, is given by

$$\boldsymbol{M}(\boldsymbol{q})\ddot{\boldsymbol{q}} + \boldsymbol{C}(\boldsymbol{q}, \dot{\boldsymbol{q}})\dot{\boldsymbol{q}} + \boldsymbol{g}(\boldsymbol{q}) = \boldsymbol{K}(\boldsymbol{\theta} - \boldsymbol{q}) + \boldsymbol{\tau}_{ext} \ , \tag{6.7}$$

$$\boldsymbol{B}\ddot{\boldsymbol{\theta}} + \boldsymbol{K}(\boldsymbol{\theta} - \boldsymbol{q}) = \boldsymbol{\tau}_m \ . \tag{6.8}$$

In order to relate this system to the triangular form from (6.1)-(6.2), respectively (6.4)-(6.5), one can perform the following change of coordinates. As coordinates $\boldsymbol{z}_1$ and $\boldsymbol{z}_2$ the joint angles $\boldsymbol{q}$ and the joint torques $\boldsymbol{\tau} = \boldsymbol{K}(\boldsymbol{\theta} - \boldsymbol{q})$ as well as their first derivatives with respect to time are chosen

$$\boldsymbol{z}_1 = \begin{pmatrix} \boldsymbol{q} \\ \dot{\boldsymbol{q}} \end{pmatrix} , \qquad \boldsymbol{z}_2 = \begin{pmatrix} \boldsymbol{\tau} \\ \dot{\boldsymbol{\tau}} \end{pmatrix} . \tag{6.9}$$

Notice that these are the same state variables as used for the singular perturbation approach in the last chapter. One may substitute $\boldsymbol{K}(\boldsymbol{\theta} - \boldsymbol{q}) = \boldsymbol{\tau}$ and $\ddot{\boldsymbol{\theta}} = \boldsymbol{K}^{-1}\ddot{\boldsymbol{\tau}} + \ddot{\boldsymbol{q}}$ in equation (6.8) in order to obtain

$$\boldsymbol{B}\boldsymbol{K}^{-1}\ddot{\boldsymbol{\tau}} + \boldsymbol{\tau} = \boldsymbol{\tau}_m - \boldsymbol{B}\ddot{\boldsymbol{q}} \ . \tag{6.10}$$

The link accelerations can be eliminated from (6.10) by using (6.7) because the inertia matrix $\boldsymbol{M}(\boldsymbol{q})$ is positive definite and therefore always invertible (Property 2.5). Therefore, one gets the new system equations

$$\boldsymbol{M}(\boldsymbol{q})\ddot{\boldsymbol{q}} + \boldsymbol{C}(\boldsymbol{q},\dot{\boldsymbol{q}})\dot{\boldsymbol{q}} + \boldsymbol{g}(\boldsymbol{q}) = \boldsymbol{\tau} + \boldsymbol{\tau}_{ext} \; , \tag{6.11}$$

$$\boldsymbol{B}\boldsymbol{K}^{-1}\ddot{\boldsymbol{\tau}} + \boldsymbol{\tau} = \boldsymbol{\tau}_m - \boldsymbol{B}\boldsymbol{M}(\boldsymbol{q})^{-1}\left(\boldsymbol{\tau} + \boldsymbol{\tau}_{ext} - \boldsymbol{C}(\boldsymbol{q},\dot{\boldsymbol{q}})\dot{\boldsymbol{q}} - \boldsymbol{g}(\boldsymbol{q})\right) \; . \tag{6.12}$$

So far, the model has been reformulated analogously to the last chapter. Clearly, the model (6.11)-(6.12) does not have a triangular structure like (6.4)-(6.5). But it can be brought into this form by the following feedback compensation

$$\boldsymbol{\tau}_m = \boldsymbol{u} + \boldsymbol{B}\boldsymbol{M}(\boldsymbol{q})^{-1}\left(\boldsymbol{\tau} + \boldsymbol{\tau}_{ext} - \boldsymbol{C}(\boldsymbol{q},\dot{\boldsymbol{q}})\dot{\boldsymbol{q}} - \boldsymbol{g}(\boldsymbol{q})\right) \; , \tag{6.13}$$

where $\boldsymbol{u} \in \mathbb{R}^n$ is an intermediate control input. The system, with decoupled torque dynamics, can then be written as

$$\boldsymbol{M}(\boldsymbol{q})\ddot{\boldsymbol{q}} + \boldsymbol{C}(\boldsymbol{q},\dot{\boldsymbol{q}})\dot{\boldsymbol{q}} + \boldsymbol{g}(\boldsymbol{q}) = \boldsymbol{\tau} + \boldsymbol{\tau}_{ext} \; , \tag{6.14}$$

$$\boldsymbol{B}\boldsymbol{K}^{-1}\ddot{\boldsymbol{\tau}} + \boldsymbol{\tau} = \boldsymbol{u} \; . \tag{6.15}$$

In order to shift the state of the torque dynamics equation (6.15) to zero, a *desired torque variable* $\boldsymbol{\tau}_d$ and a torque error variable $\boldsymbol{e}_\tau$ are introduced as

$$\boldsymbol{e}_\tau = \boldsymbol{\tau} - \boldsymbol{\tau}_d \; . \tag{6.16}$$

The desired torque $\boldsymbol{\tau}_d$ will be used later for the design of a Cartesian impedance controller. With this new state variable the system equations are written as follows:

$$\boldsymbol{M}(\boldsymbol{q})\ddot{\boldsymbol{q}} + \boldsymbol{C}(\boldsymbol{q},\dot{\boldsymbol{q}})\dot{\boldsymbol{q}} + \boldsymbol{g}(\boldsymbol{q}) = \boldsymbol{e}_\tau + \boldsymbol{\tau}_d + \boldsymbol{\tau}_{ext} \; , \tag{6.17}$$

$$\boldsymbol{B}\boldsymbol{K}^{-1}(\ddot{\boldsymbol{e}}_\tau + \ddot{\boldsymbol{\tau}}_d) + \boldsymbol{e}_\tau + \boldsymbol{\tau}_d = \boldsymbol{u} \; . \tag{6.18}$$

The intermediate input $\boldsymbol{u}$ shall be designed such that the torque error dynamics gets globally asymptotically stable. With the feedback law

$$\boldsymbol{u} = \boldsymbol{\tau}_d + \boldsymbol{B}\boldsymbol{K}^{-1}\left(\ddot{\boldsymbol{\tau}}_d - \boldsymbol{D}_\tau\dot{\boldsymbol{e}}_\tau - \boldsymbol{K}_\tau\boldsymbol{e}_\tau\right) \; , \tag{6.19}$$

where $\boldsymbol{K}_\tau$ and $\boldsymbol{D}_\tau$ are positive definite controller gain matrices, the torque error dynamics is given by the linear time-invariant system

$$\ddot{\boldsymbol{e}}_\tau + \boldsymbol{D}_\tau\dot{\boldsymbol{e}}_\tau + \left(\boldsymbol{K}_\tau + \boldsymbol{K}\boldsymbol{B}^{-1}\right)\boldsymbol{e}_\tau = \boldsymbol{0} \; , \tag{6.20}$$

which clearly is exponentially stable for positive definite controller gain matrices $\boldsymbol{D}_\tau$ and $\boldsymbol{K}_\tau$.

6.1.3 Impedance Controller Design

The closed loop system is given by equations (6.17) and (6.20) which are in triangular form, provided that the torque $\boldsymbol{\tau}_d$ is a function of the state variables

$\boldsymbol{q}$ and $\dot{\boldsymbol{q}}$ only. At the beginning of this chapter, the situation was analyzed in which the uncoupled system (6.3) was asymptotically stable. In the case of the flexible joint robot model, the system (6.3) corresponds to (6.17) with $\boldsymbol{e}_\tau = \boldsymbol{0}$. The variable $\boldsymbol{\tau}_d$ will now be used to make this system asymptotically stable.

In the following the desired impedance of Section 3.2 is considered for the non-redundant and non-singular case by restricting the considered workspace to $\bar{\mathcal{Q}}_c^p$. The same notation as in Chapter 3 is used throughout this chapter. For the ease of reference the desired dynamic behavior for the Cartesian coordinates $\boldsymbol{x} = \boldsymbol{f}(\boldsymbol{q})$ and a virtual equilibrium position $\boldsymbol{x}_d$ are recapitulated here. In the considered non-redundant and non-singular case the analytical Jacobian $\boldsymbol{J}(\boldsymbol{q}) = \partial \boldsymbol{f}(\boldsymbol{q})/\partial \boldsymbol{q}$ is assumed to be invertible. The desired impedance relation between the generalized Cartesian force $\boldsymbol{F}_{ext}$, which is related to $\boldsymbol{\tau}_{ext}$ via $\boldsymbol{\tau}_{ext} = \boldsymbol{J}(\boldsymbol{q})^T \boldsymbol{F}_{ext}$, and the Cartesian position error $\tilde{\boldsymbol{x}} = \boldsymbol{x} - \boldsymbol{x}_d$ is given by (cf. (3.16))

$$\boldsymbol{\Lambda}(\boldsymbol{x})\ddot{\tilde{\boldsymbol{x}}} + (\boldsymbol{\mu}(\boldsymbol{x}, \dot{\boldsymbol{x}}) + \boldsymbol{D}_d)\,\dot{\tilde{\boldsymbol{x}}} + \boldsymbol{K}_d \tilde{\boldsymbol{x}} = \boldsymbol{F}_{ext} \ . \tag{6.21}$$

The matrices $\boldsymbol{K}_d$ and $\boldsymbol{D}_d$ are again the symmetric and positive definite matrices of the desired stiffness and damping. The Cartesian inertia matrix $\boldsymbol{\Lambda}(\boldsymbol{x})$ and the Coriolis/centrifugal matrix $\boldsymbol{\mu}(\boldsymbol{x}, \dot{\boldsymbol{x}})$ are defined in (3.10) and (3.11). The desired trajectory $\boldsymbol{x}_d(t)$ is assumed to be four times continuously differentiable. By performing the same steps as in Section 3.1.2, it can be verified that the system equations (6.17) and (6.20) can be written in Cartesian coordinates as

$$\boldsymbol{\Lambda}(\boldsymbol{x})\ddot{\boldsymbol{x}} + \boldsymbol{\mu}(\boldsymbol{x}, \dot{\boldsymbol{x}})\dot{\boldsymbol{x}} + \boldsymbol{J}(\boldsymbol{q})^{-T}\boldsymbol{g}(\boldsymbol{q}) = \boldsymbol{F}_{ext} + \boldsymbol{J}(\boldsymbol{q})^{-T}(\boldsymbol{e}_\tau + \boldsymbol{\tau}_d) \ , \tag{6.22}$$

$$\ddot{\boldsymbol{e}}_\tau + \boldsymbol{D}_\tau \dot{\boldsymbol{e}}_\tau + \left(\boldsymbol{K}_\tau + \boldsymbol{K}\boldsymbol{B}^{-1}\right)\boldsymbol{e}_\tau = \boldsymbol{0} \ . \tag{6.23}$$

The controller from (3.18), which was designed for the rigid body model, can now be used for the desired torque $\boldsymbol{\tau}_d$. Therefore, the complete controller for the flexible joint robot model follows from equation (6.13), (6.19), and (3.18), and is given by

$$\begin{aligned}
\boldsymbol{\tau}_m &= \boldsymbol{\tau}_d + \boldsymbol{B}\boldsymbol{K}^{-1}\left(\ddot{\boldsymbol{\tau}}_d - \boldsymbol{D}_\tau(\dot{\boldsymbol{\tau}} - \dot{\boldsymbol{\tau}}_d) - \boldsymbol{K}_\tau(\boldsymbol{\tau} - \boldsymbol{\tau}_d)\right) \\
&\quad + \boldsymbol{B}\boldsymbol{M}(\boldsymbol{q})^{-1}\left(\boldsymbol{\tau} + \boldsymbol{\tau}_{ext} - \boldsymbol{C}(\boldsymbol{q}, \dot{\boldsymbol{q}})\dot{\boldsymbol{q}} - \boldsymbol{g}(\boldsymbol{q})\right) \ , \\
\boldsymbol{\tau}_d &= \boldsymbol{g}(\boldsymbol{q}) + \boldsymbol{J}(\boldsymbol{q})^T\left(\boldsymbol{\Lambda}(\boldsymbol{x})\ddot{\boldsymbol{x}}_d + \boldsymbol{\mu}(\boldsymbol{x}, \dot{\boldsymbol{x}})\dot{\boldsymbol{x}}_d - \boldsymbol{K}_d\tilde{\boldsymbol{x}} - \boldsymbol{D}_d\dot{\tilde{\boldsymbol{x}}}\right) \ .
\end{aligned}$$

Substituting $\boldsymbol{\tau}_d$ into the system equations (6.22)-(6.23) leads to the corresponding closed loop dynamics

$$\boldsymbol{\Lambda}(\boldsymbol{x})\ddot{\tilde{\boldsymbol{x}}} + (\boldsymbol{\mu}(\boldsymbol{x}, \dot{\boldsymbol{x}}) + \boldsymbol{D}_d)\dot{\tilde{\boldsymbol{x}}} + \boldsymbol{K}_d\tilde{\boldsymbol{x}} = \boldsymbol{F}_{ext} + \boldsymbol{J}(\boldsymbol{q})^{-T}\boldsymbol{e}_\tau \ , \tag{6.24}$$

$$\ddot{\boldsymbol{e}}_\tau + \boldsymbol{D}_\tau\dot{\boldsymbol{e}}_\tau + (\boldsymbol{K}_\tau + \boldsymbol{K}\boldsymbol{B}^{-1})\boldsymbol{e}_\tau = \boldsymbol{0} \ . \tag{6.25}$$

Notice that, when the coupling term $\boldsymbol{J}(\boldsymbol{q})^{-T}\boldsymbol{e}_\tau$ is neglected in (6.24), then it corresponds exactly to the desired dynamics (6.21), which, according to Proposition 3.4, is uniformly globally asymptotically stable. Notice also that the closed loop system (6.24)-(6.25) is time-varying, which is not surprising because this is also the case for the desired dynamics[2].

[2] This is due to the occurrence of $\boldsymbol{x}_d(t)$ in the equations of motion.

6.1.4 Stability Analysis

For the desired dynamics from (6.21) two important properties concerning stability and passivity were given in Proposition 3.4 and Proposition 3.5. In the following it will be shown that these statements hold also for the closed loop dynamics (6.24)-(6.25) of the flexible joint model.

Proposition 6.3. *Let the desired trajectory $\boldsymbol{x}_d(t)$ be four times continuously differentiable. Assume further that the Cartesian coordinates are globally valid, i.e. $\bar{\mathcal{Q}}_c^p = \mathbb{R}^n$. Then, for the case of free motion ($\boldsymbol{F}_{ext} = \boldsymbol{0}$) the system (6.24)-(6.25) with symmetric and positive definite matrices $\boldsymbol{K}_d$, $\boldsymbol{D}_d$, $\boldsymbol{K}_\tau$, and $\boldsymbol{D}_\tau$ is uniformly globally asymptotically stable.*

Proof. In order to rewrite the system (6.24)-(6.25) for $\boldsymbol{F}_{ext} = \boldsymbol{0}$ in the state variables $\boldsymbol{z} = (\tilde{\boldsymbol{x}}, \dot{\tilde{\boldsymbol{x}}}, \boldsymbol{e}_\tau, \dot{\boldsymbol{e}}_\tau)$ only, it is convenient to make the following substitutions: $\boldsymbol{J}(\tilde{\boldsymbol{x}}, t) := \boldsymbol{J}(\boldsymbol{f}^{-1}(\tilde{\boldsymbol{x}} + \boldsymbol{x}_d(t))) = \boldsymbol{J}(\boldsymbol{q})$, $\boldsymbol{\Lambda}(\tilde{\boldsymbol{x}}, t) := \boldsymbol{\Lambda}(\tilde{\boldsymbol{x}} + \boldsymbol{x}_d(t)) = \boldsymbol{\Lambda}(\boldsymbol{x})$, and $\boldsymbol{\mu}(\tilde{\boldsymbol{x}}, \dot{\tilde{\boldsymbol{x}}}, t) := \boldsymbol{\mu}(\tilde{\boldsymbol{x}} + \boldsymbol{x}_d(t), \dot{\tilde{\boldsymbol{x}}} + \dot{\boldsymbol{x}}_d(t)) = \boldsymbol{\mu}(\boldsymbol{x}, \dot{\boldsymbol{x}})$. Also for the linear part of the system the substitutions $\boldsymbol{w} = (\boldsymbol{w}_1^T, \boldsymbol{w}_2^T)^T = (\boldsymbol{e}_\tau^T, \dot{\boldsymbol{e}}_\tau^T)^T$ and

$$\boldsymbol{A} = \begin{bmatrix} \boldsymbol{0} & \boldsymbol{I} \\ -\boldsymbol{D}_\tau & -(\boldsymbol{K}_\tau + \boldsymbol{K}\boldsymbol{B}^{-1}) \end{bmatrix}$$

are made. This leads to the system

$$\boldsymbol{\Lambda}(\tilde{\boldsymbol{x}}, t)\ddot{\tilde{\boldsymbol{x}}} + (\boldsymbol{\mu}(\tilde{\boldsymbol{x}}, \dot{\tilde{\boldsymbol{x}}}, t) + \boldsymbol{D}_d)\dot{\tilde{\boldsymbol{x}}} + \boldsymbol{K}_d\tilde{\boldsymbol{x}} = \boldsymbol{J}(\tilde{\boldsymbol{x}}, t)^{-T}\boldsymbol{w}_1 \ , \tag{6.26}$$

$$\dot{\boldsymbol{w}} = \boldsymbol{A}\boldsymbol{w} \ . \tag{6.27}$$

Notice that this system has a cascaded structure because the linear system $\dot{\boldsymbol{w}} = \boldsymbol{A}\boldsymbol{w}$ does not depend on the state variables $\tilde{\boldsymbol{x}}$ and $\dot{\tilde{\boldsymbol{x}}}$. For a nonlinear time-invariant system in such a cascaded form to be asymptotically stable it is necessary to show that all solutions of the coupled system remain bounded and the uncoupled subsystems are asymptotically stable (Theorem 6.1). In the time-varying case the conditions of Theorem 6.2 must be fulfilled.

For the system (6.26)-(6.27) the first condition of Theorem 6.2, the existence of a nondecreasing function $H(\cdot)$ for which (6.6) holds, is fulfilled due to the assumption that the Jacobian $\boldsymbol{J}(\tilde{\boldsymbol{x}}, t)$ is non-singular. Thus, there exists a $\delta \in \mathbb{R}$, $0 < \delta < \infty$, such that

$$||\boldsymbol{J}(\tilde{\boldsymbol{x}}, t)^{-T}|| \leq \sup_{t \in [0, \infty)} \sqrt{\lambda_{max}(\boldsymbol{J}(\tilde{\boldsymbol{x}}, t)^{-1}\boldsymbol{J}(\tilde{\boldsymbol{x}}, t)^{-T})} < \delta \ ,$$

with $\lambda_{max}(\boldsymbol{H}(t))$ as the maximum eigenvalue of a matrix $\boldsymbol{H}(t)$ at the time t.

Uniform global asymptotic stability of the two uncoupled subsystems is given by Proposition 3.4 and the fact that the linear system $\dot{\boldsymbol{w}} = \boldsymbol{A}\boldsymbol{w}$ is even globally exponentially stable for positive definite matrices $\boldsymbol{D}_\tau$ and $\boldsymbol{K}_\tau$.

Hence it is sufficient to show that all solutions of the coupled system are uniformly globally bounded.

Consider therefore the following continuously differentiable, positive definite, decrescent and radially unbounded function (see Definition A.2)[3]

$$V_c(\tilde{\boldsymbol{x}}, \dot{\tilde{\boldsymbol{x}}}, \boldsymbol{w}, t) = \frac{1}{2}\dot{\tilde{\boldsymbol{x}}}^T \boldsymbol{\Lambda}(\tilde{\boldsymbol{x}}, t)\dot{\tilde{\boldsymbol{x}}} + \frac{1}{2}\tilde{\boldsymbol{x}}^T \boldsymbol{K}_d \tilde{\boldsymbol{x}} + \frac{1}{2}\boldsymbol{w}^T \boldsymbol{P}\boldsymbol{w} \tag{6.28}$$

with a positive definite matrix $\boldsymbol{P} \in \mathbb{R}^{2n\times 2n}$. At this point it is worth mentioning that $V_c(\tilde{\boldsymbol{x}}, \dot{\tilde{\boldsymbol{x}}}, \boldsymbol{w}, t)$ is bounded from above and below by some time-invariant, radially unbounded and positive definite functions[4] $W_1(\tilde{\boldsymbol{x}}, \dot{\tilde{\boldsymbol{x}}}, \boldsymbol{w})$ and $W_2(\tilde{\boldsymbol{x}}, \dot{\tilde{\boldsymbol{x}}}, \boldsymbol{w})$

$$\begin{aligned} W_1(\tilde{\boldsymbol{x}}, \dot{\tilde{\boldsymbol{x}}}, \boldsymbol{w}) &\leq V_c(\tilde{\boldsymbol{x}}, \dot{\tilde{\boldsymbol{x}}}, \boldsymbol{w}, t) \leq W_2(\tilde{\boldsymbol{x}}, \dot{\tilde{\boldsymbol{x}}}, \boldsymbol{w}) \\ W_1(\tilde{\boldsymbol{x}}, \dot{\tilde{\boldsymbol{x}}}, \boldsymbol{w}) &= \tfrac{1}{2}\lambda_1 ||\dot{\tilde{\boldsymbol{x}}}||_2^2 + \tfrac{1}{2}\tilde{\boldsymbol{x}}^T \boldsymbol{K}_d \tilde{\boldsymbol{x}} + \tfrac{1}{2}\boldsymbol{w}^T \boldsymbol{P}\boldsymbol{w} \\ W_2(\tilde{\boldsymbol{x}}, \dot{\tilde{\boldsymbol{x}}}, \boldsymbol{w}) &= \tfrac{1}{2}\lambda_2 ||\dot{\tilde{\boldsymbol{x}}}||_2^2 + \tfrac{1}{2}\tilde{\boldsymbol{x}}^T \boldsymbol{K}_d \tilde{\boldsymbol{x}} + \tfrac{1}{2}\boldsymbol{w}^T \boldsymbol{P}\boldsymbol{w} \end{aligned}$$

where

$$\begin{aligned} 0 < \lambda_1 < \inf_{t\in[0,\infty)} \lambda_{min}(\boldsymbol{\Lambda}(\tilde{\boldsymbol{x}}, t)) < \\ \sup_{t\in[0,\infty)} \lambda_{max}(\boldsymbol{\Lambda}(\tilde{\boldsymbol{x}}, t)) < \lambda_2 < \infty \end{aligned}$$

with $\lambda_{min}(\boldsymbol{\Lambda}(t))$ and $\lambda_{max}(\boldsymbol{\Lambda}(t))$ as the minimum and maximum eigenvalue of $\boldsymbol{\Lambda}(t)$ at the time t. This is ensured by Property 3.3.

From the well known skew symmetry property in Lemma 3.2 it follows that the time derivative of $V_c(\tilde{\boldsymbol{x}}, \dot{\tilde{\boldsymbol{x}}}, \boldsymbol{w}, t)$ along the solutions of (6.24)-(6.25) is given by

$$\dot{V}_c(\tilde{\boldsymbol{x}}, \dot{\tilde{\boldsymbol{x}}}, \boldsymbol{w}, t) = -\dot{\tilde{\boldsymbol{x}}}^T \boldsymbol{D}_d \dot{\tilde{\boldsymbol{x}}} - \frac{1}{2}\boldsymbol{w}^T \boldsymbol{Q}\boldsymbol{w} + \dot{\tilde{\boldsymbol{x}}}^T \boldsymbol{J}(\tilde{\boldsymbol{x}}, t)^{-T} \boldsymbol{w}_1 \ ,$$

where the matrix $\boldsymbol{Q} \in \mathbb{R}^{2n\times 2n}$ is related to $\boldsymbol{P}$ by the Lyapunov equation $\boldsymbol{Q} = -(\boldsymbol{P}\boldsymbol{A} + \boldsymbol{A}^T\boldsymbol{P})$. For the following argumentation it is important to notice that, due to Lemma A.21, one may also choose an arbitrary positive definite matrix for $\boldsymbol{Q}$ and calculate a unique positive definite matrix $\boldsymbol{P}$ for which $\boldsymbol{Q} = -(\boldsymbol{P}\boldsymbol{A} + \boldsymbol{A}^T\boldsymbol{P})$ holds. This is possible because $\boldsymbol{A}$ is Hurwitz for positive definite matrices $\boldsymbol{D}_\tau$ and $\boldsymbol{K}_\tau$ (see Lemma A.21).

Obviously, $\dot{V}_c(\tilde{\boldsymbol{x}}, \dot{\tilde{\boldsymbol{x}}}, \boldsymbol{w}, t)$ can be written in matrix form

$$\dot{V}_c(\tilde{\boldsymbol{x}}, \dot{\tilde{\boldsymbol{x}}}, \boldsymbol{w}, t) = -\begin{pmatrix} \dot{\tilde{\boldsymbol{x}}} \\ \boldsymbol{w}_1 \\ \boldsymbol{w}_2 \end{pmatrix}^T \boldsymbol{N}(\tilde{\boldsymbol{x}}, t) \begin{pmatrix} \dot{\tilde{\boldsymbol{x}}} \\ \boldsymbol{w}_1 \\ \boldsymbol{w}_2 \end{pmatrix}$$

---

[3] Notice that the eigenvalues of the matrix $\boldsymbol{\Lambda}(\tilde{\boldsymbol{x}}, t)$ are bounded from above and below by some positive constants for all $t \in \mathbb{R}$ and all $\tilde{\boldsymbol{x}} \in \mathbb{R}^n$ (Property 3.3).

[4] These properties make the function $V_c(\tilde{\boldsymbol{x}}, \dot{\tilde{\boldsymbol{x}}}, \boldsymbol{w}, t)$ positive definite, decrescent and radially unbounded.

with

$$
\boldsymbol{N}(\tilde{\boldsymbol{x}},t) = \begin{bmatrix} \boldsymbol{D}_d & \begin{bmatrix} -\frac{1}{2}\boldsymbol{J}(\tilde{\boldsymbol{x}},t)^{-T} & \mathbf{0} \end{bmatrix} \\ \begin{bmatrix} -\frac{1}{2}\boldsymbol{J}^{-1}(\tilde{\boldsymbol{x}},t) \\ \mathbf{0} \end{bmatrix} & \boldsymbol{Q} \end{bmatrix} .
$$

From Lemma A.20 it follows that a necessary and sufficient condition for $\boldsymbol{N}(\tilde{\boldsymbol{x}},t)$ to be positive definite[5] is

$$
\boldsymbol{Q} - \frac{1}{4}\boldsymbol{J}^{-1}(\tilde{\boldsymbol{x}},t)\boldsymbol{D}_d^{-1}\boldsymbol{J}^{-T}(\tilde{\boldsymbol{x}},t) > 0 \ ,
$$

which can be fulfilled for every positive definite matrix $\boldsymbol{D}_d$ because by assumption $\boldsymbol{J}(\tilde{\boldsymbol{x}},t)$ does not get singular and the matrix $\boldsymbol{Q}$ is some positive definite matrix which may be chosen arbitrarily. Hence, one can conclude that

$$
\dot{V}_c(\tilde{\boldsymbol{x}},\dot{\tilde{\boldsymbol{x}}},\boldsymbol{w},t) \le 0 \ .
$$

By using the above mentioned properties of $V_c(\tilde{\boldsymbol{x}},\dot{\tilde{\boldsymbol{x}}},\boldsymbol{w},t)$, Lemma A.8 can then be applied to show that the solutions of (6.24)-(6.25) are uniformly globally bounded. Therefore, it can be concluded that Proposition 6.3 follows from Theorem 6.2.

Notice that the need to refer to Theorem 6.2 in this stability proof results from the facts that, on the one hand, the considered system is time-varying and, on the other hand, the time derivative of the chosen function $V_c(\tilde{\boldsymbol{x}},\dot{\tilde{\boldsymbol{x}}},\boldsymbol{w},t)$ is not negative definite but only negative semi-definite. This fact, together with the remark that $\boldsymbol{Q}$ can be arbitrarily chosen, are the most important differences to the proofs in [LG96] and [LG95].

Proposition 6.4. *For $\dot{\boldsymbol{x}}_d(t) = \mathbf{0}$, the system (6.24)-(6.25) with the positive definite matrices $\boldsymbol{K}_d$, $\boldsymbol{D}_d$, $\boldsymbol{K}_\tau$, and $\boldsymbol{D}_\tau$ gets time-invariant and represents a passive map from the external force $\boldsymbol{F}_{ext}$ to the velocity error $\dot{\tilde{\boldsymbol{x}}}$.*

Proof. The function $V_c(\tilde{\boldsymbol{x}},\dot{\tilde{\boldsymbol{x}}},\boldsymbol{w},t)$ can be chosen as an appropriate storage function to show passivity. In the case of $\boldsymbol{F}_{ext} \neq \mathbf{0}$, its time derivative along the solution curve of (6.24)-(6.25) is given by

$$
\dot{V}_c(\tilde{\boldsymbol{x}},\dot{\tilde{\boldsymbol{x}}},\boldsymbol{w},t) = -\begin{pmatrix} \dot{\tilde{\boldsymbol{x}}} \\ \boldsymbol{w}_1 \\ \boldsymbol{w}_2 \end{pmatrix}^T \boldsymbol{N}(\tilde{\boldsymbol{x}},t) \begin{pmatrix} \dot{\tilde{\boldsymbol{x}}} \\ \boldsymbol{w}_1 \\ \boldsymbol{w}_2 \end{pmatrix} + \dot{\tilde{\boldsymbol{x}}}^T\boldsymbol{F}_{ext} \le \dot{\tilde{\boldsymbol{x}}}^T\boldsymbol{F}_{ext} \ . \quad (6.29)
$$

The matrix $\boldsymbol{N}(\tilde{\boldsymbol{x}},t)$ has already been shown to be positive definite. Clearly, the inequality (6.29) proves the passivity property of Proposition 6.4.

[5] Notice also that, in addition to Lemma A.20, it is also possible to show that all eigenvalues of $\boldsymbol{N}(\tilde{\boldsymbol{x}},t)$ are bounded from above and below by some positive constants, because $\boldsymbol{Q}$ can be chosen arbitrarily and the matrix $\boldsymbol{J}(\tilde{\boldsymbol{x}},t)$ is non-singular.

6.1.5 Controller Discussion

The complete controller equations are given by

$$\begin{aligned}\boldsymbol{\tau}_m &= \boldsymbol{\tau}_d + \boldsymbol{B}\boldsymbol{K}^{-1}\left(\underline{\ddot{\boldsymbol{\tau}}_d} - \boldsymbol{D}_\tau(\dot{\boldsymbol{\tau}} - \underline{\dot{\boldsymbol{\tau}}_d}) - \boldsymbol{K}_\tau(\boldsymbol{\tau} - \boldsymbol{\tau}_d)\right) \\ &\quad + \underline{\underline{\boldsymbol{B}\boldsymbol{M}(\boldsymbol{q})^{-1}\left(\boldsymbol{\tau} + \boldsymbol{\tau}_{ext} - \boldsymbol{C}(\boldsymbol{q},\dot{\boldsymbol{q}})\dot{\boldsymbol{q}} - \boldsymbol{g}(\boldsymbol{q})\right)}}\ , \\ \boldsymbol{\tau}_d &= \boldsymbol{g}(\boldsymbol{q}) + \boldsymbol{J}(\boldsymbol{q})^T\left(\boldsymbol{\Lambda}(\boldsymbol{x})\ddot{\boldsymbol{x}}_d + \boldsymbol{\mu}(\boldsymbol{x},\dot{\boldsymbol{x}})\dot{\boldsymbol{x}}_d - \boldsymbol{K}_d\tilde{\boldsymbol{x}} - \boldsymbol{D}_d\dot{\tilde{\boldsymbol{x}}}\right)\ .\end{aligned} \tag{6.30}$$

Notice that the controller structure is quite similar to the controller of the last chapter. The main advantage of this controller compared to the singular perturbation based controller is that it allows a proof of the asymptotic stability without referring to an approximate analysis. In addition to the singular perturbation based controller, this controller contains also the (underlined) *feedforward terms* $\dot{\boldsymbol{\tau}}_d$ and $\ddot{\boldsymbol{\tau}}_d$ for the inner torque control loop as well as the twice underlined term $\boldsymbol{B}\ddot{\boldsymbol{q}}$ which basically establishes a partial feedback linearization. Since $\ddot{\boldsymbol{q}}$ usually cannot be measured directly in practice it is expressed via $\boldsymbol{q}$, $\dot{\boldsymbol{q}}$, $\boldsymbol{\tau}$, and the external torques $\boldsymbol{\tau}_{ext}$ in the above controller equation. Therefore, it is necessary to measure the external torques even for the simplified desired dynamics in (6.21). This clearly is a disadvantage compared to the simpler singular perturbation based controller. Furthermore, the terms $\dot{\boldsymbol{\tau}}_d$ and $\ddot{\boldsymbol{\tau}}_d$ contain the link acceleration $\ddot{\boldsymbol{q}}$ and the jerk $\boldsymbol{q}^{(3)}$, since $\boldsymbol{\tau}_d$ is a function of $\boldsymbol{q}$ and $\dot{\boldsymbol{q}}$ (or equivalently $\boldsymbol{x}$ and $\dot{\boldsymbol{x}}$). The computation of these signals can be done for a robot with only a few joints (e.g. one to three joints) but is very problematic for a robot with six or seven joints. In principle, one can compute these signals based on the model equations via

$$\begin{aligned}\ddot{\boldsymbol{q}} &= \boldsymbol{M}(\boldsymbol{q})^{-1}\left(\boldsymbol{\tau} + \boldsymbol{\tau}_{ext} - \boldsymbol{C}(\boldsymbol{q},\dot{\boldsymbol{q}})\dot{\boldsymbol{q}} - \boldsymbol{g}(\boldsymbol{q})\right)\ , \\ \boldsymbol{q}^{(3)} &= \boldsymbol{M}(\boldsymbol{q})^{-1}\left(\dot{\boldsymbol{\tau}} - \dot{\boldsymbol{M}}(\boldsymbol{q})\ddot{\boldsymbol{q}} + \frac{\mathrm{d}}{\mathrm{d}t}\left(\boldsymbol{\tau}_{ext} - \boldsymbol{C}(\boldsymbol{q},\dot{\boldsymbol{q}})\dot{\boldsymbol{q}} - \boldsymbol{g}(\boldsymbol{q})\right)\right)\ .\end{aligned}$$

This requires additionally the computation of $\dot{\boldsymbol{\tau}}_{ext}$ (e.g. via filtering) and $\dot{\boldsymbol{C}}(\boldsymbol{q},\dot{\boldsymbol{q}})$. Another critical point is the use of the inverse inertia matrix in this computation. While the inertia matrix is always positive definite (and therefore invertible) its eigenvalues have different orders of magnitude and also can vary by several orders of magnitude depending on the joint configuration [Fea04]. The inversion can therefore become numerically ill-posed.

Another possible approach for the computation of $\ddot{\boldsymbol{q}}$ and $\boldsymbol{q}^{(3)}$ is the use of acceleration sensors[6] and/or appropriate filtering techniques. The use of joint acceleration sensors seems to be a promising approach for the implementation of the discussed controller. However, it turned out that an appropriate integration of joint acceleration sensors into the mechanics of a robot is difficult [Gei03].

In order to summarize this discussion, one can say that from an implementation point of view, this controller is indeed quite difficult. For a robot with a

---

[6] Either by a six-dof acceleration sensor at the tip in case of a six-dof robot, or even better by joint acceleration sensors.

few degrees-of-freedom, however, the implementation seems to be achievable but requires an appropriate filtering either of the external torques or of the signals from an acceleration sensor. On the other hand the discussed controller answers the question by which terms the singular perturbation based controller must be extended in order to allow a rigorous stability analysis.

6.2 Backstepping Based Approach

At first the basic design idea of (integrator) backstepping is described in Section 6.2.1. The application to the flexible joint robot model is then given in Section 6.2.2.

6.2.1 Backstepping Design Procedure

Integrator backstepping is a stepwise design procedure for the construction of Lyapunov functions which can be applied to a large class of systems. In the following the basic idea is described for a particular class of systems, in which one backstepping step is sufficient. But the procedure can readily be applied recursively to more complicated systems.

Consider a system of the form (Figure 6.2)

$$\dot{\boldsymbol{z}}_1 = \boldsymbol{f}_1(\boldsymbol{z}_1) + \boldsymbol{F}_2(\boldsymbol{z}_1)\boldsymbol{z}_2 \ , \tag{6.31}$$

$$\dot{\boldsymbol{z}}_2 = \boldsymbol{u} \ , \tag{6.32}$$

with the state variables $\boldsymbol{z}_1 \in \mathbb{R}^{n_1}$, $\boldsymbol{z}_2 \in \mathbb{R}^{n_2}$, and the input $\boldsymbol{u} \in \mathbb{R}^{n_2}$. The functions $\boldsymbol{f}_1(\boldsymbol{z}_1) : \mathbb{R}^{n_1} \to \mathbb{R}^{n_1}$ and $\boldsymbol{F}_2(\boldsymbol{z}_1) : \mathbb{R}^{n_1} \to \mathbb{R}^{n_1 \times n_2}$ are assumed to be known. Notice that the system equations for the state variables $\boldsymbol{z}_1$ are not directly influenced by the control input $\boldsymbol{u}$, but only indirectly via $\boldsymbol{z}_2$. In order to design a controller which stabilizes this system, first the variable $\boldsymbol{z}_2$ is considered as a *virtual* control input for (6.31). It is assumed now that a differentiable control law $\boldsymbol{z}_2 = \boldsymbol{k}(\boldsymbol{z}_1)$ is known, which stabilizes (6.31). Then there exists a positive definite Lyapunov function $V_1(\boldsymbol{z}_1)$ for the system

$$\dot{\boldsymbol{z}}_1 = \boldsymbol{f}_1(\boldsymbol{z}_1) + \boldsymbol{F}_2(\boldsymbol{z}_1)\boldsymbol{k}(\boldsymbol{z}_1) \tag{6.33}$$

which has a negative semi-definite derivative

$$\dot{V}_1(\boldsymbol{z}_1) = \frac{\partial V_1(\boldsymbol{z}_1)}{\partial \boldsymbol{z}_1}\left(\boldsymbol{f}_1(\boldsymbol{z}_1) + \boldsymbol{F}_2(\boldsymbol{z}_1)\boldsymbol{k}(\boldsymbol{z}_1)\right) \leq 0 \ . \tag{6.34}$$

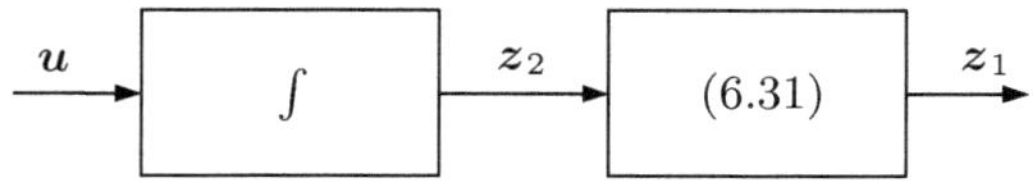

Fig. 6.2. System structure for integrator backstepping

Then the error variable $\boldsymbol{e}_2 = \boldsymbol{z}_2 - \boldsymbol{k}(\boldsymbol{z}_1)$ is introduced such that the system (6.31)-(6.32) can be written in the state $(\boldsymbol{z}_1, \boldsymbol{e}_2)$ as

$$\dot{\boldsymbol{z}}_1 = \boldsymbol{f}_1(\boldsymbol{z}_1) + \boldsymbol{F}_2(\boldsymbol{z}_1)\left(\boldsymbol{e}_2 + \boldsymbol{k}(\boldsymbol{z}_1)\right) , \tag{6.35}$$

$$\dot{\boldsymbol{e}}_2 = \boldsymbol{u} - \frac{\partial \boldsymbol{k}(\boldsymbol{z}_1)}{\partial \boldsymbol{z}_1}\left(\boldsymbol{f}_1(\boldsymbol{z}_1) + \boldsymbol{F}_2(\boldsymbol{z}_1)\left(\boldsymbol{e}_2 + \boldsymbol{k}(\boldsymbol{z}_1)\right)\right) . \tag{6.36}$$

For this system the function $V_2(\boldsymbol{z}_1, \boldsymbol{e}_2) = V_1(\boldsymbol{z}_1) + \frac{1}{2}\boldsymbol{e}_2^T\boldsymbol{e}_2$ is considered as a Lyapunov function candidate. The derivative of $V_2(\boldsymbol{z}_1, \boldsymbol{z}_2)$ along the solutions of (6.35)-(6.36) is given by

$$\begin{aligned}\dot{V}_2(\boldsymbol{z}_1, \boldsymbol{e}_2) = {} & \frac{\partial V_1(\boldsymbol{z}_1)}{\partial \boldsymbol{z}_1}\left(\boldsymbol{f}_1(\boldsymbol{z}_1) + \boldsymbol{F}_2(\boldsymbol{z}_1)\boldsymbol{k}(\boldsymbol{z}_1)\right) + \frac{\partial V_1(\boldsymbol{z}_1)}{\partial \boldsymbol{z}_1}\boldsymbol{F}_2(\boldsymbol{z}_1)\boldsymbol{e}_2 + \\ & \boldsymbol{e}_2^T\boldsymbol{u} - \boldsymbol{e}_2^T\frac{\partial \boldsymbol{k}(\boldsymbol{z}_1)}{\partial \boldsymbol{z}_1}\left(\boldsymbol{f}_1(\boldsymbol{z}_1) + \boldsymbol{F}_2(\boldsymbol{z}_1)\left(\boldsymbol{e}_2 + \boldsymbol{k}(\boldsymbol{z}_1)\right)\right) .\end{aligned}$$

Due to (6.34) the first term in $\dot{V}_2(\boldsymbol{z}_1, \boldsymbol{e}_2)$ is negative semi-definite in $\boldsymbol{z}_1$. In order to make $\dot{V}_2(\boldsymbol{z}_1, \boldsymbol{e}_2)$ negative semi-definite in $(\boldsymbol{z}_1, \boldsymbol{e}_2)$, the following control law can be used

$$\begin{aligned}\boldsymbol{u} &= -\boldsymbol{K}_e\boldsymbol{e}_2 + \frac{\partial \boldsymbol{k}(\boldsymbol{z}_1)}{\partial \boldsymbol{z}_1}\left(\boldsymbol{f}_1(\boldsymbol{z}_1) + \boldsymbol{F}_2(\boldsymbol{z}_1)\left(\boldsymbol{e}_2 + \boldsymbol{k}(\boldsymbol{z}_1)\right)\right) - \left(\frac{\partial V_1(\boldsymbol{z}_1)}{\partial \boldsymbol{z}_1}\boldsymbol{F}_2(\boldsymbol{z}_1)\right)^T \\ &= -\boldsymbol{K}_e(\boldsymbol{z}_2 - \boldsymbol{k}(\boldsymbol{z}_1)) + \frac{\partial \boldsymbol{k}(\boldsymbol{z}_1)}{\partial \boldsymbol{z}_1}\left(\boldsymbol{f}_1(\boldsymbol{z}_1) + \boldsymbol{F}_2(\boldsymbol{z}_1)\boldsymbol{z}_2\right) - \\ &\quad \left(\frac{\partial V_1(\boldsymbol{z}_1)}{\partial \boldsymbol{z}_1}\boldsymbol{F}_2(\boldsymbol{z}_1)\right)^T ,\end{aligned} \tag{6.37}$$

with an arbitrary positive definite matrix $\boldsymbol{K}_e$.

This procedure, which corresponds to one backstepping step, is formalized in the following lemma.

Lemma 6.5. *Consider the system (6.31)-(6.32). Let $\boldsymbol{k}(\boldsymbol{z}_1)$ be a differentiable state feedback law for $\boldsymbol{z}_2$, which stabilizes (6.31) (asymptotically), and $V_1(\boldsymbol{z}_1)$ be a (strict) Lyapunov function for (6.33). Then the feedback law (6.37), with a positive definite matrix $\boldsymbol{K}_e$, stabilizes the system (6.31)-(6.32) (asymptotically) and the function $V_2(\boldsymbol{z}_1, \boldsymbol{z}_2) = V_1(\boldsymbol{z}_1) + \frac{1}{2}(\boldsymbol{z}_2 - \boldsymbol{k}(\boldsymbol{z}_1))^T(\boldsymbol{z}_2 - \boldsymbol{k}(\boldsymbol{z}_1))$ can be used as a (strict) Lyapunov function.*

Notice that many extensions to this procedure are possible in order to treat a broader class of systems (see, e.g., [SJK97]). Lemma 6.5 can, for instance, readily be generalized to the time-varying case[7].

---

[7] The definitions of stability properties for time-varying systems are given in Appendix A.1.

6.2.2 Application to the Flexible Joint Model

The procedure of the last section shall now be applied to the system from Section 2.2.3:

$$\boldsymbol{M}(\boldsymbol{q})\ddot{\boldsymbol{q}} + \boldsymbol{C}(\boldsymbol{q},\dot{\boldsymbol{q}})\dot{\boldsymbol{q}} + \boldsymbol{g}(\boldsymbol{q}) = \boldsymbol{K}(\boldsymbol{\theta} - \boldsymbol{q}) + \boldsymbol{\tau}_{ext} \ , \tag{6.38}$$

$$\boldsymbol{B}\ddot{\boldsymbol{\theta}} + \boldsymbol{K}(\boldsymbol{\theta} - \boldsymbol{q}) = \boldsymbol{\tau}_m \ . \tag{6.39}$$

The structure of these system equations does not correspond identically to the particular class of systems discussed in Section 6.2.1. The following derivation in this section will therefore not directly rely on Lemma 6.5. The basic design procedure will be quite similar to the derivation which led to Lemma 6.5.

The backstepping procedure has been applied to the position control of flexible joint robots in [OL97] and in [NT92]. In this work the interest lies in the design of an impedance controller instead. Similar to the state transformation of the last section, the model can be written in the Cartesian coordinates $\boldsymbol{x} = \boldsymbol{f}(\boldsymbol{q})$ and $\dot{\boldsymbol{x}} = \boldsymbol{J}(\boldsymbol{q})\dot{\boldsymbol{q}}$ and the torque variables $\boldsymbol{\tau} = \boldsymbol{K}(\boldsymbol{\theta} - \boldsymbol{q})$ and $\dot{\boldsymbol{\tau}}$ as

$$\boldsymbol{\Lambda}(\boldsymbol{x})\ddot{\boldsymbol{x}} + \boldsymbol{\mu}(\boldsymbol{x},\dot{\boldsymbol{x}})\dot{\boldsymbol{x}} + \boldsymbol{J}(\boldsymbol{q})^{-T}\boldsymbol{g}(\boldsymbol{q}) = \boldsymbol{J}(\boldsymbol{q})^{-T}(\boldsymbol{\tau} + \boldsymbol{\tau}_{ext}) \ , \tag{6.40}$$

$$\boldsymbol{B}\boldsymbol{K}^{-1}\ddot{\boldsymbol{\tau}} + \boldsymbol{\tau} = \boldsymbol{\tau}_m - \boldsymbol{B}\ddot{\boldsymbol{q}} \ , \tag{6.41}$$

where the link acceleration $\ddot{\boldsymbol{q}}$ could also be written in the form $\ddot{\boldsymbol{q}} = \boldsymbol{M}(\boldsymbol{q})^{-1}(\boldsymbol{\tau} + \boldsymbol{\tau}_{ext} - \boldsymbol{C}(\boldsymbol{q},\dot{\boldsymbol{q}})\dot{\boldsymbol{q}} - \boldsymbol{g}(\boldsymbol{q}))$, as it was done in the last section. In this section again the non-redundant and non-singular case is treated. For the rigid body case with $\boldsymbol{\tau}_{ext} = \boldsymbol{0}$ the control law in (3.18) stabilizes the system (6.40) asymptotically. A (non-strict) Lyapunov function was given in Section 3.2 as

$$V_1(\tilde{\boldsymbol{x}},\dot{\tilde{\boldsymbol{x}}},t) = \frac{1}{2}\dot{\tilde{\boldsymbol{x}}}^T\boldsymbol{\Lambda}(\boldsymbol{x})\dot{\tilde{\boldsymbol{x}}} + \frac{1}{2}\tilde{\boldsymbol{x}}^T\boldsymbol{K}_d\tilde{\boldsymbol{x}} \ . \tag{6.42}$$

By considering the torque $\boldsymbol{\tau}$ as a virtual input vector, one can see that a feedback law of the form

$$\boldsymbol{\tau} = \boldsymbol{\tau}_d = \boldsymbol{g}(\boldsymbol{q}) + \boldsymbol{J}(\boldsymbol{q})^T(\boldsymbol{\Lambda}(\boldsymbol{x})\ddot{\boldsymbol{x}}_d + \boldsymbol{\mu}(\boldsymbol{x},\dot{\boldsymbol{x}})\dot{\boldsymbol{x}}_d - \boldsymbol{K}_d\tilde{\boldsymbol{x}} - \boldsymbol{D}_d\dot{\tilde{\boldsymbol{x}}})$$

would lead to

$$\dot{V}_1(\tilde{\boldsymbol{x}},\dot{\tilde{\boldsymbol{x}}},t) = -\dot{\tilde{\boldsymbol{x}}}^T\boldsymbol{D}_d\dot{\tilde{\boldsymbol{x}}} + \dot{\tilde{\boldsymbol{x}}}^T\boldsymbol{J}(\boldsymbol{q})^{-T}\boldsymbol{\tau}_{ext} \ .$$

This shows the stability in the case of free motion, i.e. for $\boldsymbol{\tau}_{ext} = \boldsymbol{0}$. Since $\boldsymbol{\tau}$ cannot be directly used as an input variable in the flexible joint case, one will in general have $\boldsymbol{\tau} \neq \boldsymbol{\tau}_d$ and with $\boldsymbol{e}_t = \boldsymbol{\tau} - \boldsymbol{\tau}_d$ the time derivative of $V_1(\tilde{\boldsymbol{x}},\dot{\tilde{\boldsymbol{x}}},t)$ is given by

$$\dot{V}_1(\tilde{\boldsymbol{x}},\dot{\tilde{\boldsymbol{x}}},\boldsymbol{e}_t,t) = -\dot{\tilde{\boldsymbol{x}}}^T\boldsymbol{D}_d\dot{\tilde{\boldsymbol{x}}} + \dot{\tilde{\boldsymbol{x}}}^T\boldsymbol{J}(\boldsymbol{q})^{-T}\boldsymbol{\tau}_{ext} + \underline{\dot{\tilde{\boldsymbol{x}}}^T\boldsymbol{J}(\boldsymbol{q})^{-T}\boldsymbol{e}_t}$$

instead. The backstepping procedure (Step 1) suggests to introduce an additional term in the Lyapunov function in order to cancel the cross-term. With $\boldsymbol{C}_t$ as a symmetric and positive definite matrix and

$$V_2(\tilde{\boldsymbol{x}},\dot{\tilde{\boldsymbol{x}}},\boldsymbol{e}_t,t) = \frac{1}{2}\dot{\tilde{\boldsymbol{x}}}^T\boldsymbol{\Lambda}(\boldsymbol{x})\dot{\tilde{\boldsymbol{x}}} + \frac{1}{2}\tilde{\boldsymbol{x}}^T\boldsymbol{K}_d\tilde{\boldsymbol{x}} + \underline{\frac{1}{2}\boldsymbol{e}_t^T\boldsymbol{C}_t\boldsymbol{e}_t}$$

as a new Lyapunov function candidate, the time derivative of $V_2(\tilde{\boldsymbol{x}}, \dot{\tilde{\boldsymbol{x}}}, \boldsymbol{e}_t, t)$ is given by

$$\dot{V}_2(\tilde{\boldsymbol{x}}, \dot{\tilde{\boldsymbol{x}}}, \boldsymbol{e}_t, \dot{\boldsymbol{e}}_t, t) = -\dot{\tilde{\boldsymbol{x}}}^T \boldsymbol{D}_d \dot{\tilde{\boldsymbol{x}}} + \dot{\tilde{\boldsymbol{x}}}^T \boldsymbol{J}(\boldsymbol{q})^{-T} \boldsymbol{\tau}_{ext} + \dot{\tilde{\boldsymbol{x}}}^T \boldsymbol{J}(\boldsymbol{q})^{-T} \boldsymbol{e}_t + \underline{\boldsymbol{e}_t^T \boldsymbol{C}_t \dot{\boldsymbol{e}}_t} \; .$$

Next, the error term $\dot{\boldsymbol{e}}_t$ is considered as a virtual input. By the choice

$$\dot{\boldsymbol{e}}_t = \boldsymbol{v}_t^* = -\boldsymbol{C}_t^{-1} \boldsymbol{J}(\boldsymbol{q})^{-1} \dot{\tilde{\boldsymbol{x}}} - \boldsymbol{C}_t^{-1} \boldsymbol{K}_t \boldsymbol{e}_t \; ,$$

with $\boldsymbol{K}_t$ as a positive definite controller gain matrix, one would ideally eliminate the cross-term in $\dot{V}_2(\tilde{\boldsymbol{x}}, \dot{\tilde{\boldsymbol{x}}}, \boldsymbol{e}_t, \dot{\boldsymbol{e}}_t, t)$ and obtain

$$\dot{V}_2(\tilde{\boldsymbol{x}}, \dot{\tilde{\boldsymbol{x}}}, \boldsymbol{e}_t, t) = -\dot{\tilde{\boldsymbol{x}}}^T \boldsymbol{D}_d \dot{\tilde{\boldsymbol{x}}} + \dot{\tilde{\boldsymbol{x}}}^T \boldsymbol{J}(\boldsymbol{q})^{-T} \boldsymbol{\tau}_{ext} - \underline{\boldsymbol{e}_t^T \boldsymbol{K}_t \boldsymbol{e}_t} \; .$$

But since in general $\dot{\boldsymbol{e}}_t \neq \boldsymbol{v}_t^*$, one actually gets, by introducing the error variable $\boldsymbol{e}_s = \dot{\boldsymbol{e}}_t - \boldsymbol{v}_t^*$, the following change of the Lyapunov function along the solution curves

$$\dot{V}_2(\tilde{\boldsymbol{x}}, \dot{\tilde{\boldsymbol{x}}}, \boldsymbol{e}_t, \boldsymbol{e}_s, t) = -\dot{\tilde{\boldsymbol{x}}}^T \boldsymbol{D}_d \dot{\tilde{\boldsymbol{x}}} + \dot{\tilde{\boldsymbol{x}}}^T \boldsymbol{J}(\boldsymbol{q})^{-T} \boldsymbol{\tau}_{ext} - \boldsymbol{e}_t^T \boldsymbol{K}_t \boldsymbol{e}_t + \underline{\boldsymbol{e}_t^T \boldsymbol{C}_t \boldsymbol{e}_s} \; .$$

In order to eliminate also the remaining cross-term, one has to perform an additional backstepping step (Step 2), by first introducing a new term in the Lyapunov function with a symmetric and positive definite matrix $\boldsymbol{C}_s$

$$V_3(\tilde{\boldsymbol{x}}, \dot{\tilde{\boldsymbol{x}}}, \boldsymbol{e}_t, \boldsymbol{e}_s, t) = \frac{1}{2} \dot{\tilde{\boldsymbol{x}}}^T \boldsymbol{\Lambda}(\boldsymbol{x}) \dot{\tilde{\boldsymbol{x}}} + \frac{1}{2} \tilde{\boldsymbol{x}}^T \boldsymbol{K}_d \tilde{\boldsymbol{x}} + \frac{1}{2} \boldsymbol{e}_t^T \boldsymbol{C}_t \boldsymbol{e}_t + \underline{\frac{1}{2} \boldsymbol{e}_s^T \boldsymbol{C}_s \boldsymbol{e}_s} \; ,$$

such that the time derivative reads as

$$\dot{V}_3(\tilde{\boldsymbol{x}}, \dot{\tilde{\boldsymbol{x}}}, \boldsymbol{e}_t, \boldsymbol{e}_s, \dot{\boldsymbol{e}}_s, t) = -\dot{\tilde{\boldsymbol{x}}}^T \boldsymbol{D}_d \dot{\tilde{\boldsymbol{x}}} + \dot{\tilde{\boldsymbol{x}}}^T \boldsymbol{J}(\boldsymbol{q})^{-T} \boldsymbol{\tau}_{ext} - \boldsymbol{e}_t^T \boldsymbol{K}_t \boldsymbol{e}_t + \boldsymbol{e}_t^T \boldsymbol{C}_t \boldsymbol{e}_s + \underline{\boldsymbol{e}_s^T \boldsymbol{C}_s \dot{\boldsymbol{e}}_s} \; .$$

With $\dot{\boldsymbol{e}}_s = \ddot{\boldsymbol{\tau}} - \ddot{\boldsymbol{\tau}}_d - \dot{\boldsymbol{v}}_t^*$ and by substituting the torque dynamics equation (6.41), one obtains

$$\dot{V}_3(\tilde{\boldsymbol{x}}, \dot{\tilde{\boldsymbol{x}}}, \boldsymbol{e}_t, \boldsymbol{e}_s, t) = -\dot{\tilde{\boldsymbol{x}}}^T \boldsymbol{D}_d \dot{\tilde{\boldsymbol{x}}} + \dot{\tilde{\boldsymbol{x}}}^T \boldsymbol{J}(\boldsymbol{q})^{-T} \boldsymbol{\tau}_{ext} - \boldsymbol{e}_t^T \boldsymbol{K}_t \boldsymbol{e}_t + \boldsymbol{e}_t^T \boldsymbol{C}_t \boldsymbol{e}_s + \boldsymbol{e}_s^T \boldsymbol{C}_s (\underline{\boldsymbol{K}\boldsymbol{B}^{-1}(\boldsymbol{\tau}_m - \boldsymbol{B}\ddot{\boldsymbol{q}} - \boldsymbol{\tau})} - \ddot{\boldsymbol{\tau}}_d - \dot{\boldsymbol{v}}_t^*) \; .$$

Controller Design: By completing the backstepping step the controller can now be chosen as

$$\boldsymbol{\tau}_m = \boldsymbol{B}\ddot{\boldsymbol{q}} + \boldsymbol{\tau} + \boldsymbol{B}\boldsymbol{K}^{-1}(\ddot{\boldsymbol{\tau}}_d + \dot{\boldsymbol{v}}_t^* - \boldsymbol{C}_s^{-1} \boldsymbol{C}_t \boldsymbol{e}_t - \boldsymbol{C}_s^{-1} \boldsymbol{K}_s \boldsymbol{e}_s) \tag{6.43}$$

with a positive definite matrix $\boldsymbol{K}_s$. This leads to

$$\dot{V}_3(\tilde{\boldsymbol{x}}, \dot{\tilde{\boldsymbol{x}}}, \boldsymbol{e}_t, \boldsymbol{e}_s, t) = -\dot{\tilde{\boldsymbol{x}}}^T \boldsymbol{D}_d \dot{\tilde{\boldsymbol{x}}} + \dot{\tilde{\boldsymbol{x}}}^T \boldsymbol{J}(\boldsymbol{q})^{-T} \boldsymbol{\tau}_{ext} - \boldsymbol{e}_t^T \boldsymbol{K}_t \boldsymbol{e}_t - \boldsymbol{e}_s^T \boldsymbol{K}_s \boldsymbol{e}_s \; ,$$

and hence $\dot{V}_3(\tilde{\boldsymbol{x}}, \dot{\tilde{\boldsymbol{x}}}, \boldsymbol{e}_t, \boldsymbol{e}_s, t)$ is negative semi-definite for the case of free motion. Thus, one may conclude that the equilibrium is stable in the sense of Lyapunov.

The controller in (6.43) can be brought into a more handy form by substituting the expressions $\boldsymbol{e}_s = \dot{\boldsymbol{e}}_t - \boldsymbol{v}_t^*$ and $\boldsymbol{v}_t^* = -\boldsymbol{C}_t^{-1}\boldsymbol{J}(\boldsymbol{q})^{-1}\dot{\tilde{\boldsymbol{x}}} - \boldsymbol{C}_t^{-1}\boldsymbol{K}_t\boldsymbol{e}_t$ to obtain

$$\begin{aligned}\boldsymbol{\tau}_m &= \boldsymbol{B}\ddot{\boldsymbol{q}} + \boldsymbol{\tau} + \boldsymbol{B}\boldsymbol{K}^{-1}(\ddot{\boldsymbol{\tau}}_d - \boldsymbol{C}_t^{-1}\frac{\mathrm{d}}{\mathrm{d}t}(\boldsymbol{J}(\boldsymbol{q})^{-1}\dot{\tilde{\boldsymbol{x}}}) - \boldsymbol{C}_t^{-1}\boldsymbol{K}_t\dot{\boldsymbol{e}}_t - \boldsymbol{C}_s^{-1}\boldsymbol{C}_t\boldsymbol{e}_t)\\ &\quad -\boldsymbol{B}\boldsymbol{K}^{-1}\boldsymbol{C}_s^{-1}\boldsymbol{K}_s(\dot{\boldsymbol{e}}_t + \boldsymbol{C}_t^{-1}\boldsymbol{J}(\boldsymbol{q})^{-1}\dot{\tilde{\boldsymbol{x}}} + \boldsymbol{C}_t^{-1}\boldsymbol{K}_t\boldsymbol{e}_t)\ .\end{aligned}$$

By rearranging the terms, and with the substitutions $\bar{\boldsymbol{K}}_t = \boldsymbol{C}_t^{-1}\boldsymbol{K}_t$ and $\bar{\boldsymbol{K}}_s = \boldsymbol{C}_s^{-1}\boldsymbol{K}_s$, the controller can finally be written as

$$\begin{aligned}\boldsymbol{\tau}_m &= \boldsymbol{B}\ddot{\boldsymbol{q}} + \boldsymbol{\tau} + \boldsymbol{B}\boldsymbol{K}^{-1}\left(\ddot{\boldsymbol{\tau}}_d - (\bar{\boldsymbol{K}}_s\bar{\boldsymbol{K}}_t + \boldsymbol{C}_s^{-1}\boldsymbol{C}_t)\boldsymbol{e}_t - (\bar{\boldsymbol{K}}_s + \bar{\boldsymbol{K}}_t)\dot{\boldsymbol{e}}_t\right)\\ &\quad -\boldsymbol{B}\boldsymbol{K}^{-1}(\boldsymbol{C}_t^{-1}\frac{\mathrm{d}}{\mathrm{d}t}(\boldsymbol{J}(\boldsymbol{q})^{-1}\dot{\tilde{\boldsymbol{x}}}) + \bar{\boldsymbol{K}}_s\boldsymbol{C}_t^{-1}\boldsymbol{J}(\boldsymbol{q})^{-1}\dot{\tilde{\boldsymbol{x}}})\ ,\end{aligned} \tag{6.44}$$

with $\boldsymbol{\tau}_d$ given by

$$\boldsymbol{\tau}_d = \boldsymbol{g}(\boldsymbol{q}) + \boldsymbol{J}(\boldsymbol{q})^T(\boldsymbol{\Lambda}(\boldsymbol{x})\ddot{\boldsymbol{x}}_d + \boldsymbol{\mu}(\boldsymbol{x},\dot{\boldsymbol{x}})\dot{\boldsymbol{x}}_d - \boldsymbol{K}_d\tilde{\boldsymbol{x}} - \boldsymbol{D}_d\dot{\tilde{\boldsymbol{x}}})\ .$$

The result can be summarized in the following proposition.

Proposition 6.6. *Let the desired trajectory $\boldsymbol{x}_d(t)$ be four times continuously differentiable. Assume further that the Cartesian coordinates are globally valid, i.e. $\bar{\mathcal{Q}}_c^p = \mathbb{R}^n$. Then, for the case of free motion ($\boldsymbol{F}_{ext} = \boldsymbol{0}$) the controller (6.44) with symmetric and positive definite matrices $\boldsymbol{K}_d$, $\boldsymbol{D}_d$, $\boldsymbol{C}_t$, $\boldsymbol{C}_s$, $\boldsymbol{K}_t$, and $\boldsymbol{K}_s$ stabilizes the system (6.40)-(6.41) globally.*

Proof. (Sketch) The proof can be easily done by showing that the function $V_3(\boldsymbol{e}_x, \dot{\boldsymbol{e}}_x, \boldsymbol{e}_t, \boldsymbol{e}_s, t)$ is a (time-varying) Lyapunov function with a negative semi-definite derivative along the solution curves of the closed loop system. This can be done by following the steps of the above backstepping based controller derivation.

In addition to this stability statement, a passivity property similar to Proposition 3.5, follows easily.

Proposition 6.7. *Consider a closed loop system according to the control law (6.44) with symmetric and positive definite matrices $\boldsymbol{K}_d$, $\boldsymbol{D}_d$, $\boldsymbol{C}_t$, $\boldsymbol{C}_s$, $\boldsymbol{K}_t$, and $\boldsymbol{K}_s$ applied to the system (6.40)-(6.41). For $\dot{\boldsymbol{x}}_d(t) = \boldsymbol{0}$, this closed loop system gets time-invariant and represents a passive map from the external force $\boldsymbol{F}_{ext}$ to the velocity error $\dot{\tilde{\boldsymbol{x}}}$.*

Proof. This can be proven by considering $V_3(\boldsymbol{e}_x, \dot{\boldsymbol{e}}_x, \boldsymbol{e}_t, \boldsymbol{e}_s)$, which for $\dot{\boldsymbol{x}}_d(t) = \boldsymbol{0}$ gets time-invariant, as a storage function for the closed loop system.

6.2.3 Controller Discussion

Notice the similarity of the controller (6.44) to the controller from the last section (6.30). With the correspondence $\boldsymbol{K}_\tau = \bar{\boldsymbol{K}}_s\bar{\boldsymbol{K}}_t + \boldsymbol{C}_s^{-1}\boldsymbol{C}_t - \boldsymbol{K}\boldsymbol{B}^{-1}$ and $\boldsymbol{D}_\tau =$

$\bar{\boldsymbol{K}}_s + \bar{\boldsymbol{K}}_t$ and by substituting $\ddot{\boldsymbol{q}} = \boldsymbol{M}(\boldsymbol{q})^{-1}(\boldsymbol{\tau} + \boldsymbol{\tau}_{ext} - \boldsymbol{C}(\boldsymbol{q}, \dot{\boldsymbol{q}})\dot{\boldsymbol{q}} - \boldsymbol{g}(\boldsymbol{q}))$ the controller (6.44) can be written as

$$\begin{aligned}\boldsymbol{\tau}_m = \boldsymbol{\tau}_d + \boldsymbol{B}\boldsymbol{K}^{-1}\left(\ddot{\boldsymbol{\tau}}_d - \boldsymbol{K}_\tau \boldsymbol{e}_t - \boldsymbol{D}_\tau \dot{\boldsymbol{e}}_t\right) \\ \boldsymbol{B}\boldsymbol{M}(\boldsymbol{q})^{-1}\left(\boldsymbol{\tau} + \boldsymbol{\tau}_{ext} - \boldsymbol{C}(\boldsymbol{q}, \dot{\boldsymbol{q}})\dot{\boldsymbol{q}} - \boldsymbol{g}(\boldsymbol{q})\right) \\ -\boldsymbol{B}\boldsymbol{K}^{-1}(\boldsymbol{C}_t^{-1}\frac{\mathrm{d}}{\mathrm{d}t}(\boldsymbol{J}(\boldsymbol{q})^{-1}\dot{\tilde{\boldsymbol{x}}}) + \bar{\boldsymbol{K}}_s \boldsymbol{C}_t^{-1}\boldsymbol{J}(\boldsymbol{q})^{-1}\dot{\tilde{\boldsymbol{x}}}) \ .\end{aligned}$$

The first part of this controller thus has the same form as in (6.30). But it should be noticed that in general the controller gains $\boldsymbol{K}_\tau$ and $\boldsymbol{D}_\tau$ must be chosen differently. For the decoupling based controller these matrices must be positive definite only, but in the backstepping based controller they must be chosen such that the matrices $\boldsymbol{K}_t, \boldsymbol{K}_s, \boldsymbol{C}_t$, and $\boldsymbol{C}_s$ are all symmetric and positive definite.

From a computational point of view the two controllers from this chapter are comparable. The main difficulty arises from the fact that for the implementation the link accelerations $\ddot{\boldsymbol{q}}$ and the jerks $\boldsymbol{q}^{(3)}$ have to be determined (see also Section 6.1.5).

The parameter choice for the backstepping based controller is more difficult. Clearly, the design of the matrices $\boldsymbol{K}_t, \boldsymbol{K}_s, \boldsymbol{C}_t$, and $\boldsymbol{C}_s$, which must be chosen additionally to the given impedance parameters $\boldsymbol{K}_d$ and $\boldsymbol{D}_d$, is not very intuitive. It can be simplified, when two of the parameters are fixed. In the simulations of this controller (see also Chapter 8) it turned out that the choice $\boldsymbol{C}_s = \boldsymbol{C}_t = \boldsymbol{I}$ leads to good results. Thereby, the design is reduced to the choice of positive definite matrices $\boldsymbol{K}_t$ and $\boldsymbol{K}_s$.

Notice that the above analysis allows to give a proof of stability, but nothing has been said about asymptotic stability. In principle, the same procedure can be applied to a strict Lyapunov function for the rigid body part of the system. Santibanez and Kelly proposed such a strict Lyapunov function for the rigid body part in [SK97b]. This would enable to treat the case of asymptotic stability but would further increase the complexity of the controller.

6.3 Summary

In this chapter two Cartesian impedance controllers were proposed which take advantage of the special structure of the reduced flexible joint robot model.

In the first approach an exact decoupling of the torque dynamics from the link side dynamics is established. In combination with an impedance control law for the link side dynamics, this allows to give a stability proof based on the control theory for time-varying cascaded systems. In contrast to the singular perturbation approach from the last chapter no further simplifications or approximations of the reduced flexible joint robot model are required.

In the second part of this chapter a constructive approach, namely the backstepping approach, was applied. In this approach the Lyapunov function for a rigid body impedance controller was used as a basis to construct a Lyapunov function for the flexible joint robot model. This is only possible by additional feedback of the joint torques.

The approaches of this chapter are based on the particular structure of the system dynamics. Undesired coupling terms between the torque dynamics and the link side dynamics are eliminated by appropriate feedback. A rather different approach, based on the passivity properties of the robot model, will be discussed in the next chapter.

7 A Passivity Based Approach

In this chapter a passivity based approach for the impedance control of a flexible joint robot is described. The basic idea behind the controller formulation is a physical interpretation of the torque feedback. This allows to analyze the stability based on the passivity properties of the system.

The energy shaping methodology for flexible joint robots was first introduced in [Tom91], where it was proven that a motor position based PD-controller leads to a stable closed loop system similar to the rigid body case. The generalization to the Cartesian case was treated by Goldsmith et al. in [GFG99], where the stability analysis for a Cartesian hybrid position/force controller for a flexible joint robot model without gravitational effects was presented. The gravity compensation problem for PD-like regulation controllers was addressed in [DLSZ05, ZSL+03, ZLS04, ZSL+05].

However, it has been shown that in practice only quite unsatisfactory results can be achieved with a restriction to purely motor position (and velocity) based feedback controllers (without additional non-collocated feedback) for the case of a flexible joint robot. In some works a controller structure based on a feedback of the joint torques as well as the link side positions was considered and it was shown that this can indeed lead to better results (see, e.g., [Spo89]). This was also verified experimentally with the DLR lightweight robots [ASOFH03]. From a theoretical point of view the use of an inner torque feedback loop usually is justified (for a sufficiently high joint stiffness) by an approximate analysis based on the singular perturbation theory (see Chapter 5).

In this chapter a physical interpretation of the torque feedback is given instead. Therefore, a stability proof based on passivity properties of the system can be given. It is important to note that the described controller itself is not passive due to the feedback of the joint torque, but it will be shown that the controlled motor dynamics in combination with the torque feedback are passive [OASK+05, OASK+04a]. Together with the passive (link side) rigid body dynamics the closed loop system can, therefore, be represented as a feedback interconnection of two passive systems.

C. Ott: Cartesian Imped. Cntrl. of Redund. & Flexible-Joint Robots, STAR 49, pp. 93–121, 2008.
springerlink.com

In prior work at the DLR, a controller with a complete static state feedback (position and torque as well as their first derivatives) was introduced in [ASH00], for which asymptotic stability was shown based on the passivity properties of the controller. Analogously to [Tom91], a gravity compensation term based on the desired configuration was used in [ASH00]. For an impedance controller this is not appropriate, since for low desired stiffness a large deviation from the desired configuration may occur. In Section 7.3 a gravity compensation term will be designed which is based on the measurement of the motor position. This is far better for the impedance control problem.

A similar passivity based controller structure, which can be implemented without measurement of the joint velocities, is given by the *IPC*[1] from [Str01]. It should be mentioned that the original idea of the presented work from this chapter was the result of considerations about how such a passivity based controller designed for rigid body robots (like a simple PD controller or the more involved IPC) could be best implemented for a robot with flexible joints.

The presented controller is also strongly related to the state feedback controller from [ASH00] as described further in Section 7.8.

This chapter is organized as follows: Section 7.1 describes the design idea which is based on some simple physical observations. The application of the design idea to the flexible joint robot model is presented in Section 7.2. In Section 7.3 the problem of gravity compensation is treated. A detailed stability analysis of the controller based on the passivity properties is given in Section 7.4. Furthermore, in Section 7.5 the stiffness design is treated in more detail. Some straightforward generalizations of the controller are given in Section 7.6. The tracking case is considered in more detail in Section 7.7. Finally, in Section 7.8, the relations of the controller to some other design methods are shortly discussed.

7.1 Design Idea

In this section the basic idea of the controller design is described. It is motivated by some simple considerations for a one-dimensional model.

Consider at first the model of a single flexible joint as it is sketched in Figure 7.1 for the second joint of the DLR-Lightweight-Robot-III. The motor torque τ_m acts here on the rotor inertia B of the motor[2]. The elasticity of the transmission between the rotor and the following link of the robot[3] is modeled in form of a linear spring with stiffness K.

The goal of the impedance controller is to achieve a desired dynamical behavior with respect to an external force F_{ext} acting on the link side. In the following it is assumed that this dynamical behavior is given by a differential equation of second order representing a mass-spring-damper system with mass M, desired stiffness K_θ, and desired damping D_θ. For a robot with rigid

[1] *Intrinsically Passive Controller.*

[2] The motors are modeled as ideal torque sources without considering the dynamics of the electrical drives.

[3] In Figure 7.1 represented in a simplified form with a constant mass M.

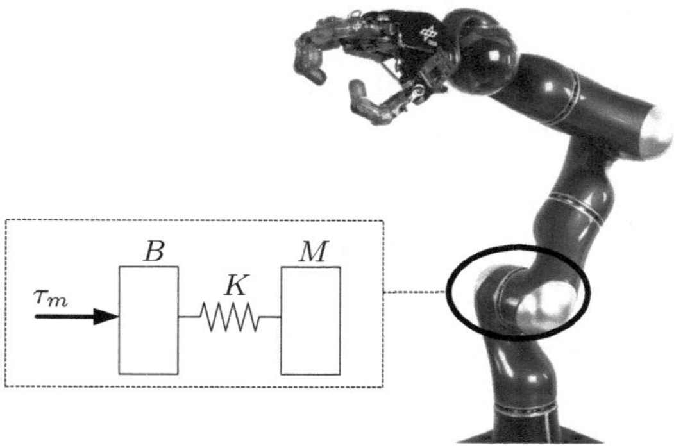

Fig. 7.1. Sketch of the model for a flexible joint robot

(i.e. non-elastic) joints this behavior could be realized by a simple PD-controller. If one uses a motor position based PD-controller in case of a robot with elastic joints as shown in Figure 7.2 for the one-dimensional case, then the resulting dynamics will clearly be influenced also by the joint elasticity and the motor inertia. Intuitively speaking, the deviation from the desired behavior will be less significant when the rotor mass B gets smaller and the joint stiffness K gets larger.

At this point it should be mentioned that the joint stiffness values of a *typical* flexible joint robot are indeed quite large[4] but not negligible. By a negative feedback of the joint torque τ the apparent inertia (of the rotor) can now be scaled down, which means that the closed loop system reacts to external forces as if the rotor inertia were smaller. The desired dynamical behavior will be approximated the better, the smaller the apparent rotor inertia is. This approach, as suggested in Figure 7.2 intuitively, will be put in concrete terms in the following section for the model of a flexible joint robot. Furthermore, in Section 7.5 a method for compensating the influence of the spring K will also be presented.

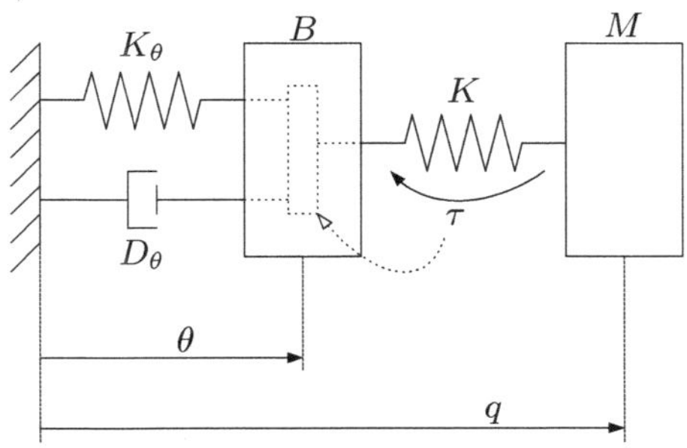

Fig. 7.2. Motor position based PD-control of a single joint

[4] For the lower joints of the DLR lightweight robots these values lie in the range 10.000 – 15.000 Nm/rad.

7.2 Application to the Flexible Joint Model

In Section 2.2.3 the reduced model of a flexible joint manipulator was derived. For the ease of reference the model equations are recapitulated here

$$\boldsymbol{M}(\boldsymbol{q})\ddot{\boldsymbol{q}} + \boldsymbol{C}(\boldsymbol{q},\dot{\boldsymbol{q}})\dot{\boldsymbol{q}} + \boldsymbol{g}(\boldsymbol{q}) = \boldsymbol{K}(\boldsymbol{\theta} - \boldsymbol{q}) + \boldsymbol{\tau}_{ext} \ , \tag{7.1}$$

$$\boldsymbol{B}\ddot{\boldsymbol{\theta}} + \boldsymbol{K}(\boldsymbol{\theta} - \boldsymbol{q}) = \boldsymbol{\tau}_m \ , \tag{7.2}$$

where the same notation is used as in Chapter 2. As already mentioned in the last section, the apparent motor inertia can be reduced from $\boldsymbol{B}$ to $\boldsymbol{B}_\theta$ by feeding back the joint torque $\boldsymbol{\tau} = \boldsymbol{K}(\boldsymbol{\theta} - \boldsymbol{q})$. This is realized by the feedback law

$$\boldsymbol{\tau}_m = \boldsymbol{B}\boldsymbol{B}_\theta^{-1}\boldsymbol{u} + (\boldsymbol{I} - \boldsymbol{B}\boldsymbol{B}_\theta^{-1})\boldsymbol{\tau} \ , \tag{7.3}$$

wherein $\boldsymbol{u}$ is a new input variable. The resulting system dynamics is given by

$$\boldsymbol{M}(\boldsymbol{q})\ddot{\boldsymbol{q}} + \boldsymbol{C}(\boldsymbol{q},\dot{\boldsymbol{q}})\dot{\boldsymbol{q}} + \boldsymbol{g}(\boldsymbol{q}) = \boldsymbol{K}(\boldsymbol{\theta} - \boldsymbol{q}) + \boldsymbol{\tau}_{ext} \ , \tag{7.4}$$

$$\boldsymbol{B}_\theta\ddot{\boldsymbol{\theta}} + \boldsymbol{K}(\boldsymbol{\theta} - \boldsymbol{q}) = \boldsymbol{u} \ . \tag{7.5}$$

Clearly, (7.4) and (7.5) have the same form as the original equations but with scaled motor inertia. The input variable $\boldsymbol{u}$ is further split up into one term $\boldsymbol{u}_{imp}$, which actually implements the stiffness and damping, and another term $\boldsymbol{u}_g$, which acts as a gravity compensation

$$\boldsymbol{u} = \boldsymbol{u}_{imp} + \boldsymbol{u}_g \ . \tag{7.6}$$

In the remaining part of this section the term $\boldsymbol{u}_{imp}$ is designed. The construction of an appropriate gravity compensation term $\boldsymbol{u}_g$ will be treated in Section 7.3.2. In case that the desired impedance behavior is defined with respect to joint coordinates in form of positive definite stiffness and damping matrices $\boldsymbol{K}_\theta$ and $\boldsymbol{D}_\theta$, respectively, and by a (constant) virtual equilibrium configuration $\boldsymbol{\theta}_d$, a motor position based PD-controller

$$\boldsymbol{u}_{imp} = -\boldsymbol{K}_\theta(\boldsymbol{\theta} - \boldsymbol{\theta}_d) - \boldsymbol{D}_\theta\dot{\boldsymbol{\theta}} \tag{7.7}$$

can be used. Therefore, one gets the following closed loop equations

$$\boldsymbol{M}(\boldsymbol{q})\ddot{\boldsymbol{q}} + \boldsymbol{C}(\boldsymbol{q},\dot{\boldsymbol{q}})\dot{\boldsymbol{q}} + \boldsymbol{g}(\boldsymbol{q}) = \boldsymbol{\tau} + \boldsymbol{\tau}_{ext} \ , \tag{7.8}$$

$$\boldsymbol{B}_\theta\ddot{\boldsymbol{\theta}} + \boldsymbol{D}_\theta\dot{\boldsymbol{\theta}} + \boldsymbol{K}_\theta(\boldsymbol{\theta} - \boldsymbol{\theta}_d) + \boldsymbol{\tau} = \boldsymbol{u}_g \ . \tag{7.9}$$

7.2.1 Generalization to Cartesian Coordinates

Usually, the desired impedance behavior is defined with respect to Cartesian coordinates $\boldsymbol{x} \in \mathbb{R}^m$, which describe the position and orientation of the robot end-effector, rather than in joint coordinates. In the following it is assumed that the forward kinematics mapping from the joint space coordinates $\boldsymbol{q}$ to the Cartesian coordinates $\boldsymbol{x} = \boldsymbol{f}(\boldsymbol{q})$ as well as the Jacobian matrix $\boldsymbol{J}(\boldsymbol{q}) = \frac{\partial \boldsymbol{f}(\boldsymbol{q})}{\partial \boldsymbol{q}}$ are known.

The controller (7.7) can then be easily generalized to Cartesian coordinates, by using the motor angles $\boldsymbol{\theta}$ instead of the link side angles $\boldsymbol{q}$ in the forward kinematics, i.e. $\boldsymbol{x}_\theta = \boldsymbol{f}(\boldsymbol{\theta})$. The desired stiffness and damping matrices are given by the positive definite matrices $\boldsymbol{K}_d$ and $\boldsymbol{D}_d$. Then the feedback law

$$\boldsymbol{u}_{imp} = -\boldsymbol{J}(\boldsymbol{\theta})^T(\boldsymbol{K}_d\tilde{\boldsymbol{x}}_\theta + \boldsymbol{D}_d\dot{\boldsymbol{x}}_\theta) \ , \tag{7.10}$$

$$\tilde{\boldsymbol{x}}_\theta = \boldsymbol{f}(\boldsymbol{\theta}) - \boldsymbol{x}_{\theta,d} \ , \tag{7.11}$$

$$\dot{\boldsymbol{x}}_\theta = \boldsymbol{J}(\boldsymbol{\theta})\dot{\boldsymbol{\theta}} \tag{7.12}$$

generalizes (7.7) to Cartesian coordinates. Herein, $\boldsymbol{x}_{\theta,d}$ is the *virtual* motor side position in Cartesian coordinates. Notice that in the design of $\boldsymbol{x}_{\theta,d}$ the static (i.e. in equilibrium state for $\boldsymbol{\tau}_{ext} = \boldsymbol{0}$) difference of the motor and link side angles due to gravity should be considered. This means that for a given link side position $\boldsymbol{q}_d$ which corresponds to the desired Cartesian position $\boldsymbol{x}_d = \boldsymbol{f}(\boldsymbol{q}_d)$, $\boldsymbol{x}_{\theta,d}$ should be chosen as $\boldsymbol{x}_{\theta,d} = \boldsymbol{f}(\boldsymbol{q}_d + \boldsymbol{K}^{-1}\boldsymbol{g}(\boldsymbol{q}_d))$.

The controller in (7.10) leads then, together with (7.3), to the closed loop system

$$\boldsymbol{M}(\boldsymbol{q})\ddot{\boldsymbol{q}} + \boldsymbol{C}(\boldsymbol{q}, \dot{\boldsymbol{q}})\dot{\boldsymbol{q}} + \boldsymbol{g}(\boldsymbol{q}) = \boldsymbol{\tau} + \boldsymbol{\tau}_{ext} \ , \tag{7.13}$$

$$\boldsymbol{B}_\theta\ddot{\boldsymbol{\theta}} + \boldsymbol{J}(\boldsymbol{\theta})^T(\boldsymbol{K}_d\tilde{\boldsymbol{x}}_\theta + \boldsymbol{D}_d\dot{\boldsymbol{x}}_\theta) + \boldsymbol{\tau} = \boldsymbol{u}_g \ . \tag{7.14}$$

7.3 Gravity Compensation

In the control law (7.10) the effects of the gravity torque $\boldsymbol{g}(\boldsymbol{q})$ were not considered. It has been shown in [Tom91] that for a motor position based PD-controller a feedforward term of the gravity torques in the desired steady state can be used. This indeed leads in the case of a position controller usually to a good performance because the deviations from the desired position can be kept small. For an impedance controller, however, this is not true. Here a pure feedforward action for the gravity compensation does not give satisfactory results because large deviations from the steady state positions may occur in the case of a small desired stiffness $\boldsymbol{K}_d$.

The problem of constructing an online gravity compensation for a flexible joint robot based on the motor position was first treated in [ZSL+03]. The solution in [ZSL+03], however, still leads to lower bounds on $\boldsymbol{K}_d$, limiting the generality of the impedance controller. In contrast to this the solution presented herein does not require such additional constraints [OASK+04b, ASOH04b, OASKH08]. In the following some further details on the gravity term $\boldsymbol{g}(\boldsymbol{q})$ are analyzed in Section 7.3.1. The actual gravity compensation will then be formulated in Section 7.3.2 and a derivation of the corresponding potential function is given in Section 7.3.3.

7.3.1 On the Boundedness of the Gravity Hessian

According to its derivation in Chapter 2 the gravity term $\boldsymbol{g}(\boldsymbol{q})$ corresponds to the differential of the gravity potential $V_g(\boldsymbol{q})$, i.e. $\boldsymbol{g}(\boldsymbol{q}) = (\partial V_g(\boldsymbol{q})/\partial \boldsymbol{q})^T$. Property 2.8

ensures the boundedness of the Hessian of the gravity potential for an arbitrary norm. In case that the manipulator under consideration has both rotational and prismatic joints, the bound α from Property 2.8 is, from a physical point of view, not well defined, since it clearly depends on the chosen physical units for the translational and rotational coordinates. In order to overcome this problem particular matrix and vector norms are defined in the following by scaling with the joint stiffness matrix[5].

Let therefore $\boldsymbol{R} \in \mathbb{R}^{n\times n}$ be the square root of the joint stiffness matrix $\boldsymbol{K}$, i.e.

$$\boldsymbol{K} = \boldsymbol{R}^T\boldsymbol{R} \,. \tag{7.15}$$

Since $\boldsymbol{K}$ is a diagonal matrix (see Chapter 2), the matrix $\boldsymbol{R}$ is given by $\boldsymbol{R} = \mathrm{diag}(\sqrt{K_i})$. Then a vector norm $||\cdot||_K : \mathbb{R}^n \rightarrow \mathbb{R}^+$ for a vector $\boldsymbol{v} \in \mathbb{R}^n$ can be defined via the Euclidean vector norm $||\cdot||_2$ as

$$||\boldsymbol{v}||_K := \left(\sum_{i=1}^{n} K_i v_i^2\right)^{1/2} = ||\boldsymbol{R}\boldsymbol{v}||_2 = \left(\boldsymbol{v}^T\boldsymbol{K}\boldsymbol{v}\right)^{1/2} \,. \tag{7.16}$$

The matrix $\boldsymbol{R}$, respectively $\boldsymbol{K}$, is used herein as a normalization of the chosen physical units. Corresponding to this vector norm, a matrix norm $||\cdot||_K : \mathbb{R}^{n\times n} \rightarrow \mathbb{R}^+$ for a matrix $\boldsymbol{A} \in \mathbb{R}^{n\times n}$ is defined in the following via the spectral norm[6] $||\cdot||_{i2}$. Since the interest of this section lies on the Hessian of the gravity potential, it is reasonable[7] to consider the quadratic form $\boldsymbol{v}^T\boldsymbol{A}\boldsymbol{v}$ for a matrix $\boldsymbol{A}$. Notice that the inequality

$$|\boldsymbol{v}^T\boldsymbol{A}\boldsymbol{v}| \leq ||\boldsymbol{A}\boldsymbol{v}||_2||\boldsymbol{v}||_2 \leq ||\boldsymbol{A}||_{i2}||\boldsymbol{v}||_2^2 \tag{7.17}$$

implies

$$|\boldsymbol{v}^T\boldsymbol{A}\boldsymbol{v}| = |\boldsymbol{v}^T\boldsymbol{R}^T\boldsymbol{R}^{-T}\boldsymbol{A}\boldsymbol{R}^{-1}\boldsymbol{R}\boldsymbol{v}| \leq ||\boldsymbol{R}^{-T}\boldsymbol{A}\boldsymbol{R}^{-1}||_{i2}||\boldsymbol{v}||_K^2 \,, \tag{7.18}$$

with the *K-norm* for the vector $\boldsymbol{v}$ as defined in (7.16). This motivates the choice[8]

$$||\boldsymbol{A}||_K := ||\boldsymbol{R}^{-T}\boldsymbol{A}\boldsymbol{R}^{-1}||_{i2} \,.$$

for the definition of the matrix norm $||\cdot||_K$. Notice that, when this norm is applied to the joint stiffness matrix $\boldsymbol{K}$, one obtains $||\boldsymbol{K}||_K = 1$.

---

[5] Notice that the following design of the gravity compensation does not involve the complete dynamics of the manipulator. Therefore, the stiffness matrix is the appropriate metric for the considered problem rather than the inertia matrix.

[6] The spectral norm is the matrix norm induced by the Euclidean vector norm.

[7] The reason for considering quadratic forms at this point will become clear later in the proof of Proposition 7.1.

[8] Notice that the term $\boldsymbol{R}^{-T}\boldsymbol{A}\boldsymbol{R}^{-1}$ corresponds to the coordinate transformation of a covariant tensor $\boldsymbol{A}$ of rank two when $\boldsymbol{R}$ is the Jacobian of the coordinate transformation. A linear transformation (i.e. a mixed tensor), instead, would be transformed as $\boldsymbol{R}\boldsymbol{A}\boldsymbol{R}^{-1}$.

Let $\boldsymbol{H}(\boldsymbol{q})$ be the Hessian of the gravity potential

$$\boldsymbol{H}(\boldsymbol{q}) := \frac{\partial^2 V_g(\boldsymbol{q})}{\partial \boldsymbol{q}^2} \ . \tag{7.19}$$

Let further α_g be an upper bound for the K-norm of this Hessian in[9] $\mathcal{Q}^p$, i.e.

$$\alpha_g := \sup_{\forall \boldsymbol{q} \in \mathcal{Q}^p} ||\boldsymbol{H}(\boldsymbol{q})||_K \ . \tag{7.20}$$

The existence of this bound $\alpha_g < \infty$ is ensured by Property 2.8 and it implies the following statement for the gravity potential.

Property 7.1. *Let α_g, as defined in (7.20), be an upper bound for the Hessian of the gravity potential $V_g(\boldsymbol{q})$ with respect to the K-norm. Then the inequality*

$$|V_g(\boldsymbol{q}_1) - V_g(\boldsymbol{q}_2) + \boldsymbol{g}(\boldsymbol{q}_1)^T(\boldsymbol{q}_2 - \boldsymbol{q}_1)| \leq \frac{1}{2}\alpha_g ||\boldsymbol{q}_2 - \boldsymbol{q}_1||_K^2 \tag{7.21}$$

holds for all $\boldsymbol{q}_1, \boldsymbol{q}_2 \in \mathcal{Q}^p$.

Proof. The gravity potential $V_g(\boldsymbol{q})$ at the point $\boldsymbol{q}_2$ can be written as the integral of the differential $\boldsymbol{g}(\boldsymbol{q})$ starting from $\boldsymbol{q}_1$

$$\begin{aligned} V_g(\boldsymbol{q}_2) &= V_g(\boldsymbol{q}_1) + \int_{q_1}^{q_2} \frac{\partial V_g(\boldsymbol{q})}{\partial \boldsymbol{q}} \mathrm{d}\boldsymbol{q} \\ &= V_g(\boldsymbol{q}_1) + \int_{q_1}^{q_2} \boldsymbol{g}(\boldsymbol{q})^T \mathrm{d}\boldsymbol{q} \ . \end{aligned} \tag{7.22}$$

Analogously, one can write the differential $\boldsymbol{g}(\boldsymbol{q})$ as

$$\boldsymbol{g}(\boldsymbol{q}) = \boldsymbol{g}(\boldsymbol{q}_1) + \int_{q_1}^{q} \boldsymbol{H}(\boldsymbol{\xi}) \mathrm{d}\boldsymbol{\xi} \ , \tag{7.23}$$

where $\boldsymbol{H}(\boldsymbol{\xi})$ is the Hessian matrix $\boldsymbol{H}(\boldsymbol{\xi}) := (\frac{\partial \boldsymbol{g}(\boldsymbol{q})}{\partial \boldsymbol{q}})_{\boldsymbol{q}=\boldsymbol{\xi}} = (\frac{\partial^2 V_g(\boldsymbol{q})}{\partial \boldsymbol{q}^2})_{\boldsymbol{q}=\boldsymbol{\xi}}$. Combining (7.22) and (7.23) leads to

$$V_g(\boldsymbol{q}_2) = V_g(\boldsymbol{q}_1) + \int_{q_1}^{q_2} \left(\boldsymbol{g}(\boldsymbol{q}_1) + \int_{q_1}^{q} \boldsymbol{H}(\boldsymbol{\xi}) \mathrm{d}\boldsymbol{\xi} \right)^T \mathrm{d}\boldsymbol{q} \ ,$$

from which one can follow

$$V_g(\boldsymbol{q}_2) - V_g(\boldsymbol{q}_1) - \boldsymbol{g}(\boldsymbol{q}_1)^T(\boldsymbol{q}_2 - \boldsymbol{q}_1) = \int_{q_1}^{q_2} \left(\int_{q_1}^{q} \boldsymbol{H}(\boldsymbol{\xi}) \mathrm{d}\boldsymbol{\xi} \right)^T \mathrm{d}\boldsymbol{q} \ .$$

[9] Remember that $\mathcal{Q}^p$ is an area in the configuration space wherein all the prismatic joint variables keep bounded (see (2.36)).

By using the transformations $\boldsymbol{z} = \boldsymbol{R}\boldsymbol{q}$ and $\boldsymbol{z}_\xi = \boldsymbol{R}\boldsymbol{\xi}$ the integral on the right hand side can be written as

$$\int_{q_1}^{q_2} \left(\int_{q_1}^{q} \boldsymbol{H}(\boldsymbol{\xi}) \mathrm{d}\boldsymbol{\xi} \right)^T \mathrm{d}\boldsymbol{q} = \int_{z_1}^{z_2} \left(\int_{z_1}^{z} \boldsymbol{H}(\boldsymbol{R}^{-1}\boldsymbol{z}_\xi)\boldsymbol{R}^{-1} \mathrm{d}\boldsymbol{z}_\xi \right)^T \boldsymbol{R}^{-1} \mathrm{d}\boldsymbol{z}$$
$$= \int_{z_1}^{z_2} \left(\int_{z_1}^{z} \boldsymbol{R}^{-T}\boldsymbol{H}(\boldsymbol{R}^{-1}\boldsymbol{z}_\xi)\boldsymbol{R}^{-1} \mathrm{d}\boldsymbol{z}_\xi \right)^T \mathrm{d}\boldsymbol{z} \, .$$

By using Lemma A.23, one can then conclude

$$|V_g(\boldsymbol{q}_2) - V_g(\boldsymbol{q}_1) - \boldsymbol{g}(\boldsymbol{q}_1)^T(\boldsymbol{q}_2 - \boldsymbol{q}_1)| \leq \frac{1}{2} \sup_{\forall \boldsymbol{q} \in \mathbb{R}^n} ||\boldsymbol{R}^{-T}\boldsymbol{H}(\boldsymbol{q})\boldsymbol{R}^{-1}||_{i2} ||\boldsymbol{z}_2 - \boldsymbol{z}_1||_2^2$$
$$= \frac{1}{2}\alpha_g ||\boldsymbol{q}_2 - \boldsymbol{q}_1||_K^2 \, .$$

Property 7.1 is a direct consequence of Property 2.8 and will be of interest later in the stability analysis. Additionally, one further assumption on the gravity potential will be needed.

Assumption 7.2. *The Hessian* $\boldsymbol{H}(\boldsymbol{q}) = \frac{\partial^2 V_g(\boldsymbol{q})}{\partial \boldsymbol{q}^2}$ *of the gravity potential* $V_g(\boldsymbol{q})$ *satisfies the condition*

$$\alpha_g = \sup_{\forall \boldsymbol{q} \in \mathcal{Q}^p} ||\boldsymbol{H}(\boldsymbol{q})||_K < ||\boldsymbol{K}||_K = 1 \, . \tag{7.24}$$

Notice that this assumption is not restrictive at all. Intuitively speaking it states nothing else than the fact that the manipulator should be designed properly. In particular, this means that the joint stiffness is sufficiently high such that it can prevent the manipulator from falling down under the load of its own weight. For a real manipulator the area in which $||\boldsymbol{H}(\boldsymbol{q})||_K < 1$ holds will be much larger than $\mathcal{Q}^p$.

It should also be mentioned that the quantity α_g is dimensionless since it is defined via the norm $||\cdot||_K$.

7.3.2 Construction of the Gravity Compensation Term

In the following a compensation for the static effects of the gravity term $\boldsymbol{g}(\boldsymbol{q})$ is constructed. This compensation is solely based on the motor position and can compensate for the link side gravity torques (in a *quasi-stationary* fashion). Consider first the set $\Omega := \{(\boldsymbol{q}, \boldsymbol{\theta}) \mid \boldsymbol{K}(\boldsymbol{\theta} - \boldsymbol{q}) = \boldsymbol{g}(\boldsymbol{q})\}$ of stationary points (for $\boldsymbol{\tau}_{ext} = \boldsymbol{0}$) for which the torque due to the joint elasticity counterbalances the link side gravity torque. The goal of the gravity compensation is now to construct a compensation term $\bar{\boldsymbol{g}}(\boldsymbol{\theta})$ such that in Ω the equilibrium condition

$$\bar{\boldsymbol{g}}(\boldsymbol{\theta}) = \boldsymbol{g}(\boldsymbol{q}) \quad \forall (\boldsymbol{q}, \boldsymbol{\theta}) \in \Omega \tag{7.25}$$

holds. This means that the gravity compensation term counterbalances the link side gravity torque in all stationary points.

Notice that for any point $(\boldsymbol{q}_0, \boldsymbol{\theta}_0) \in \Omega$ the motor position can be uniquely expressed as a function of the link side position

$$\boldsymbol{\theta}_0 = \boldsymbol{q}_0 + \boldsymbol{K}^{-1}\boldsymbol{g}(\boldsymbol{q}_0) =: \boldsymbol{h}_g(\boldsymbol{q}_0) \ . \tag{7.26}$$

Furthermore, by the use of the contraction mapping theorem (see the proof of Proposition 7.3 below for more details on this) it can be shown that the inverse function to $\boldsymbol{h}_g(\boldsymbol{q}_0)$ exists. Then

$$\boldsymbol{q}_0 = \boldsymbol{h}_g^{-1}(\boldsymbol{\theta}_0) =: \bar{\boldsymbol{q}}(\boldsymbol{\theta}_0) \tag{7.27}$$

can be used for the construction of a gravity compensation term of the form

$$\boldsymbol{u}_g = \bar{\boldsymbol{g}}(\boldsymbol{\theta}) := \boldsymbol{g}(\bar{\boldsymbol{q}}(\boldsymbol{\theta})) \ , \tag{7.28}$$

which clearly fulfills (7.25). Finally, the question about the existence of the function $\bar{\boldsymbol{q}}(\boldsymbol{\theta})$ is answered by the following proposition.

Proposition 7.3. *If (7.24) from Assumption 7.2 holds globally (i.e. for* $\mathcal{Q}^p = \mathbb{R}^n$*), the inverse function* $\boldsymbol{h}_g^{-1}(\boldsymbol{\theta}) = \bar{\boldsymbol{q}}(\boldsymbol{\theta})$ *of* $\boldsymbol{h}_g(\boldsymbol{q}) = \boldsymbol{q} + \boldsymbol{K}^{-1}\boldsymbol{g}(\boldsymbol{q}) : \mathbb{R}^n \to \mathbb{R}^n$ *exists globally. Moreover, the iteration*

$$\hat{\boldsymbol{q}}_{n+1} = \boldsymbol{T}_g(\hat{\boldsymbol{q}}_n) \tag{7.29}$$

with $\boldsymbol{T}_g(\boldsymbol{q}) := \boldsymbol{\theta} - \boldsymbol{K}^{-1}\boldsymbol{g}(\boldsymbol{q})$ *converges for every fixed* $\boldsymbol{\theta}$ *and for every starting point* $\hat{\boldsymbol{q}}_0$ *to* $\bar{\boldsymbol{q}}(\boldsymbol{\theta})$.

Proof. The proposition can be proven by showing first that the mapping $\boldsymbol{T}_g(\boldsymbol{q}) : \mathbb{R}^n \to \mathbb{R}^n$ is a global *contraction* (see [Vid93]) for the vector norm $||\cdot||_K$. Since the vector space $\mathbb{R}^n$ together with the norm $||\cdot||_K$ indubitably is a Banach space[10], one must only show that there exists a $\rho < 1$ such that $\boldsymbol{T}_g(\boldsymbol{q})$ satisfies the condition

$$||\boldsymbol{T}_g(\boldsymbol{q}_2) - \boldsymbol{T}_g(\boldsymbol{q}_1)||_K \leq \rho||\boldsymbol{q}_2 - \boldsymbol{q}_1||_K \quad \forall \boldsymbol{q}_1, \boldsymbol{q}_2 \in \mathbb{R}^n \ .$$

The function $\boldsymbol{T}_g(\boldsymbol{q})$ at $\boldsymbol{q}_2$ can be written as the integral

$$\begin{aligned}\boldsymbol{T}_g(\boldsymbol{q}_2) &= \boldsymbol{T}_g(\boldsymbol{q}_1) + \int_{q_1}^{q_2} \frac{\partial \boldsymbol{T}_g(\boldsymbol{q})}{\partial \boldsymbol{q}} \mathrm{d}\boldsymbol{q} \\ &= \boldsymbol{T}_g(\boldsymbol{q}_1) - \int_{q_1}^{q_2} \boldsymbol{K}^{-1}\boldsymbol{H}(\boldsymbol{q})\mathrm{d}\boldsymbol{q} \ ,\end{aligned}$$

[10] A Banach space is a normed vector space which is *complete* in the sense that every *Cauchy sequence* converges to an element of the space (see [Vid93]).

from which one can conclude (by using the transformation $\boldsymbol{z} = \boldsymbol{R}\boldsymbol{q}$ and Lemma A.23)

$$\begin{aligned}
\boldsymbol{T}_g(\boldsymbol{q}_2) - \boldsymbol{T}_g(\boldsymbol{q}_1) &= -\int_{q_1}^{q_2} \boldsymbol{K}^{-1}\boldsymbol{H}(\boldsymbol{q})\mathrm{d}\boldsymbol{q} \\
\boldsymbol{R}(\boldsymbol{T}_g(\boldsymbol{q}_2) - \boldsymbol{T}_g(\boldsymbol{q}_1)) &= -\int_{q_1}^{q_2} \boldsymbol{R}\boldsymbol{K}^{-1}\boldsymbol{H}(\boldsymbol{q})\mathrm{d}\boldsymbol{q} \\
\boldsymbol{R}(\boldsymbol{T}_g(\boldsymbol{q}_2) - \boldsymbol{T}_g(\boldsymbol{q}_1)) &= -\int_{q_1}^{q_2} \boldsymbol{R}^{-T}\boldsymbol{H}(\boldsymbol{q})\mathrm{d}\boldsymbol{q} \\
\boldsymbol{R}(\boldsymbol{T}_g(\boldsymbol{q}_2) - \boldsymbol{T}_g(\boldsymbol{q}_1)) &= -\int_{z_1}^{z_2} \boldsymbol{R}^{-T}\boldsymbol{H}(\boldsymbol{R}^{-1}\boldsymbol{z})\boldsymbol{R}^{-1}\mathrm{d}\boldsymbol{z} \\
||\boldsymbol{R}(\boldsymbol{T}_g(\boldsymbol{q}_2) - \boldsymbol{T}_g(\boldsymbol{q}_1))||_2 &\leq \sup_{\forall \boldsymbol{q}\in\mathbb{R}^n} ||\boldsymbol{R}^{-T}\boldsymbol{H}(\boldsymbol{q})\boldsymbol{R}^{-1}||_{i2}||\boldsymbol{z}_2 - \boldsymbol{z}_1||_2 \\
||\boldsymbol{T}_g(\boldsymbol{q}_2) - \boldsymbol{T}_g(\boldsymbol{q}_1)||_K &\leq \sup_{\forall \boldsymbol{q}\in\mathbb{R}^n} ||\boldsymbol{H}(\boldsymbol{q})||_K||\boldsymbol{q}_2 - \boldsymbol{q}_1||_K \ .
\end{aligned}$$

From Assumption 7.2 it follows that $\boldsymbol{T}_g(\boldsymbol{q})$ is a (global) contraction. By the *contraction mapping theorem*[11] (see, e.g., [Vid93]) one can therefore conclude that the mapping $\boldsymbol{T}_g(\boldsymbol{q})$ has a unique fixed point $\boldsymbol{q}^* = \boldsymbol{T}_g(\boldsymbol{q}^*)$ and that the iteration of (7.29) converges to this fixed point

$$\lim_{n\to\infty} \hat{\boldsymbol{q}}_n = \boldsymbol{q}^* \ .$$

By comparing $\boldsymbol{T}_g(\boldsymbol{q})$ with $\boldsymbol{h}_g(\boldsymbol{q})$, one can easily see that (for each particular value of $\boldsymbol{\theta}$) this fixed point $\boldsymbol{q}^*$ corresponds to $\bar{\boldsymbol{q}}(\boldsymbol{\theta})$.

While in general the inverse function $\boldsymbol{h}_g^{-1}(\boldsymbol{\theta})$ cannot be computed directly in practice, it is thus possible to approximate it with arbitrary accuracy by iteration. From a practical point of view one or two iteration steps already lead to quite satisfactory results. Notice also that by a first order approximation with $\hat{\boldsymbol{q}}_0 = \boldsymbol{q}_d$ one would obtain the online gravity compensation term of [ZSL+03].

In the following analysis it is therefore assumed that the inverse function $\boldsymbol{h}_g^{-1}(\boldsymbol{\theta})$ is known exactly, although it can only be approximated in practice.

Another remark about the range in which Property 7.3 holds is important. The assumption $\mathcal{Q}^p = \mathbb{R}^n$, which holds for instance when the robot has only rotational joints, was needed to ensure that $\boldsymbol{T}_g(\boldsymbol{q})$ is a global contraction. If instead $\mathcal{Q}^p \subset \mathbb{R}^n$, then one must additionally ensure that the points $\hat{\boldsymbol{q}}_i$ of the iteration (7.29) stay in an area in which $||\boldsymbol{H}(\boldsymbol{q})||_K < ||\boldsymbol{K}||_K = 1$ holds. While this is not a critical issue from a practical point of view[12], it is difficult to be proven in general.

Since $\bar{\boldsymbol{g}}(\boldsymbol{\theta})$ is the motor torque needed (statically) to prevent the robot from falling down under the action of its own weight, one can see that $\bar{\boldsymbol{g}}(\boldsymbol{\theta})$ must

[11] Also called *Banach fixed point theorem.*

[12] Notice that for a real robot the area in which $||\boldsymbol{H}(\boldsymbol{q})||_K < ||\boldsymbol{K}||_K = 1$ holds will be much larger than $\mathcal{Q}^p$.

be given as the differential of a potential function $V_{\bar{g}}(\boldsymbol{\theta})$, which is related to the potential energy (gravity plus joint stiffness) of the robot. This potential function will be of interest for the passivity and stability analysis in Section 7.4. A detailed derivation of $V_{\bar{g}}(\boldsymbol{\theta})$ is therefore given in the next section.

7.3.3 Derivation of the Gravity Compensation Potential

In this section the potential function $V_{\bar{g}}(\boldsymbol{\theta})$ for the gravity compensation term $\bar{\boldsymbol{g}}(\boldsymbol{\theta})$ is derived, such that $\bar{\boldsymbol{g}}(\boldsymbol{\theta}) = (\partial V_{\bar{g}}(\boldsymbol{\theta})/\partial\boldsymbol{\theta})^T$ holds. Remember that for the construction of $\bar{\boldsymbol{g}}(\boldsymbol{\theta}) = \boldsymbol{g}(\bar{\boldsymbol{q}}(\boldsymbol{\theta}))$ in the previous section the function $\bar{\boldsymbol{q}}(\boldsymbol{\theta}) = \boldsymbol{h}_g^{-1}(\boldsymbol{\theta})$, i.e. the inverse of the function $\boldsymbol{h}_g(\boldsymbol{q}) = \boldsymbol{q}+\boldsymbol{K}^{-1}\boldsymbol{g}(\boldsymbol{q})$, was used. Existence and uniqueness of $\boldsymbol{h}_g^{-1}(\boldsymbol{\theta})$ were established in Proposition 7.1 by the use of Assumption 7.2.

In the following the Jacobian matrix $\partial\bar{\boldsymbol{q}}(\boldsymbol{\theta})/\partial\boldsymbol{\theta}$ will be needed. Consider first the Jacobian matrix of the function $\boldsymbol{h}_g(\boldsymbol{q})$

$$\frac{\partial \boldsymbol{h}_g(\boldsymbol{q})}{\partial \boldsymbol{q}} = \left(\boldsymbol{I} + \boldsymbol{K}^{-1}\frac{\partial \boldsymbol{g}(\boldsymbol{q})}{\partial \boldsymbol{q}}\right) . \tag{7.30}$$

Due to $\boldsymbol{h}_g(\bar{\boldsymbol{q}}(\boldsymbol{\theta})) = \boldsymbol{\theta}$ one has

$$\frac{\partial \boldsymbol{h}_g(\bar{\boldsymbol{q}}(\boldsymbol{\theta}))}{\partial \boldsymbol{\theta}} = \frac{\partial \boldsymbol{h}_g(\bar{\boldsymbol{q}})}{\partial \bar{\boldsymbol{q}}}\frac{\partial \bar{\boldsymbol{q}}(\boldsymbol{\theta})}{\partial \boldsymbol{\theta}} = \boldsymbol{I} \ ,$$

and therefore the Jacobian matrix $\frac{\partial \bar{\boldsymbol{q}}(\boldsymbol{\theta})}{\partial \boldsymbol{\theta}}$ is given by

$$\frac{\partial \bar{\boldsymbol{q}}(\boldsymbol{\theta})}{\partial \boldsymbol{\theta}} = \left(\boldsymbol{I} + \boldsymbol{K}^{-1}\frac{\partial \boldsymbol{g}(\bar{\boldsymbol{q}})}{\partial \bar{\boldsymbol{q}}}\right)^{-1}_{\bar{\boldsymbol{q}}=\bar{\boldsymbol{q}}(\boldsymbol{\theta})} . \tag{7.31}$$

The potential function $V_{\bar{g}}(\boldsymbol{\theta})$ clearly can be written in the form

$$V_{\bar{g}}(\boldsymbol{\theta}) = V_{\bar{g}}(\boldsymbol{h}_g(\bar{\boldsymbol{q}}(\boldsymbol{\theta}))) =: V_{\bar{g}h}(\bar{\boldsymbol{q}}(\boldsymbol{\theta})) \ .$$

For the differential $\partial V_{\bar{g}}(\boldsymbol{\theta})/\partial\boldsymbol{\theta}$ one obtains

$$\frac{\partial V_{\bar{g}}(\boldsymbol{\theta})}{\partial \boldsymbol{\theta}} = \left(\frac{\partial V_{\bar{g}h}(\bar{\boldsymbol{q}})}{\partial \bar{\boldsymbol{q}}}\right)_{\bar{\boldsymbol{q}}=\bar{\boldsymbol{q}}(\boldsymbol{\theta})}\frac{\partial \bar{\boldsymbol{q}}(\boldsymbol{\theta})}{\partial \boldsymbol{\theta}} \ .$$

By substituting $\frac{\partial V_{\bar{g}}(\boldsymbol{\theta})}{\partial \boldsymbol{\theta}} = \bar{\boldsymbol{g}}(\boldsymbol{\theta}) = \boldsymbol{g}(\bar{\boldsymbol{q}}(\boldsymbol{\theta}))^T$ and $\frac{\partial \bar{\boldsymbol{q}}(\boldsymbol{\theta})}{\partial \boldsymbol{\theta}}$ from (7.31), one gets

$$\begin{aligned}\frac{\partial V_{\bar{g}h}(\bar{\boldsymbol{q}})}{\partial \bar{\boldsymbol{q}}} &= \boldsymbol{g}(\bar{\boldsymbol{q}})^T\left(\boldsymbol{I} + \boldsymbol{K}^{-1}\frac{\partial \boldsymbol{g}(\bar{\boldsymbol{q}})}{\partial \bar{\boldsymbol{q}}}\right) , \\ &= \boldsymbol{g}(\bar{\boldsymbol{q}})^T + \boldsymbol{g}(\bar{\boldsymbol{q}})^T\boldsymbol{K}^{-1}\frac{\partial \boldsymbol{g}(\bar{\boldsymbol{q}})}{\partial \bar{\boldsymbol{q}}} \ .\end{aligned}$$

This differential can then be integrated to $V_{\bar{g}h}(\bar{\boldsymbol{q}}) = V_g(\bar{\boldsymbol{q}})+\frac{1}{2}\boldsymbol{g}(\bar{\boldsymbol{q}})^T\boldsymbol{K}^{-1}\boldsymbol{g}(\bar{\boldsymbol{q}})+\boldsymbol{c}$, with an arbitrary constant $\boldsymbol{c} \in \mathbb{R}^n$. Setting $\boldsymbol{c} = \boldsymbol{0}$ leads to the gravity compensation potential

$$V_{\bar{g}}(\boldsymbol{\theta}) = V_{\bar{g}h}(\bar{\boldsymbol{q}}(\boldsymbol{\theta})) = V_g(\bar{\boldsymbol{q}}(\boldsymbol{\theta})) + \frac{1}{2}\boldsymbol{g}(\bar{\boldsymbol{q}}(\boldsymbol{\theta}))^T\boldsymbol{K}^{-1}\boldsymbol{g}(\bar{\boldsymbol{q}}(\boldsymbol{\theta})) \ . \tag{7.32}$$

Notice also that the potential energy of the manipulator $V_{pot}(\boldsymbol{q}, \boldsymbol{\theta})$ from (2.24) is identical to the gravity compensation potential for all stationary points, i.e.

$$V_{pot}(\boldsymbol{q}, \boldsymbol{\theta}) = V_{\bar{g}}(\boldsymbol{\theta}) \quad \forall\ (\boldsymbol{q}, \boldsymbol{\theta}) \in \Omega\ . \tag{7.33}$$

From this it follows that $V_{\bar{g}}(\boldsymbol{\theta})$ can also be written as

$$V_{\bar{g}}(\boldsymbol{\theta}) = V_{pot}(\bar{\boldsymbol{q}}(\boldsymbol{\theta}), \boldsymbol{\theta}) = V_g(\bar{\boldsymbol{q}}(\boldsymbol{\theta})) + V_k(\bar{\boldsymbol{q}}(\boldsymbol{\theta}), \boldsymbol{\theta})\ . \tag{7.34}$$

7.4 Analysis

The complete control law with gravity compensation is given by

$$\boldsymbol{\tau}_m = \boldsymbol{B}\boldsymbol{B}_\theta^{-1}\boldsymbol{u} + (\boldsymbol{I} - \boldsymbol{B}\boldsymbol{B}_\theta^{-1})\boldsymbol{\tau}\ , \tag{7.35}$$

$$\boldsymbol{u} = \boldsymbol{u}_{imp} + \boldsymbol{u}_g = -\boldsymbol{J}(\boldsymbol{\theta})^T(\boldsymbol{K}_d\tilde{\boldsymbol{x}}_\theta + \boldsymbol{D}_d\dot{\boldsymbol{x}}_\theta) + \bar{\boldsymbol{g}}(\boldsymbol{\theta})\ , \tag{7.36}$$

and leads to the closed loop system equations

$$\boldsymbol{M}(\boldsymbol{q})\ddot{\boldsymbol{q}} + \boldsymbol{C}(\boldsymbol{q}, \dot{\boldsymbol{q}})\dot{\boldsymbol{q}} + \boldsymbol{g}(\boldsymbol{q}) = \boldsymbol{\tau} + \boldsymbol{\tau}_{ext}\ , \tag{7.37}$$

$$\boldsymbol{B}_\theta\ddot{\boldsymbol{\theta}} + \boldsymbol{J}(\boldsymbol{\theta})^T(\boldsymbol{K}_d\tilde{\boldsymbol{x}}_\theta + \boldsymbol{D}_d\dot{\boldsymbol{x}}_\theta) + \boldsymbol{\tau} = \bar{\boldsymbol{g}}(\boldsymbol{\theta})\ . \tag{7.38}$$

In this section it will first be shown that in case of a globally bounded potential function $V_g(\boldsymbol{q})$ the closed loop system can be written as the feedback interconnection of two passive subsystems. Notice that this implies the passivity of the complete system [vdS00]. Additionally, a proof of the asymptotic stability is given in Section 7.4.2 which holds also without the assumption of a bounded gravity potential[13].

7.4.1 Passivity

For the passivity analysis of this section it will be assumed that there exists a real $\beta > 0$, such that

$$|V_g(\boldsymbol{q})| < \beta \qquad \forall \boldsymbol{q} \in \mathbb{R}^n \tag{7.39}$$

holds. This is for instance satisfied for all robots with solely rotational joints (i.e. without prismatic joints). Due to Property 2.8 also the gravity torque vector $\boldsymbol{g}(\boldsymbol{q})$ will then be globally bounded. Furthermore, (7.39) also implies the boundedness of $V_{\bar{g}}(\boldsymbol{\theta})$ and $\bar{\boldsymbol{g}}(\boldsymbol{\theta})$. Notice that the requirement of a bounded gravity potential is only needed for the passivity analysis, while the proof of the asymptotic stability in Section 7.4.2 will also be valid for a general potential.

According to Definition A.17, a sufficient condition for a system (with input $\boldsymbol{u}$ and output $\boldsymbol{y}$) to be passive is given by the existence of a continuous storage function S [vdS00, Kug01] which is bounded from below and for which the

[13] Notice that this is relevant only for robots with prismatic joints.

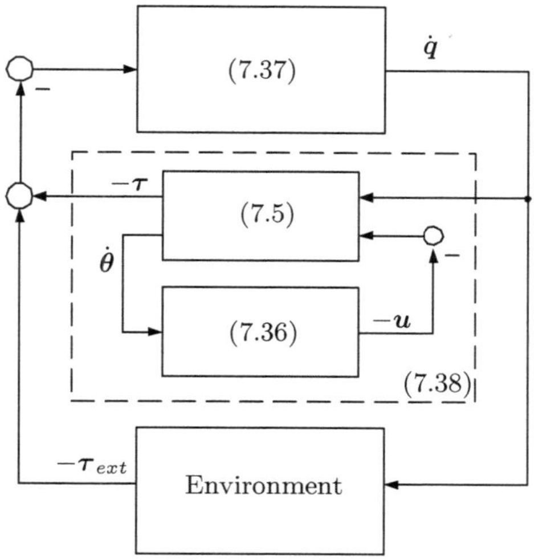

Fig. 7.3. System representation as an interconnection of passive subsystems

derivative with respect to time along the solutions of the system satisfies the inequality $\dot{S} \leq \boldsymbol{y}^T \boldsymbol{u}$.

In the following it will be shown that the system (7.37)-(7.38), as outlined in Figure 7.3, consists of two passive subsystems in feedback interconnection. It is often assumed that also the environment of the robot can be described by a passive mapping $(\dot{\boldsymbol{q}} \rightarrow -\boldsymbol{\tau}_{ext})$. The passivity of (7.37), as a mapping $(\boldsymbol{\tau}+\boldsymbol{\tau}_{ext}) \rightarrow \dot{\boldsymbol{q}}$, is well known due to purely physical reasons and can be shown with the storage function

$$S_q(\boldsymbol{q}, \dot{\boldsymbol{q}}) = \frac{1}{2}\dot{\boldsymbol{q}}^T \boldsymbol{M}(\boldsymbol{q})\dot{\boldsymbol{q}} + V_g(\boldsymbol{q}) \tag{7.40}$$

for which the derivative along the solutions of (7.37) is given by[14]

$$\dot{S}_q(\boldsymbol{q}, \dot{\boldsymbol{q}}) = \dot{\boldsymbol{q}}^T(\boldsymbol{\tau} + \boldsymbol{\tau}_{ext}) \ . \tag{7.41}$$

In a similar way the passivity of (7.38), as a mapping $\dot{\boldsymbol{q}} \rightarrow -\boldsymbol{\tau}$, can be shown with the storage function

$$\begin{aligned} S_\theta(\boldsymbol{q}, \boldsymbol{\theta}, \dot{\boldsymbol{\theta}}) = &\frac{1}{2}\dot{\boldsymbol{\theta}}^T \boldsymbol{B}_\theta \dot{\boldsymbol{\theta}} + \frac{1}{2}(\boldsymbol{\theta} - \boldsymbol{q})^T \boldsymbol{K}(\boldsymbol{\theta} - \boldsymbol{q}) \\ &+ \frac{1}{2}\tilde{\boldsymbol{x}}_\theta^T \boldsymbol{K}_d \tilde{\boldsymbol{x}}_\theta - V_{\bar{g}}(\boldsymbol{\theta}) \ . \end{aligned} \tag{7.42}$$

The derivative of $S_\theta(\boldsymbol{q}, \boldsymbol{\theta}, \dot{\boldsymbol{\theta}})$ along the solutions of (7.38) is then given by

$$\dot{S}_\theta(\boldsymbol{q}, \boldsymbol{\theta}, \dot{\boldsymbol{\theta}}) = -\dot{\boldsymbol{x}}_\theta^T \boldsymbol{D}_d \dot{\boldsymbol{x}}_\theta - \dot{\boldsymbol{q}}^T \boldsymbol{\tau} \ . \tag{7.43}$$

[14] This is due to Property 2.6.

The desired passivity property follows directly from (7.41) and (7.43) and from the fact that the feedback interconnection of passive systems is again passive. It should also be mentioned that these passivity properties are still valid if the PD-controller in (7.36) is replaced by any other passive (with respect to $\dot{\boldsymbol{\theta}} \rightarrow -\boldsymbol{u}$) controller. Notice that the structure of the closed loop system in form of a feedback interconnection of passive subsystems, as depicted in Figure 7.3, brings along very advantageous robustness properties.

7.4.2 Stability

The following stability analysis is restricted to the non-redundant case ($m = n$). In addition, singular configurations have to be avoided. Thus, the further analysis is restricted to an area in the workspace in which the Jacobian $\boldsymbol{J}(\boldsymbol{\theta})$ is non-singular and in which the inverse mapping of $\boldsymbol{x}_\theta = \boldsymbol{f}(\boldsymbol{\theta})$ can be uniquely solved.

In the following it will be shown that the closed loop system is asymptotically stable for the case of free motion (i.e. $\boldsymbol{\tau}_{ext} = \boldsymbol{0}$).

Determination of the steady state

The steady state conditions of the system (7.37)-(7.38) are given by

$$\boldsymbol{K}(\boldsymbol{\theta}_0 - \boldsymbol{q}_0) = \boldsymbol{g}(\boldsymbol{q}_0) \ , \tag{7.44}$$

$$\boldsymbol{K}(\boldsymbol{\theta}_0 - \boldsymbol{q}_0) + \boldsymbol{J}(\boldsymbol{\theta}_0)^T \boldsymbol{K}_d(\boldsymbol{f}(\boldsymbol{\theta}_0) - \boldsymbol{x}_{\theta,d}) = \bar{\boldsymbol{g}}(\boldsymbol{\theta}_0) \ . \tag{7.45}$$

Herein the matrix $\boldsymbol{K}_d$ is positive definite. Due to (7.25) it follows that

$$\boldsymbol{J}(\boldsymbol{\theta}_0)^T \boldsymbol{K}_d(\boldsymbol{f}(\boldsymbol{\theta}_0) - \boldsymbol{x}_{\theta,d}) = \boldsymbol{0} \tag{7.46}$$

must be satisfied in the steady state. As already mentioned above, the stability analysis is restricted to an area in which this condition can be solved uniquely for $\boldsymbol{\theta}_0$. The steady state is then given by

$$\boldsymbol{\theta}_0 = \boldsymbol{f}^{-1}(\boldsymbol{x}_{\theta,d}) \ , \tag{7.47}$$

$$\boldsymbol{q}_0 = \boldsymbol{h}_g^{-1}(\boldsymbol{\theta}_0) \ , \tag{7.48}$$

$$\dot{\boldsymbol{q}}_0 = \boldsymbol{0} \tag{7.49}$$

$$\dot{\boldsymbol{\theta}}_0 = \boldsymbol{0} \ , \tag{7.50}$$

and $\boldsymbol{\tau}_{m,0} = \boldsymbol{\tau}_0 = \bar{\boldsymbol{g}}(\boldsymbol{\theta}_0)$ is the steady state input torque.

Lyapunov-Function

Consider the following function $V(\boldsymbol{q}, \dot{\boldsymbol{q}}, \boldsymbol{\theta}, \dot{\boldsymbol{\theta}})$ as a Lyapunov function candidate

$$V(\boldsymbol{q}, \dot{\boldsymbol{q}}, \boldsymbol{\theta}, \dot{\boldsymbol{\theta}}) = S_q(\boldsymbol{q}, \dot{\boldsymbol{q}}) + S_\theta(\boldsymbol{q}, \boldsymbol{\theta}, \dot{\boldsymbol{\theta}}) \ . \tag{7.51}$$

In the steady state the following holds (due to (7.34))[15]

$$V(\boldsymbol{q}_0, \boldsymbol{0}, \boldsymbol{\theta}_0, \boldsymbol{0}) = V_g(\boldsymbol{q}_0) - V_{\bar{g}}(\boldsymbol{\theta}_0) + \frac{1}{2}(\boldsymbol{\theta}_0 - \boldsymbol{q}_0)^T \boldsymbol{K}(\boldsymbol{\theta}_0 - \boldsymbol{q}_0) = 0 \ .$$

[15] Remember that in steady state $\boldsymbol{q}_0 = \bar{\boldsymbol{q}}(\boldsymbol{\theta}_0)$.

The kinetic part of $V(\boldsymbol{q}, \dot{\boldsymbol{q}}, \boldsymbol{\theta}, \dot{\boldsymbol{\theta}})$

$$V_{kin}(\boldsymbol{q}, \dot{\boldsymbol{q}}, \dot{\boldsymbol{\theta}}) = \frac{1}{2}\dot{\boldsymbol{q}}^T \boldsymbol{M}(\boldsymbol{q})\dot{\boldsymbol{q}} + \frac{1}{2}\dot{\boldsymbol{\theta}}^T \boldsymbol{B}_\theta \dot{\boldsymbol{\theta}}$$

is positive definite with respect to $\dot{\boldsymbol{q}}$ and $\dot{\boldsymbol{\theta}}$ because the inertia matrices are positive definite. In order to show that $V(\boldsymbol{q}, \dot{\boldsymbol{q}}, \boldsymbol{\theta}, \dot{\boldsymbol{\theta}})$ is positive definite, it is then sufficient to show that the potential part

$$V_{pot}(\boldsymbol{q}, \boldsymbol{\theta}) = V(\boldsymbol{q}, \dot{\boldsymbol{q}}, \boldsymbol{\theta}, \dot{\boldsymbol{\theta}}) - V_{kin}(\boldsymbol{q}, \dot{\boldsymbol{q}}, \dot{\boldsymbol{\theta}}) \tag{7.52}$$

is positive definite with respect to $\boldsymbol{q}$ and $\boldsymbol{\theta}$.

Consider at first only the part of the potential energy due to $\boldsymbol{K}$. In the remaining part of this section $\bar{\boldsymbol{q}}$ is written instead of $\bar{\boldsymbol{q}}(\boldsymbol{\theta})$ in order to simplify the notation.

$$\begin{aligned} V_k(\boldsymbol{q}, \boldsymbol{\theta}) &= \frac{1}{2}(\boldsymbol{\theta} - \boldsymbol{q})^T \boldsymbol{K}(\boldsymbol{\theta} - \boldsymbol{q}) \\ &= \frac{1}{2}(\boldsymbol{\theta} - \bar{\boldsymbol{q}} + \bar{\boldsymbol{q}} - \boldsymbol{q})^T \boldsymbol{K}(\boldsymbol{\theta} - \bar{\boldsymbol{q}} + \bar{\boldsymbol{q}} - \boldsymbol{q}) \\ &= \frac{1}{2}\boldsymbol{g}(\bar{\boldsymbol{q}})^T \boldsymbol{K}^{-1}\boldsymbol{g}(\bar{\boldsymbol{q}}) + \frac{1}{2}(\bar{\boldsymbol{q}} - \boldsymbol{q})^T \boldsymbol{K}(\bar{\boldsymbol{q}} - \boldsymbol{q}) + (\bar{\boldsymbol{q}} - \boldsymbol{q})^T \boldsymbol{g}(\bar{\boldsymbol{q}}) \end{aligned}$$

Herein the relationship $\boldsymbol{K}(\boldsymbol{\theta} - \bar{\boldsymbol{q}}) = \boldsymbol{g}(\bar{\boldsymbol{q}})$ was used which (contrary to (7.25)) holds everywhere (i.e. not only in Ω). The potential energy can then be written (with (7.34)) as follows

$$\begin{aligned} V_{pot}(\boldsymbol{q}, \boldsymbol{\theta}) &= V_k(\boldsymbol{q}, \boldsymbol{\theta}) + \frac{1}{2}\tilde{\boldsymbol{x}}_\theta^T \boldsymbol{K}_d \tilde{\boldsymbol{x}}_\theta + V_g(\boldsymbol{q}) - V_{\bar{g}}(\boldsymbol{\theta}) \\ &= V_k(\boldsymbol{q}, \boldsymbol{\theta}) + \frac{1}{2}\tilde{\boldsymbol{x}}_\theta^T \boldsymbol{K}_d \tilde{\boldsymbol{x}}_\theta + V_g(\boldsymbol{q}) - V_g(\bar{\boldsymbol{q}}) - \frac{1}{2}\boldsymbol{g}(\bar{\boldsymbol{q}})^T \boldsymbol{K}^{-1}\boldsymbol{g}(\bar{\boldsymbol{q}}) \end{aligned}$$

Due to Property 7.1 the following inequality holds

$$\begin{aligned} V_{pot}(\boldsymbol{q}, \boldsymbol{\theta}) &\geq \frac{1}{2}(\bar{\boldsymbol{q}} - \boldsymbol{q})^T \boldsymbol{K}(\bar{\boldsymbol{q}} - \boldsymbol{q}) + \frac{1}{2}\tilde{\boldsymbol{x}}_\theta^T \boldsymbol{K}_d \tilde{\boldsymbol{x}}_\theta \\ &\quad - |V_g(\boldsymbol{q}) - V_g(\bar{\boldsymbol{q}}) + (\bar{\boldsymbol{q}} - \boldsymbol{q})^T \boldsymbol{g}(\bar{\boldsymbol{q}})| \\ &\geq \frac{1}{2}(\bar{\boldsymbol{q}} - \boldsymbol{q})^T \boldsymbol{K}(\bar{\boldsymbol{q}} - \boldsymbol{q}) - \frac{1}{2}\alpha_g ||\bar{\boldsymbol{q}} - \boldsymbol{q}||_K^2 + \frac{1}{2}\tilde{\boldsymbol{x}}_\theta^T \boldsymbol{K}_d \tilde{\boldsymbol{x}}_\theta \\ &= \frac{1}{2}(1 - \alpha_g)||\bar{\boldsymbol{q}} - \boldsymbol{q}||_K^2 + \frac{1}{2}\tilde{\boldsymbol{x}}_\theta^T \boldsymbol{K}_d \tilde{\boldsymbol{x}}_\theta \; . \end{aligned}$$

The right hand side of the last inequality is nonnegative for all $(\boldsymbol{q}, \boldsymbol{\theta})$, since by Assumption 7.2 the bound α_g satisfies $\alpha_g < 1$. The area in which the term $\tilde{\boldsymbol{x}}_\theta^T \boldsymbol{K}_d \tilde{\boldsymbol{x}}_\theta$ is positive definite (in $\boldsymbol{\theta}$) finally determines the area in which the Lyapunov function is positive definite. For the case of a general forward kinematics mapping only local statements can be made therefore.

Derivative of the Lyapunov-Function

The derivative of $V(\boldsymbol{q}, \dot{\boldsymbol{q}}, \boldsymbol{\theta}, \dot{\boldsymbol{\theta}})$ along the solutions of the system (7.37)-(7.38) (for $\boldsymbol{\tau}_{ext} = \mathbf{0}$) is given by

$$\dot{V}(\boldsymbol{q}, \dot{\boldsymbol{q}}, \boldsymbol{\theta}, \dot{\boldsymbol{\theta}}) = \dot{S}_q(\boldsymbol{q}, \dot{\boldsymbol{q}}) + \dot{S}_\theta(\boldsymbol{q}, \boldsymbol{\theta}, \dot{\boldsymbol{\theta}}) = -\dot{\boldsymbol{x}}_\theta^T \boldsymbol{D}_d \dot{\boldsymbol{x}}_\theta \ . \tag{7.53}$$

Due to the fact that the matrix $\boldsymbol{D}_d$ is positive definite, it can be concluded that the equilibrium point (7.47)-(7.50) is stable. Furthermore, asymptotic stability can be shown by the use of the invariance principle of LaSalle. According to this the system state will converge to the largest positively invariant set for which $\dot{\boldsymbol{x}}_\theta = \mathbf{0}$ holds. From the system equations it follows that there does not exist any trajectory for which $\dot{\boldsymbol{x}}_\theta = \mathbf{0}$ holds except for the restriction to the equilibrium point[16]. Therefore, asymptotic stability can be concluded.

7.4.3 Controller Discussion

The passivity analysis in Section 7.4.1 showed that the closed loop system can be seen as a feedback interconnection of passive subsystems. In many applications the environment can also be treated as a passive system with respect to the input $\dot{\boldsymbol{q}}$ and the output $-\boldsymbol{\tau}_{ext}$. Therefore, one can conclude very advantageous robustness properties of the whole system. Stability is for instance guaranteed for *arbitrary*[17] errors in the dynamical parameters of the mass matrices $\boldsymbol{M}(\boldsymbol{q})$ and $\boldsymbol{B}$.

Concerning the formulation of the gravity compensation term it should be mentioned that, in contrast to any related previous works, no restrictions are imposed on the positive definite matrix $\boldsymbol{K}_d$ for stability, meaning that the stiffness can be commanded arbitrarily close to zero.

Notice that by the same argumentation as above one can in principle also show convergence of the Cartesian error $\tilde{\boldsymbol{x}} \to \mathbf{0}$ in the redundant case $m < n$ (as long as singular configurations are avoided). However, then of course also a nullspace damping term is needed to ensure $\dot{\boldsymbol{\theta}} \to \mathbf{0}$ (see, e.g., [Kha87]).

It should also be mentioned that the stability analysis would have led to a global statement for a joint space impedance controller. The Cartesian impedance controller defines the impedance in local Cartesian coordinates. This is the reason that only local stability can be proven for the Cartesian controller. In the case of rotational joints any set of Cartesian coordinates which describe the position and orientation of the end-effector must necessarily be periodic in $\boldsymbol{\theta}$. Furthermore, it is well known that, due to topological reasons[18], it is not possible to design a potential function in $SE(3)$ which has only a single global extremum.

---

[16] Notice again that the analysis is restricted to a workspace in which the Jacobian is non-singular.

[17] Notice that the motor inertia matrix $\boldsymbol{B}$ must be diagonal.

[18] And Morse's theory [Mil63].

7.5 Exact Stiffness Design

In the previous sections the impedance controller was designed basically as a serial interconnection of the joint stiffness $\boldsymbol{K}$ and a desired stiffness. In the following it is shown, how one can take account of the joint stiffness in the design of $\boldsymbol{K}_\theta$ (respectively $\boldsymbol{K}_d$) in order to achieve a particular value for the overall external stiffness in the case of a joint level impedance and a Cartesian impedance.

7.5.1 Design of an Exact Joint Level Stiffness

Consider first the case of a joint level impedance with a stiffness $\boldsymbol{K}_\theta$, cf. (7.7). The resulting stiffness $\boldsymbol{K}_{\theta,ext}$, i.e. the static relation between $\boldsymbol{\tau}_{ext}$ and $\boldsymbol{q} - \boldsymbol{q}_d$, is given by

$$\boldsymbol{K}_{\theta,ext} = \left(\boldsymbol{K}_\theta^{-1} + \boldsymbol{K}^{-1}\right)^{-1} .$$

When the joint stiffness $\boldsymbol{K}$ is much larger than $\boldsymbol{K}_\theta$, the stiffness $\boldsymbol{K}_{\theta,ext}$ will be close to $\boldsymbol{K}_\theta$. But the relationship between $\boldsymbol{K}_{\theta,ext}$ and $\boldsymbol{K}_\theta$ can of course be used also explicitly for the design of $\boldsymbol{K}_\theta$. To exactly realize a stiffness of $\boldsymbol{K}_{\theta,ext}$, the controller stiffness in (7.7) must be set to

$$\boldsymbol{K}_\theta = \left(\boldsymbol{K}_{\theta,ext}^{-1} - \boldsymbol{K}^{-1}\right)^{-1} .$$

Since $\boldsymbol{K}_\theta$ must be positive definite, $\boldsymbol{K}$ is the maximum achievable stiffness for a controller of the form (7.7).

For the Cartesian impedance controller from (7.10) the same can be done only locally at $\boldsymbol{q}_d$. It can easily be verified that at $\boldsymbol{q}_d$ the local stiffness from (7.10) with respect to joint coordinates is given by $\boldsymbol{J}(\boldsymbol{q}_d)^T \boldsymbol{K}_d \boldsymbol{J}(\boldsymbol{q}_d)$. For a desired external stiffness $\boldsymbol{K}_{d,ext}$, which must be chosen such that the matrix $\boldsymbol{K}_{d,ext}^{-1} - \boldsymbol{J}(\boldsymbol{q}_d)\boldsymbol{K}^{-1}\boldsymbol{J}(\boldsymbol{q}_d)^T$ is positive definite, the matrix $\boldsymbol{K}_d$ reads as

$$\boldsymbol{K}_d = \left(\boldsymbol{K}_{d,ext}^{-1} - \boldsymbol{J}(\boldsymbol{q}_d)\boldsymbol{K}^{-1}\boldsymbol{J}(\boldsymbol{q}_d)^T\right)^{-1} .$$

But this simple design law matches the desired stiffness only locally in contrast to the joint level case. This can indeed be practically relevant for applications that require a high stiffness in one Cartesian direction and a low stiffness in another direction. In order to overcome this problem, in the following a different controller structure than in (7.36) is formulated.

7.5.2 Design of an Exact Cartesian Stiffness

For the derivation of the gravity compensation term $\bar{\boldsymbol{g}}(\boldsymbol{\theta})$ only the free ($\boldsymbol{\tau}_{ext} = \boldsymbol{0}$) steady state condition has been used. In [ASOH04a] and [OASKH08] it was shown, how to implement an exact Cartesian link side stiffness together with an exact gravity compensation by extending the design ideas of [OASK+04b, ASOH04b].

Consider the case that a constant generalized external force $\boldsymbol{F}_{ext}$ acts on the robot. In the steady state at a position $\boldsymbol{q}_0$ the generalized external force $\boldsymbol{F}_{ext}$ is related to the external torques $\boldsymbol{\tau}_{ext}$ via $\boldsymbol{\tau}_{ext} = \boldsymbol{J}(\boldsymbol{q}_0)^T \boldsymbol{F}_{ext}$. The equilibrium condition for this case is

$$\boldsymbol{K}(\boldsymbol{\theta}_0 - \boldsymbol{q}_0) = \boldsymbol{g}(\boldsymbol{q}_0) - \boldsymbol{J}(\boldsymbol{q}_0)^T \boldsymbol{F}_{ext} \ , \tag{7.54}$$

$$\boldsymbol{K}(\boldsymbol{\theta}_0 - \boldsymbol{q}_0) = \boldsymbol{u}_0 \ , \tag{7.55}$$

where $\boldsymbol{u}_0$ is the static value of $\boldsymbol{u}$. In the following the desired stiffness relation

$$\boldsymbol{K}_d(\boldsymbol{x}(\boldsymbol{q}_0) - \boldsymbol{x}_d) = \boldsymbol{F}_{ext} \tag{7.56}$$

shall be achieved statically. By combining (7.56) with (7.54), one gets the condition

$$\boldsymbol{K}(\boldsymbol{\theta}_0 - \boldsymbol{q}_0) = \boldsymbol{g}(\boldsymbol{q}_0) - \boldsymbol{J}(\boldsymbol{q}_0)^T \boldsymbol{K}_d(\boldsymbol{x}(\boldsymbol{q}_0) - \boldsymbol{x}_d) \ . \tag{7.57}$$

This condition can be seen as a relationship between the static motor side position $\boldsymbol{\theta}_0$ and the static link side position $\boldsymbol{q}_0$. In order to stress the similarity of the following derivation to the derivation of the gravity compensation term in Section 7.3.2 the function $\boldsymbol{l}(\boldsymbol{q})$ is defined as

$$\boldsymbol{l}(\boldsymbol{q}) := \boldsymbol{g}(\boldsymbol{q}) - \boldsymbol{J}(\boldsymbol{q})^T \boldsymbol{K}_d(\boldsymbol{x}(\boldsymbol{q}) - \boldsymbol{x}_d) \ . \tag{7.58}$$

The following procedure is then completely analogous to the design of the gravity compensation term in Section 7.3.2. The function $\boldsymbol{l}(\boldsymbol{q})$ plays now the same role as the gravity function $\boldsymbol{g}(\boldsymbol{q})$ in Section 7.3.2. Notice that the equation (7.57) can also be written as $\boldsymbol{K}(\boldsymbol{\theta}_0 - \boldsymbol{q}_0) = \boldsymbol{l}(\boldsymbol{q}_0)$.

By defining the function

$$\boldsymbol{h}_l(\boldsymbol{q}) := \boldsymbol{q} + \boldsymbol{K}^{-1}\boldsymbol{l}(\boldsymbol{q}) \ , \tag{7.59}$$

the static motor side position $\boldsymbol{\theta}_0$ can be expressed as $\boldsymbol{\theta}_0 = \boldsymbol{h}_l(\boldsymbol{q}_0)$. At this point it is assumed that the inverse function of $\boldsymbol{h}_l(\boldsymbol{q})$ exists and it will be denoted by

$$\bar{\boldsymbol{q}}_l(\boldsymbol{\theta}) := \boldsymbol{h}_l^{-1}(\boldsymbol{\theta}) \ . \tag{7.60}$$

A sufficient condition for the existence of this inverse function, as well as an iterative computation procedure, will be given later in Proposition 7.5. By means of $\bar{\boldsymbol{q}}_l(\boldsymbol{\theta})$ a control law combining the gravity compensation with a statically exact stiffness can be designed in the form

$$\begin{aligned} \boldsymbol{u} &= \boldsymbol{l}(\bar{\boldsymbol{q}}_l(\boldsymbol{\theta})) - \boldsymbol{J}(\bar{\boldsymbol{q}}_l(\boldsymbol{\theta}))^T \boldsymbol{D}_d \boldsymbol{J}(\bar{\boldsymbol{q}}_l(\boldsymbol{\theta}))\dot{\boldsymbol{\theta}} \\ &= \boldsymbol{g}(\bar{\boldsymbol{q}}_l(\boldsymbol{\theta})) - \boldsymbol{J}(\bar{\boldsymbol{q}}_l(\boldsymbol{\theta}))^T \left(\boldsymbol{K}_d(\boldsymbol{x}(\bar{\boldsymbol{q}}_l(\boldsymbol{\theta})) - \boldsymbol{x}_d) + \boldsymbol{D}_d \boldsymbol{J}(\bar{\boldsymbol{q}}_l(\boldsymbol{\theta}))\dot{\boldsymbol{\theta}}\right) \ . \end{aligned} \tag{7.61}$$

The function $\boldsymbol{l}(\boldsymbol{q})$, as defined in (7.58), is the differential of the potential function

$$V_l(\boldsymbol{q}) = V_g(\boldsymbol{q}) - \frac{1}{2}(\boldsymbol{x}(\boldsymbol{q}) - \boldsymbol{x}_d)^T \boldsymbol{K}_d(\boldsymbol{x}(\boldsymbol{q}) - \boldsymbol{x}_d) \ , \tag{7.62}$$

i.e. $\boldsymbol{l}(\boldsymbol{q}) = (\partial V_l(\boldsymbol{q})/\partial \boldsymbol{q})^T$. Analogously to Assumption 7.2 the following assumption is needed now.

Assumption 7.4. *The Hessian* $\boldsymbol{H}_l(\boldsymbol{q}) = \frac{\partial^2 V_l(\boldsymbol{q})}{\partial \boldsymbol{q}^2}$ *of the potential function* $V_l(\boldsymbol{q})$ *satisfies the condition*

$$\alpha_l := \sup_{\forall \boldsymbol{q} \in \mathbb{R}^n} ||\boldsymbol{H}_l(\boldsymbol{q})||_K < ||\boldsymbol{K}||_K = 1 \ . \tag{7.63}$$

Notice that this assumption implicitly contains an upper bound on the desired stiffness $\boldsymbol{K}_d$. This is not surprising since the controller basically implements a stiffness, which is in serial interconnection to the joint stiffness $\boldsymbol{K}$. The combined stiffness $\boldsymbol{K}_d$ therefore must be *smaller* than $\boldsymbol{K}$. Assumption 7.4, however, ensures the existence of the inverse function $\boldsymbol{h}_l^{-1}(\boldsymbol{\theta})$ as formulated in the following proposition which is analogous to Proposition 7.3.

Proposition 7.5. *Under the Assumption 7.4 the inverse function* $\boldsymbol{h}_l^{-1}(\boldsymbol{\theta}) := \bar{\boldsymbol{q}}_l(\boldsymbol{\theta})$ *of* $\boldsymbol{h}_l(\boldsymbol{q}) = \boldsymbol{q} + \boldsymbol{K}^{-1}\boldsymbol{l}(\boldsymbol{q}) : \mathbb{R}^n \to \mathbb{R}^n$ *exists. Moreover, the iteration*

$$\hat{\boldsymbol{q}}_{l,n+1} = \boldsymbol{T}_l(\hat{\boldsymbol{q}}_{l,n}) \tag{7.64}$$

with $\boldsymbol{T}_l(\boldsymbol{q}) := \boldsymbol{\theta} - \boldsymbol{K}^{-1}\boldsymbol{l}(\boldsymbol{q})$ *converges for every fixed* $\boldsymbol{\theta}$ *and for every starting point* $\hat{\boldsymbol{q}}_{l,0}$ *to* $\bar{\boldsymbol{q}}_l(\boldsymbol{\theta})$.

Proof. The proposition can be proven exactly the same way as Proposition 7.3 when $\boldsymbol{g}(\boldsymbol{q})$ is replaced by $\boldsymbol{l}(\boldsymbol{q})$.

Furthermore, by following the same derivation as in Section 7.3.3 (with $\boldsymbol{l}(\boldsymbol{q})$ instead of $\boldsymbol{g}(\boldsymbol{q})$), one can show that the controller term $\boldsymbol{l}(\bar{\boldsymbol{q}}_l(\boldsymbol{\theta}))$ can be written as the differential of the potential function

$$V_{\bar{l}}(\boldsymbol{\theta}) = V_l(\bar{\boldsymbol{q}}_l(\boldsymbol{\theta})) + \frac{1}{2}\boldsymbol{l}(\bar{\boldsymbol{q}}_l(\boldsymbol{\theta}))^T\boldsymbol{K}^{-1}\boldsymbol{l}(\bar{\boldsymbol{q}}_l(\boldsymbol{\theta})) \ , \tag{7.65}$$

i.e. $\boldsymbol{l}(\bar{\boldsymbol{q}}_l(\boldsymbol{\theta})) = (\partial V_{\bar{l}}(\boldsymbol{\theta})/\partial \boldsymbol{\theta})^T$. For the following stability analysis also a statement analogously to Property 7.1 is of interest.

Property 7.6. *Let* α_l, *as defined in Assumption 7.4, be an upper bound for the Hessian of the potential function* $V_l(\boldsymbol{q})$. *Then the inequality*

$$|V_l(\boldsymbol{q}_1) - V_l(\boldsymbol{q}_2) + \boldsymbol{l}(\boldsymbol{q}_1)^T(\boldsymbol{q}_2 - \boldsymbol{q}_1)| \leq \frac{1}{2}\alpha_l||\boldsymbol{q}_2 - \boldsymbol{q}_1||_K^2 \tag{7.66}$$

holds for all $\boldsymbol{q}_1, \boldsymbol{q}_2 \in \mathbb{R}^n$.

Proof. This property can be proven exactly the same way as Property 7.1.

The passivity and stability statements from Section 7.4 can also be shown for the controller from this section in an analogous manner. In the following only the stability proof is sketched. With (7.61) the closed loop system is given by

$$\boldsymbol{M}(\boldsymbol{q})\ddot{\boldsymbol{q}} + \boldsymbol{C}(\boldsymbol{q},\dot{\boldsymbol{q}})\dot{\boldsymbol{q}} + \boldsymbol{g}(\boldsymbol{q}) = \boldsymbol{K}(\boldsymbol{\theta} - \boldsymbol{q}) + \boldsymbol{\tau}_{ext} \tag{7.67}$$

$$\boldsymbol{B}_\theta\ddot{\boldsymbol{\theta}} + \boldsymbol{K}(\boldsymbol{\theta} - \boldsymbol{q}) = \boldsymbol{l}(\bar{\boldsymbol{q}}_l(\boldsymbol{\theta})) - \boldsymbol{J}(\bar{\boldsymbol{q}}_l(\boldsymbol{\theta}))^T\boldsymbol{D}_d\boldsymbol{J}(\bar{\boldsymbol{q}}_l(\boldsymbol{\theta}))\dot{\boldsymbol{\theta}} \ . \tag{7.68}$$

Stability for the case of free motion ($\boldsymbol{\tau}_{ext} = \mathbf{0}$) can be shown by considering the Lyapunov function candidate

$$V_e(\boldsymbol{q}, \dot{\boldsymbol{q}}, \boldsymbol{\theta}, \dot{\boldsymbol{\theta}}) = \frac{1}{2}\dot{\boldsymbol{q}}^T \boldsymbol{M}(\boldsymbol{q})\dot{\boldsymbol{q}} + \frac{1}{2}\dot{\boldsymbol{\theta}}^T \boldsymbol{B}_\theta \dot{\boldsymbol{\theta}} + V_g(\boldsymbol{q}) + V_k(\boldsymbol{q}, \boldsymbol{\theta}) - V_{\bar{l}}(\boldsymbol{\theta}) \ . \quad (7.69)$$

In order to show that this function is positive definite one can substitute

$$V_g(\boldsymbol{q}) = V_l(\boldsymbol{q}) + \frac{1}{2}(\boldsymbol{x}(\boldsymbol{q}) - \boldsymbol{x}_d)^T \boldsymbol{K}_d (\boldsymbol{x}(\boldsymbol{q}) - \boldsymbol{x}_d)$$

from (7.62) and

$$\begin{aligned} V_k(\boldsymbol{q}, \boldsymbol{\theta}) &= \frac{1}{2}(\boldsymbol{\theta} - \boldsymbol{q})^T \boldsymbol{K}(\boldsymbol{\theta} - \boldsymbol{q}) \\ &= \frac{1}{2}(\bar{\boldsymbol{q}}_l(\boldsymbol{\theta}) - \boldsymbol{q})^T \boldsymbol{K}(\bar{\boldsymbol{q}}_l(\boldsymbol{\theta}) - \boldsymbol{q}) + \frac{1}{2}\boldsymbol{l}(\bar{\boldsymbol{q}}_l(\boldsymbol{\theta}))^T \boldsymbol{K}^{-1} \boldsymbol{l}(\bar{\boldsymbol{q}}_l(\boldsymbol{\theta})) \\ &\quad + \boldsymbol{l}(\bar{\boldsymbol{q}}_l(\boldsymbol{\theta}))^T (\bar{\boldsymbol{q}}_l - \boldsymbol{q}) \ , \end{aligned}$$

which follows from the definition of $\boldsymbol{l}(\boldsymbol{q})$, into the expression for $V_e(\boldsymbol{q}, \dot{\boldsymbol{q}}, \boldsymbol{\theta}, \dot{\boldsymbol{\theta}})$. Together with the expression for $V_{\bar{l}}(\boldsymbol{\theta})$ from (7.65) and by using Proposition 7.6, one can show that this Lyapunov function fulfills the inequality

$$\begin{aligned} V_e(\boldsymbol{q}, \dot{\boldsymbol{q}}, \boldsymbol{\theta}, \dot{\boldsymbol{\theta}}) \geq & \frac{1}{2}\dot{\boldsymbol{q}}^T \boldsymbol{M}(\boldsymbol{q})\dot{\boldsymbol{q}} + \frac{1}{2}\dot{\boldsymbol{\theta}}^T \boldsymbol{B}_\theta \dot{\boldsymbol{\theta}} \\ & + \frac{1}{2}(1 - \alpha_l)||\boldsymbol{q} - \bar{\boldsymbol{q}}_l(\boldsymbol{\theta})||_K^2 + \frac{1}{2}(\boldsymbol{x}(\boldsymbol{q}) - \boldsymbol{x}_d)^T \boldsymbol{K}_d (\boldsymbol{x}(\boldsymbol{q}) - \boldsymbol{x}_d) \end{aligned}$$

from which one can follow that $V_e(\boldsymbol{q}, \dot{\boldsymbol{q}}, \boldsymbol{\theta}, \dot{\boldsymbol{\theta}})$ is positive definite. Its time derivative along the solution curves of (7.67)-(7.68) can be computed as

$$\dot{V}_e(\boldsymbol{q}, \dot{\boldsymbol{q}}, \boldsymbol{\theta}, \dot{\boldsymbol{\theta}}) = -\dot{\boldsymbol{\theta}}^T \boldsymbol{J}(\bar{\boldsymbol{q}}_l(\boldsymbol{\theta}))^T \boldsymbol{D}_d \boldsymbol{J}(\bar{\boldsymbol{q}}_l(\boldsymbol{\theta}))\dot{\boldsymbol{\theta}} \quad (7.70)$$

and therefore is negative semi-definite. This ensures stability. Asymptotic stability can be shown again by invoking LaSalle's invariance principle.

7.6 Further Generalizations

In the previous sections the basic setting of the proposed impedance controller was described in detail. Several extensions of the controller are possible. Some of them will be discussed in this section. First, in Section 7.6.1 it is shown, how gear damping can be included in the model, respectively in the controller. After that a more general controller structure is considered in Section 7.6.2 in which also a feedback of the torque derivative $\dot{\boldsymbol{\tau}}$ is included. Finally, in Section 7.6.3 and Section 7.6.4 some comments on the use of filtered sensor signals and on the problems concerning shaping of the inertia matrix are given.

7.6.1 Including Joint Damping

Since the analysis of the controller was based on a physical interpretation of the torque feedback it is also possible to include joint damping, i.e. gear damping, very easily. The considered model with joint damping was given in (2.39)-(2.40). The equations are repeated here for the ease of reference

$$\boldsymbol{M}(\boldsymbol{q})\ddot{\boldsymbol{q}} + \boldsymbol{C}(\boldsymbol{q},\dot{\boldsymbol{q}})\dot{\boldsymbol{q}} + \boldsymbol{g}(\boldsymbol{q}) = \boldsymbol{K}(\boldsymbol{\theta}-\boldsymbol{q}) + \boldsymbol{D}(\dot{\boldsymbol{\theta}}-\dot{\boldsymbol{q}}) + \boldsymbol{\tau}_{ext} \ , \tag{7.71}$$

$$\boldsymbol{B}\ddot{\boldsymbol{\theta}} + \boldsymbol{K}(\boldsymbol{\theta}-\boldsymbol{q}) + \boldsymbol{D}(\dot{\boldsymbol{\theta}}-\dot{\boldsymbol{q}}) = \boldsymbol{\tau}_m \ , \tag{7.72}$$

$$\boldsymbol{\tau} = \boldsymbol{K}(\boldsymbol{\theta}-\boldsymbol{q}) \ .$$

The matrix $\boldsymbol{D} \in \mathbb{R}^{n\times n}$ is a diagonal and positive definite damping matrix. For this model the same type of controller as in the last section can be used, when instead of (7.3) the control law

$$\boldsymbol{\tau}_m = \boldsymbol{B}\boldsymbol{B}_\theta^{-1}\boldsymbol{u} + (\boldsymbol{I} - \boldsymbol{B}\boldsymbol{B}_\theta^{-1})\left(\boldsymbol{\tau} + \boldsymbol{D}\boldsymbol{K}^{-1}\dot{\boldsymbol{\tau}}\right) \tag{7.73}$$

is applied. This leads to the closed loop system

$$\boldsymbol{M}(\boldsymbol{q})\ddot{\boldsymbol{q}} + \boldsymbol{C}(\boldsymbol{q},\dot{\boldsymbol{q}})\dot{\boldsymbol{q}} + \boldsymbol{g}(\boldsymbol{q}) = \boldsymbol{K}(\boldsymbol{\theta}-\boldsymbol{q}) + \boldsymbol{D}(\dot{\boldsymbol{\theta}}-\dot{\boldsymbol{q}}) + \boldsymbol{\tau}_{ext} \tag{7.74}$$

$$\boldsymbol{B}_\theta\ddot{\boldsymbol{\theta}} + \boldsymbol{K}(\boldsymbol{\theta}-\boldsymbol{q}) + \boldsymbol{D}(\dot{\boldsymbol{\theta}}-\dot{\boldsymbol{q}}) = \boldsymbol{u} \tag{7.75}$$

for which the intermediate control input $\boldsymbol{u}$ can be chosen in the same way as in the last sections. In particular, the control law from (7.36) is considered in more detail in the following. Similar passivity and stability properties can be shown under the same conditions as in Section 7.4. In the following only the passivity properties will be considered. It is shown that the same storage functions $S_q(\boldsymbol{q},\dot{\boldsymbol{q}})$ and $S_\theta(\boldsymbol{q},\boldsymbol{\theta},\dot{\boldsymbol{\theta}})$ can be used as in (7.40) and (7.42) from Section 7.4.1. For the derivative of the storage function $S_q(\boldsymbol{q},\dot{\boldsymbol{q}})$ along (7.74) one obtains now

$$\dot{S}_q(\boldsymbol{q},\dot{\boldsymbol{q}}) = \dot{\boldsymbol{q}}^T(\boldsymbol{\tau} + \boldsymbol{D}\boldsymbol{K}^{-1}\dot{\boldsymbol{\tau}} + \boldsymbol{\tau}_{ext}) \ ,$$

from which one can follow that the system (7.74) represents a passive mapping $\left(\boldsymbol{\tau} + \boldsymbol{D}\boldsymbol{K}^{-1}\dot{\boldsymbol{\tau}} + \boldsymbol{\tau}_{ext}\right) \to \dot{\boldsymbol{q}}$ if the gravity potential $V_g(\boldsymbol{q})$ is bounded. Moreover, the derivative of $S_\theta(\boldsymbol{q},\boldsymbol{\theta},\dot{\boldsymbol{\theta}})$ along (7.75) with the control law (7.36) is given by

$$\dot{S}_\theta(\boldsymbol{q},\boldsymbol{\theta},\dot{\boldsymbol{\theta}}) = -\dot{\boldsymbol{x}}_\theta^T\boldsymbol{D}_d\dot{\boldsymbol{x}}_\theta - (\dot{\boldsymbol{\theta}}-\dot{\boldsymbol{q}})^T\boldsymbol{D}(\dot{\boldsymbol{\theta}}-\dot{\boldsymbol{q}}) - \dot{\boldsymbol{q}}^T(\boldsymbol{\tau} + \boldsymbol{D}\boldsymbol{K}^{-1}\dot{\boldsymbol{\tau}}) \ .$$

Therefore, it follows that (7.75) is a passive mapping $\dot{\boldsymbol{q}} \to -\left(\boldsymbol{\tau} + \boldsymbol{D}\boldsymbol{K}^{-1}\dot{\boldsymbol{\tau}}\right)$. The closed loop system can thus be seen again as the feedback interconnection of two passive subsystems, with the only difference that the port variable $\boldsymbol{\tau}$ is replaced by the variable $\boldsymbol{\tau} + \boldsymbol{D}\boldsymbol{K}^{-1}\dot{\boldsymbol{\tau}}$.

Furthermore, asymptotic stability for the case of free motion can be shown by considering $S_q(\boldsymbol{q},\dot{\boldsymbol{q}}) + S_\theta(\boldsymbol{q},\boldsymbol{\theta},\dot{\boldsymbol{\theta}})$ as a Lyapunov function. Since the proof can be done completely similar to Section 7.4.2 it is omitted here.

7.6.2 Generalization to Full State Feedback

Instead of using the control law (7.73) one can consider also the more general controller

$$\boldsymbol{\tau}_m = \boldsymbol{B}\boldsymbol{B}_\theta^{-1}\boldsymbol{u} + (\boldsymbol{I} - \boldsymbol{B}\boldsymbol{B}_\theta^{-1})\left(\boldsymbol{\tau} + \boldsymbol{K}_s\boldsymbol{K}^{-1}\dot{\boldsymbol{\tau}}\right) , \tag{7.76}$$

where $\boldsymbol{K}_s$ is a positive definite controller gain matrix, which may be different from $\boldsymbol{D}$. Together with (7.36) the closed loop system is then given by

$$\begin{aligned} \boldsymbol{M}(\boldsymbol{q})\ddot{\boldsymbol{q}} + \boldsymbol{C}(\boldsymbol{q},\dot{\boldsymbol{q}})\dot{\boldsymbol{q}} + \boldsymbol{g}(\boldsymbol{q}) &= \boldsymbol{\tau} + \boldsymbol{D}\boldsymbol{K}^{-1}\dot{\boldsymbol{\tau}} + \boldsymbol{\tau}_{ext} , \\ \boldsymbol{B}_\theta\ddot{\boldsymbol{\theta}} + \boldsymbol{\tau} + (\boldsymbol{B}_\theta\boldsymbol{B}^{-1}\boldsymbol{D} - \boldsymbol{B}_\theta\boldsymbol{B}^{-1}\boldsymbol{K}_s + \boldsymbol{K}_s)\boldsymbol{K}^{-1}\dot{\boldsymbol{\tau}} &= \boldsymbol{u} , \\ \boldsymbol{u} &= \bar{\boldsymbol{g}}(\boldsymbol{\theta}) - \boldsymbol{J}(\boldsymbol{\theta})^T(\boldsymbol{K}_d\tilde{\boldsymbol{x}}_\theta + \boldsymbol{D}_d\dot{\boldsymbol{x}}_\theta) . \end{aligned}$$

In order to analyze the stability of this system, one can consider again the same Lyapunov function

$$V(\boldsymbol{q},\dot{\boldsymbol{q}},\boldsymbol{\theta},\dot{\boldsymbol{\theta}}) = \frac{1}{2}\dot{\boldsymbol{q}}^T\boldsymbol{M}(\boldsymbol{q})\dot{\boldsymbol{q}} + \frac{1}{2}\dot{\boldsymbol{\theta}}^T\boldsymbol{B}_\theta\dot{\boldsymbol{\theta}} + V_{pot}(\boldsymbol{q},\boldsymbol{\theta})$$

as in Section 7.4.2. By a lengthy calculation one can derive that the time derivative of $V(\boldsymbol{q},\dot{\boldsymbol{q}},\boldsymbol{\theta},\dot{\boldsymbol{\theta}})$ can now be written as

$$\dot{V}(\boldsymbol{q},\dot{\boldsymbol{q}},\boldsymbol{\theta},\dot{\boldsymbol{\theta}}) = -\begin{pmatrix}\dot{\boldsymbol{\theta}}-\dot{\boldsymbol{q}}\\ \dot{\boldsymbol{\theta}}\end{pmatrix}^T \boldsymbol{Q}(\boldsymbol{\theta}) \begin{pmatrix}\dot{\boldsymbol{\theta}}-\dot{\boldsymbol{q}}\\ \dot{\boldsymbol{\theta}}\end{pmatrix}$$

$$\boldsymbol{Q}(\boldsymbol{\theta}) = \begin{bmatrix} \boldsymbol{D} & \frac{1}{2}(\boldsymbol{B}_\theta\boldsymbol{B}^{-1}-\boldsymbol{I})(\boldsymbol{D}-\boldsymbol{K}_s) \\ \frac{1}{2}(\boldsymbol{B}_\theta\boldsymbol{B}^{-1}-\boldsymbol{I})(\boldsymbol{D}-\boldsymbol{K}_s) & \boldsymbol{J}(\boldsymbol{\theta})^T\boldsymbol{D}_d\boldsymbol{J}(\boldsymbol{\theta}) \end{bmatrix}$$

From Lemma A.20 one can see that $\dot{V}(\boldsymbol{q},\dot{\boldsymbol{q}},\boldsymbol{\theta},\dot{\boldsymbol{\theta}})$ is negative semi-definite, if the condition

$$\boldsymbol{J}(\boldsymbol{\theta})^T\boldsymbol{D}_d\boldsymbol{J}(\boldsymbol{\theta}) - \frac{1}{4}(\boldsymbol{D}-\boldsymbol{K}_s)(\boldsymbol{B}_\theta\boldsymbol{B}^{-1}-\boldsymbol{I})\boldsymbol{D}^{-1}(\boldsymbol{B}_\theta\boldsymbol{B}^{-1}-\boldsymbol{I})(\boldsymbol{D}-\boldsymbol{K}_s) > 0$$

holds for all $\boldsymbol{\theta} \in \mathbb{R}^n$, which is an implicit condition on $\boldsymbol{K}_s$ and $\boldsymbol{D}_d$.

7.6.3 On the Filtering of the Velocities

Another generalization of the controller from this chapter can be done by using a filter for the implementation of the damping term. For the DLR lightweight robots, for instance, only the motor positions and the joint torques can be measured, while the time derivatives of these signals must be computed via numerical differentiation or filtering[19]. In [KS98] a large class of linear position filters, including e.g. simple nonmodel-based first and second order filters, is proposed.

[19] In the literature a filtered version of such a time derivative is often called a *dirty derivative*.

Therein it is shown that the use of these filters for the estimation of $\dot{\boldsymbol{\theta}}$ does not jeopardize the stability properties of an energy shaping based controller for flexible joint robots. Due to the physical interpretation of the torque feedback according to Section 7.1 this result can indeed be applied also to the controllers presented in this chapter. More details on such velocity observers can be found in [KS98] and the references cited therein.

7.6.4 Inertia Shaping

One disadvantage of the presented controller is its limitation for inertia shaping. Torque feedback, on the one hand, can be seen as a shaping of the motor inertia, as was discussed in more detail in the introduction of this chapter. The possibility to shape the apparent link side inertia, on the other hand, seems limited. At least it is possible to approximate a system behavior with a bigger inertia by referring to a controller structure according to Figure 7.4. By choosing a high (i.e. $K_c >> K_\theta$) stiffness value for K_c (and with an appropriate damping factor D_c), the closed loop behavior will be dominated in the low frequency domain by the behavior of a spring-mass-damper-system with the mass $M_d + B_\theta + M$.

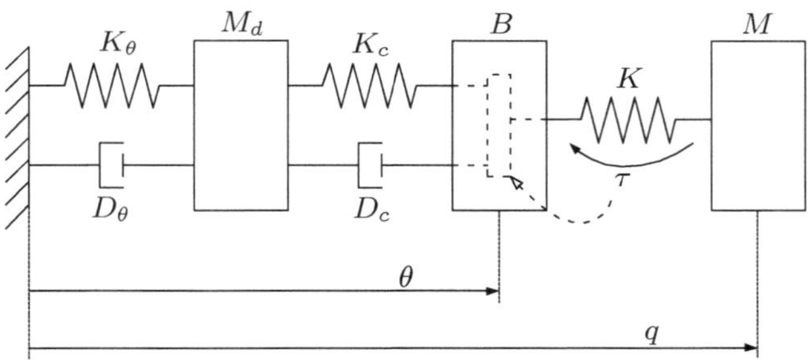

Fig. 7.4. Control structure for simulating a virtual mass

7.7 Tracking

The controller from Section 7.2 with the gravity compensation term from Section 7.3.2 was designed especially for regulation, where the desired virtual equilibrium position $\boldsymbol{x}_d = \boldsymbol{f}(\boldsymbol{q}_d)$ (respectively $\boldsymbol{x}_{\theta,d}$) is constant. A direct extension of this controller to the tracking case (with a time-varying virtual equilibrium position), however, is not clear.

For the gravity torques a compensation term was designed which compensates the gravity effects in all stationary points. This compensation term $\boldsymbol{u}_g = \bar{\boldsymbol{g}}(\boldsymbol{\theta})$ is based on quasi-static considerations such that it depends on the motor side position $\boldsymbol{\theta}$ only. This allows to perform a passivity based stability analysis for the regulation case. Apart from passivity considerations another question that arises in the context of gravity compensation is, if also the compensation term $\boldsymbol{u}_g = \boldsymbol{g}(\boldsymbol{q})$ leads to a stable closed loop system. This question is still open.

In the following a different passivity based approach is discussed. Contrary to the previous sections of this chapter, the design will now concentrate on the tracking case. But it is desired that, when reduced to the regulation case, a controller similar to the controller of Section 7.2 should be obtained. The relation of the resulting controller to the passivity based tracking controller from [BOL95] will be discussed later in Section 7.7.3. Moreover, the resulting controller will also give a hint - not an answer - concerning the above mentioned question about the usability of the gravity compensation term $\boldsymbol{u}_g = \boldsymbol{g}(\boldsymbol{q})$ for the regulation case.

For ease of presentation, this section will only consider the case of a joint level impedance. The matrices $\boldsymbol{K}_\theta$ and $\boldsymbol{D}_\theta$ are the desired symmetric and positive definite stiffness and damping matrices. An extension of the presented controller to the Cartesian case can be done in exactly the same way as in Section 7.2. The virtual equilibrium position $\boldsymbol{q}_d$ is assumed to be continuously differentiable up to the fourth time derivative.

7.7.1 The Gravity Free Case

Consider first a gravity free version of the dynamics from (7.1)-(7.2)

$$\boldsymbol{M}(\boldsymbol{q})\ddot{\boldsymbol{q}} + \boldsymbol{C}(\boldsymbol{q},\dot{\boldsymbol{q}})\dot{\boldsymbol{q}} = \boldsymbol{\tau} + \boldsymbol{\tau}_{ext} \ , \tag{7.77}$$

$$\boldsymbol{B}\ddot{\boldsymbol{\theta}} + \boldsymbol{\tau} = \boldsymbol{\tau}_m \ , \tag{7.78}$$

where $\boldsymbol{\tau}$ is again given by $\boldsymbol{\tau} = \boldsymbol{K}(\boldsymbol{\theta} - \boldsymbol{q})$. Analogously to Section 7.2 a torque feedback of the form

$$\boldsymbol{\tau}_m = \boldsymbol{B}\boldsymbol{B}_\theta^{-1}\boldsymbol{u} + (\boldsymbol{I} - \boldsymbol{B}\boldsymbol{B}_\theta^{-1})\boldsymbol{\tau} \ ,$$

with $\boldsymbol{u}$ as a new input variable, is applied in order to scale the apparent motor inertia to $\boldsymbol{B}_\theta$. This leads to

$$\boldsymbol{M}(\boldsymbol{q})\ddot{\boldsymbol{q}} + \boldsymbol{C}(\boldsymbol{q},\dot{\boldsymbol{q}})\dot{\boldsymbol{q}} = \boldsymbol{\tau} + \boldsymbol{\tau}_{ext} \ , \tag{7.79}$$

$$\boldsymbol{B}_\theta\ddot{\boldsymbol{\theta}} + \boldsymbol{\tau} = \boldsymbol{u} \ . \tag{7.80}$$

The passivity based controller design in Section 7.2 was then based on the fact that (7.79) represents a passive mapping $(\boldsymbol{\tau} + \boldsymbol{\tau}_{ext}) \rightarrow \dot{\boldsymbol{q}}$ and that (7.80) represents a passive mapping $\dot{\boldsymbol{q}} \rightarrow -\boldsymbol{\tau}$. Passivity of (7.79) can easily be shown with the storage function $S_1 = 1/2\dot{\boldsymbol{q}}^T\boldsymbol{M}(\boldsymbol{q})\dot{\boldsymbol{q}}$. For the tracking case the velocity error $\dot{\tilde{\boldsymbol{q}}} = \dot{\boldsymbol{q}} - \dot{\boldsymbol{q}}_d$ is more relevant than the velocity $\dot{\boldsymbol{q}}$. Therefore, it is interesting to consider

$$\bar{S}_q(\tilde{\boldsymbol{q}},\dot{\tilde{\boldsymbol{q}}},t) = \frac{1}{2}\dot{\tilde{\boldsymbol{q}}}^T\boldsymbol{M}(\boldsymbol{q})\dot{\tilde{\boldsymbol{q}}} \tag{7.81}$$

as a (time-varying) storage function for (7.79). Due to

$$\dot{\bar{S}}_q(\tilde{\boldsymbol{q}},\dot{\tilde{\boldsymbol{q}}},t) = \dot{\tilde{\boldsymbol{q}}}^T\left(\boldsymbol{\tau} - \boldsymbol{M}(\boldsymbol{q})\ddot{\boldsymbol{q}}_d - \boldsymbol{C}(\boldsymbol{q},\dot{\boldsymbol{q}})\dot{\boldsymbol{q}}_d + \boldsymbol{\tau}_{ext}\right) \tag{7.82}$$

one can see that the rigid body system (7.79) represents a passive mapping from the input $(\boldsymbol{\tau} - \boldsymbol{M}(\boldsymbol{q})\ddot{\boldsymbol{q}}_d - \boldsymbol{C}(\boldsymbol{q},\dot{\boldsymbol{q}})\dot{\boldsymbol{q}}_d + \boldsymbol{\tau}_{ext})$ to the output $\dot{\tilde{\boldsymbol{q}}}$. In order to transform the motor dynamics (7.80) into a passive mapping for the input $\dot{\tilde{\boldsymbol{q}}}$ and the output $-(\boldsymbol{\tau}-\boldsymbol{M}(\boldsymbol{q})\ddot{\boldsymbol{q}}_d-\boldsymbol{C}(\boldsymbol{q},\dot{\boldsymbol{q}})\dot{\boldsymbol{q}}_d)$, i.e. in order to achieve *port matching*, the storage function

$$\bar{S}_\theta(\tilde{\boldsymbol{q}},\tilde{\boldsymbol{\theta}},\dot{\tilde{\boldsymbol{\theta}}})=\frac{1}{2}\dot{\tilde{\boldsymbol{\theta}}}^T\boldsymbol{B}_\theta\dot{\tilde{\boldsymbol{\theta}}}+\frac{1}{2}(\tilde{\boldsymbol{\theta}}-\tilde{\boldsymbol{q}})^T\boldsymbol{K}(\tilde{\boldsymbol{\theta}}-\tilde{\boldsymbol{q}})+\frac{1}{2}\tilde{\boldsymbol{\theta}}^T\boldsymbol{K}_\theta\tilde{\boldsymbol{\theta}} \tag{7.83}$$

is considered, with the motor position error $\tilde{\boldsymbol{\theta}}=\boldsymbol{\theta}-\boldsymbol{\theta}_s$. The *desired motor position* $\boldsymbol{\theta}_s$ is determined later. Then the time derivative of the storage function $\bar{S}_\theta(\tilde{\boldsymbol{q}},\tilde{\boldsymbol{\theta}},\dot{\tilde{\boldsymbol{\theta}}})$ along (7.80) is given by

$$\begin{aligned}\dot{\bar{S}}_\theta(\tilde{\boldsymbol{q}},\tilde{\boldsymbol{\theta}},\dot{\tilde{\boldsymbol{\theta}}})&=\dot{\tilde{\boldsymbol{\theta}}}^T\boldsymbol{B}_\theta\ddot{\tilde{\boldsymbol{\theta}}}+(\tilde{\boldsymbol{\theta}}-\tilde{\boldsymbol{q}})^T\boldsymbol{K}(\dot{\tilde{\boldsymbol{\theta}}}-\dot{\tilde{\boldsymbol{q}}})+\tilde{\boldsymbol{\theta}}^T\boldsymbol{K}_\theta\dot{\tilde{\boldsymbol{\theta}}}\\&=\dot{\tilde{\boldsymbol{\theta}}}^T\left(\boldsymbol{u}-\boldsymbol{K}(\boldsymbol{\theta}-\boldsymbol{q})-\boldsymbol{B}_\theta\ddot{\boldsymbol{\theta}}_s\right)+(\tilde{\boldsymbol{\theta}}-\tilde{\boldsymbol{q}})^T\boldsymbol{K}(\dot{\tilde{\boldsymbol{\theta}}}-\dot{\tilde{\boldsymbol{q}}})+\tilde{\boldsymbol{\theta}}^T\boldsymbol{K}_\theta\dot{\tilde{\boldsymbol{\theta}}}\ .\end{aligned}$$

A controller of the form

$$\boldsymbol{u}=\boldsymbol{B}_\theta\ddot{\boldsymbol{\theta}}_s+\boldsymbol{K}(\boldsymbol{\theta}_s-\boldsymbol{q}_d)-\boldsymbol{K}_\theta\tilde{\boldsymbol{\theta}}-\boldsymbol{D}_\theta\dot{\tilde{\boldsymbol{\theta}}} \tag{7.84}$$

then leads to

$$\dot{\bar{S}}_\theta(\tilde{\boldsymbol{q}},\tilde{\boldsymbol{\theta}},\dot{\tilde{\boldsymbol{\theta}}})=-\dot{\tilde{\boldsymbol{\theta}}}^T\boldsymbol{D}_\theta\dot{\tilde{\boldsymbol{\theta}}}-\dot{\tilde{\boldsymbol{\theta}}}^T\boldsymbol{K}(\tilde{\boldsymbol{\theta}}-\tilde{\boldsymbol{q}})+(\dot{\tilde{\boldsymbol{\theta}}}-\dot{\tilde{\boldsymbol{q}}})^T\boldsymbol{K}(\tilde{\boldsymbol{\theta}}-\tilde{\boldsymbol{q}}) \tag{7.85}$$

$$=-\dot{\tilde{\boldsymbol{\theta}}}^T\boldsymbol{D}_\theta\dot{\tilde{\boldsymbol{\theta}}}-\dot{\tilde{\boldsymbol{q}}}^T\boldsymbol{K}(\tilde{\boldsymbol{\theta}}-\tilde{\boldsymbol{q}}) \tag{7.86}$$

$$=-\dot{\tilde{\boldsymbol{\theta}}}^T\boldsymbol{D}_\theta\dot{\tilde{\boldsymbol{\theta}}}-\dot{\tilde{\boldsymbol{q}}}^T\left(\boldsymbol{\tau}-\boldsymbol{K}(\boldsymbol{\theta}_s-\boldsymbol{q}_d)\right)\ . \tag{7.87}$$

In order to ensure the desired passivity property the desired motor side position $\boldsymbol{\theta}_s$ is defined now via the relationship

$$\boldsymbol{M}(\boldsymbol{q})\ddot{\boldsymbol{q}}_d+\boldsymbol{C}(\boldsymbol{q},\dot{\boldsymbol{q}})\dot{\boldsymbol{q}}_d=\boldsymbol{K}(\boldsymbol{\theta}_s-\boldsymbol{q}_d) \tag{7.88}$$

as

$$\boldsymbol{\theta}_s=\boldsymbol{q}_d+\boldsymbol{K}^{-1}\left(\boldsymbol{M}(\boldsymbol{q})\ddot{\boldsymbol{q}}_d+\boldsymbol{C}(\boldsymbol{q},\dot{\boldsymbol{q}})\dot{\boldsymbol{q}}_d\right) \tag{7.89}$$

and therefore is a function of $\boldsymbol{q}$ and $\dot{\boldsymbol{q}}$, i.e. $\boldsymbol{\theta}_s=\boldsymbol{\theta}_s(\boldsymbol{q},\dot{\boldsymbol{q}},t)$. Thereby, the required passivity property is obtained. It can easily be verified that the controller leads to the time-varying closed loop system

$$\boldsymbol{M}(\boldsymbol{q})\ddot{\tilde{\boldsymbol{q}}}+\boldsymbol{C}(\boldsymbol{q},\dot{\boldsymbol{q}})\dot{\tilde{\boldsymbol{q}}}=\boldsymbol{K}(\tilde{\boldsymbol{\theta}}-\tilde{\boldsymbol{q}})+\boldsymbol{\tau}_{ext}\ , \tag{7.90}$$

$$\boldsymbol{B}_\theta\ddot{\tilde{\boldsymbol{\theta}}}+\boldsymbol{K}(\tilde{\boldsymbol{\theta}}-\tilde{\boldsymbol{q}})+\boldsymbol{D}_\theta\dot{\tilde{\boldsymbol{\theta}}}+\boldsymbol{K}_\theta\tilde{\boldsymbol{\theta}}=\boldsymbol{0}\ . \tag{7.91}$$

Notice that, due to the particular choice of $\boldsymbol{\theta}_s$ in (7.89), the term $-\boldsymbol{K}(\tilde{\boldsymbol{\theta}}-\tilde{\boldsymbol{q}})$ corresponds to the *port output variable* $-(\boldsymbol{\tau}-\boldsymbol{M}(\boldsymbol{q})\ddot{\boldsymbol{q}}_d-\boldsymbol{C}(\boldsymbol{q},\dot{\boldsymbol{q}})\dot{\boldsymbol{q}}_d)$ of the motor dynamics, see also Figure 7.5. Moreover, by considering $\bar{S}_q(\tilde{\boldsymbol{q}},\dot{\tilde{\boldsymbol{q}}},t)+\bar{S}_\theta(\tilde{\boldsymbol{q}},\tilde{\boldsymbol{\theta}},\dot{\tilde{\boldsymbol{\theta}}})$ as a time-varying Lyapunov function it can easily be shown that, for free motion ($\boldsymbol{\tau}_{ext}=\boldsymbol{0}$), the origin $\tilde{\boldsymbol{\theta}}=\tilde{\boldsymbol{q}}=\dot{\tilde{\boldsymbol{\theta}}}=\dot{\tilde{\boldsymbol{q}}}=\boldsymbol{0}$ is a stable equilibrium point.

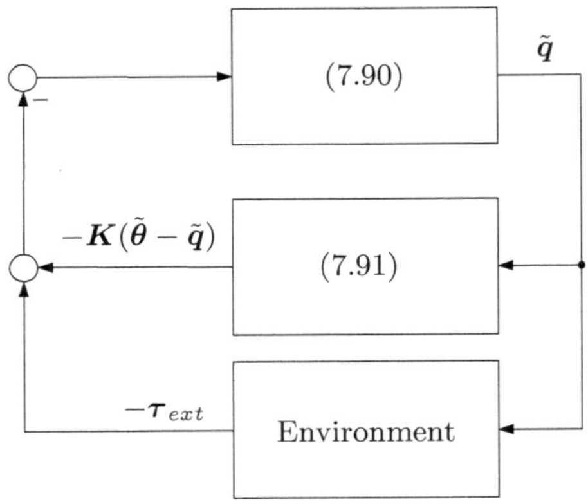

Fig. 7.5. Closed loop interconnection structure for the tracking case

7.7.2 Including Gravity

Consider now the link side dynamics with gravity

$$\boldsymbol{M}(\boldsymbol{q})\ddot{\boldsymbol{q}} + \boldsymbol{C}(\boldsymbol{q}, \dot{\boldsymbol{q}})\dot{\boldsymbol{q}} + \boldsymbol{g}(\boldsymbol{q}) = \boldsymbol{\tau} + \boldsymbol{\tau}_{ext} \tag{7.92}$$

instead of (7.79). Then $\dot{\tilde{S}}_q(\tilde{\boldsymbol{q}}, \dot{\tilde{\boldsymbol{q}}}, t)$ is given by

$$\dot{\tilde{S}}_q(\tilde{\boldsymbol{q}}, \dot{\tilde{\boldsymbol{q}}}, t) = \dot{\tilde{\boldsymbol{q}}}^T \left(\boldsymbol{\tau} - \boldsymbol{M}(\boldsymbol{q})\ddot{\boldsymbol{q}}_d - \boldsymbol{C}(\boldsymbol{q}, \dot{\boldsymbol{q}})\dot{\boldsymbol{q}}_d - \boldsymbol{g}(\boldsymbol{q}) + \boldsymbol{\tau}_{ext}\right) . \tag{7.93}$$

One can see that the link side rigid body dynamics (7.92) can be considered as a passive mapping $(\boldsymbol{\tau} - \boldsymbol{M}(\boldsymbol{q})\ddot{\boldsymbol{q}}_d - \boldsymbol{C}(\boldsymbol{q}, \dot{\boldsymbol{q}})\dot{\boldsymbol{q}}_d - \boldsymbol{g}(\boldsymbol{q})) \rightarrow \dot{\tilde{\boldsymbol{q}}}$. A gravity compensation term can therefore be included in the controller by choosing $\boldsymbol{\theta}_s$ according to

$$\boldsymbol{M}(\boldsymbol{q})\ddot{\boldsymbol{q}}_d + \boldsymbol{C}(\boldsymbol{q}, \dot{\boldsymbol{q}})\dot{\boldsymbol{q}}_d + \boldsymbol{g}(\boldsymbol{q}) = \boldsymbol{K}(\boldsymbol{\theta}_s - \boldsymbol{q}_d) . \tag{7.94}$$

instead of (7.88). The complete tracking controller is then given by

$$\boldsymbol{\tau}_m = \boldsymbol{B}\boldsymbol{B}_\theta^{-1}\boldsymbol{u} + (\boldsymbol{I} - \boldsymbol{B}\boldsymbol{B}_\theta^{-1})\boldsymbol{\tau} , \tag{7.95}$$

$$\boldsymbol{u} = \boldsymbol{B}_\theta\ddot{\boldsymbol{\theta}}_s + \boldsymbol{K}(\boldsymbol{\theta}_s - \boldsymbol{q}_d) - \boldsymbol{K}_\theta\tilde{\boldsymbol{\theta}} - \boldsymbol{D}_\theta\dot{\tilde{\boldsymbol{\theta}}} , \tag{7.96}$$

$$\boldsymbol{\theta}_s = \boldsymbol{q}_d + \boldsymbol{K}^{-1}\left(\boldsymbol{M}(\boldsymbol{q})\ddot{\boldsymbol{q}}_d + \boldsymbol{C}(\boldsymbol{q}, \dot{\boldsymbol{q}})\dot{\boldsymbol{q}}_d + \boldsymbol{g}(\boldsymbol{q})\right) , \tag{7.97}$$

and leads to a closed loop system which has the same form as (7.90)-(7.91).

7.7.3 Controller Discussion

Notice that the gravity free version in Section 7.7.1 is a direct extension of the regulation controller from Section 7.2, since it reduces to the same feedback law for $\dot{\boldsymbol{q}}_d = \boldsymbol{0}$. The solution with gravity from Section 7.7.2, however, is different. For $\dot{\boldsymbol{q}}_d = \boldsymbol{0}$ the controller (7.96)-(7.97) is simplified to

$$\boldsymbol{\tau}_m = \boldsymbol{B}\boldsymbol{B}_\theta^{-1}\boldsymbol{u} + (\boldsymbol{I} - \boldsymbol{B}\boldsymbol{B}_\theta^{-1})\boldsymbol{\tau} \ , \tag{7.98}$$

$$\boldsymbol{u} = \underline{\boldsymbol{B}_\theta \boldsymbol{K}^{-1}\ddot{g}(\boldsymbol{q})} + \boldsymbol{g}(\boldsymbol{q}) - \boldsymbol{K}_\theta(\boldsymbol{\theta} - \boldsymbol{\theta}_s) - \boldsymbol{D}_\theta(\dot{\boldsymbol{\theta}} - \underline{\dot{\boldsymbol{\theta}}_s}) \ , \tag{7.99}$$

$$\boldsymbol{\theta}_s = \boldsymbol{q}_d + \boldsymbol{K}^{-1}\boldsymbol{g}(\boldsymbol{q}) \ . \tag{7.100}$$

This suggests[20] that the use of $\boldsymbol{u}_g = \boldsymbol{g}(\boldsymbol{q})$ instead of the gravity compensation term from Section 7.3.2 is insufficient to ensure stability in the regulation case. In (7.99) additionally the second derivative $\ddot{\boldsymbol{g}}(\boldsymbol{q})$ is needed.

A combination of the motor position based online gravity compensation from Section 7.3.2 with the tracking controller from this section seems difficult because the design of the gravity compensation term from Section 7.3.2 is based on quasi-static considerations.

Notice also that the extension to the case of a Cartesian impedance controller can be done analogously to Section 7.2.

It should also be mentioned that the presented tracking controller is related to the passivity based tracking controllers in [BOL95] and [LO95]. In [BOL95] a simplified version of the controller from [LB92] is presented, in which basically the rigid body tracking control law from Slotine and Li [SL87] is used instead of the relationship in (7.94). Similar to the presentation herein, the physical interpretation of the torque feedback from Section 7.1 allows to augment these controllers by an inner torque feedback loop, which is not included in [BOL95, LO95, LB92], and thus to enhance the control performance. Moreover, the controller in [BOL95] also uses a slightly different controller equation than (7.96) for the motor dynamics by again applying the controller structure from [SL87].

7.8 Relation to Other Methods

In this section the relation of the presented passivity based impedance controller to some other controller design methods from the literature are shortly discussed.

7.8.1 Comparison to the Singular Perturbation Controller

The physical interpretation of the torque feedback can also be used in order to give an interpretation of the effective inertia of the steady state system from the singular perturbation analysis in Section 5.4.1. The effective inertia in (5.26) is given by $\boldsymbol{M}(\boldsymbol{q}) + (\boldsymbol{I} + \boldsymbol{K}_\tau)^{-1}\boldsymbol{B}$. The torque feedback gain $-\boldsymbol{K}_\tau$ from Section 5.4.1 clearly corresponds to the gain $\boldsymbol{I} - \boldsymbol{B}\boldsymbol{B}_\theta^{-1}$ from (7.3), by which the scaling of the motor inertia from $\boldsymbol{B}$ to $\boldsymbol{B}_\theta$ was achieved. The new effective inertia of the singular perturbation approach can therefore be seen as the sum of the link side inertia matrix and the scaled motor inertia $\boldsymbol{M}(\boldsymbol{q}) + \boldsymbol{B}_\theta$.

[20] This clearly is not a proof but merely a conjecture.

7.8.2 Comparison to the Controller from Albu-Schäffer

The Cartesian impedance controller presented in this chapter has a structure quite similar to the controllers from [AS01, ASH01a, ASH00, ASH01b]. In these works a position controller for a flexible joint robot model was developed which consists of a feedback of the joint torques and of the motor side joint position errors. For the gravity torques a compensation term based on the desired link side position was used, similar to [Tom91]. Contrary to this, in this work an online gravity compensation term was developed which is more appropriate for impedance control. Moreover, the physical interpretation of the torque feedback from Section 7.2 subsequently allows to generalize a joint level impedance controller, which actually is quite similar to [AS01], to the Cartesian case (see also [ASmOH07]).

7.8.3 Relation to the IDA-PBC and the Method of Controlled Lagrangians

The *IDA-PBC*[21] [OvdSME02, OSGEB02] and the *Method of Controlled Lagrangians* [BLM00, BCLM01] which have been presented recently in the control literature are two extensions of the energy-shaping methodology. An excellent comparison of these two methods is given in [BOvdS02]. It is interesting to notice the relation of the controllers presented in this chapter to those methods. Consider therefore the Cartesian impedance controller from Section 7.2 which was analyzed in Section 7.4 in detail. Since the stability analysis of Section 7.4 was based on the passivity properties of the system one can easily use the storage functions of Section 7.4.1 in order to set up the energy functions for the IDA-PBC or the Method of Controlled Lagrangians. The Lyapunov function from 7.4.2 for instance is exactly the target Lagrangian which leads to the controller from Section 7.2 in the framework of the controlled Lagrangians. Notice that the storage functions used in this chapter were implicitly designed such that they satisfy the matching conditions of the Method of Controlled Lagrangians for underactuated systems. An additional degree-of-freedom of the controller design based on the IDA-PBC framework is the use of an additional skew-symmetric matrix in order to shape the internal interconnection structure of the closed loop system [OvdSME02, OSGEB02, BOvdS02]. Without going into details here, it should be mentioned that an interesting application of this is for example the nullspace decoupling in (4.48).

7.9 Summary

In this chapter a passivity based control approach is treated. Motivated by the discussion in Chapter 3 for the rigid body case, an impedance behavior without inertia shaping is considered. The basic idea behind the proposed controller is

[21] Interconnection and Damping Assignment Passivity Based Control.

the physical interpretation of the torque feedback as a scaling of the motor inertia. In addition to the inner torque feedback loop an outer impedance controller implements the desired Cartesian stiffness and damping. Moreover, a compensation term for the gravity torques was designed based on the measurements of the motor side position only.

It was shown that the closed loop system can be seen as the feedback interconnection of passive subsystems. This brings along very advantageous robustness properties. A proof of the asymptotic stability was presented, which was based on the passivity properties of the system.

After discussing the basic controller, several extensions concerning full state feedback, the stiffness design, and the tracking case were presented.

8 Evaluation

This chapter reports the simulations and experiments for validating and comparing the different control laws from the previous chapters. First the controllers are compared in a simulation study in Section 8.1. The Cartesian impedance controllers based on the results of Chapter 5 and 7 are evaluated in more detail by experiments with the DLR-Lightweight-Robot-II. Contrary to typical industrial robots, this robot is equipped with joint torque sensors and is thus very well suited for the implementation of the presented controllers.

8.1 Simulation Results

In this section some simulation results are reported, by which the different controllers from this book are compared. For the simulations a flexible joint robot model according to the first six joints of the DLR-Lightweight-Robot-II was considered. The kinematic structure of this robot is given in the Appendix A.6. For the relevant Cartesian coordinates the error variable

$$\tilde{\boldsymbol{x}} = \begin{pmatrix} {}_t\boldsymbol{p}_{td}(\boldsymbol{q},t) \\ {}_t\boldsymbol{\phi}_{td}(\boldsymbol{q},t) \end{pmatrix} =: \begin{pmatrix} e_x \\ e_y \\ e_z \\ \phi_x \\ \phi_y \\ \phi_z \end{pmatrix} \tag{8.1}$$

was chosen (see also Section 3.5) with a roll-pitch-yaw representation for the Euler angles ${}_t\boldsymbol{\phi}_{td}(\boldsymbol{q},t)$. The desired stiffness and damping matrix thus are defined with respect to the tool frame $\mathcal{T}$.

The desired Cartesian impedance is then given by

$$\boldsymbol{\Lambda}(\boldsymbol{x})\ddot{\tilde{\boldsymbol{x}}} + (\boldsymbol{\mu}(\boldsymbol{x},\dot{\boldsymbol{x}}) + \boldsymbol{D}_d)\,\dot{\tilde{\boldsymbol{x}}} + \boldsymbol{K}_d\tilde{\boldsymbol{x}} = \boldsymbol{F}_{ext}\ , \tag{8.2}$$

where the inertia matrix $\boldsymbol{\Lambda}(\boldsymbol{x})$ and the Coriolis/centrifugal matrix $\boldsymbol{\mu}(\boldsymbol{x},\dot{\boldsymbol{x}})$ are the relevant matrices from the link side rigid body model of the DLR-Lightweight-Robot-II.

C. Ott: Cartesian Imped. Cntrl. of Redund. & Flexible-Joint Robots, STAR 49, pp. 123–147, 2008.
springerlink.com

In order to evaluate the *nominal performance* of the controllers it is assumed for the subsequent simulation studies that all required measurement signals are available and they are not corrupted by noise. A step response is evaluated for a step of 5 cm in x-direction for the virtual equilibrium position. After the step the virtual equilibrium pose $\boldsymbol{x}_d$ is constant, i.e. $\dot{\boldsymbol{x}}_d = \mathbf{0}$, which leads to a regulation problem. The starting configuration is shown in Figure 8.1. For the desired stiffness and damping matrices diagonal matrices with the entries from Table 8.1 are used. These values remain the same for all simulations.

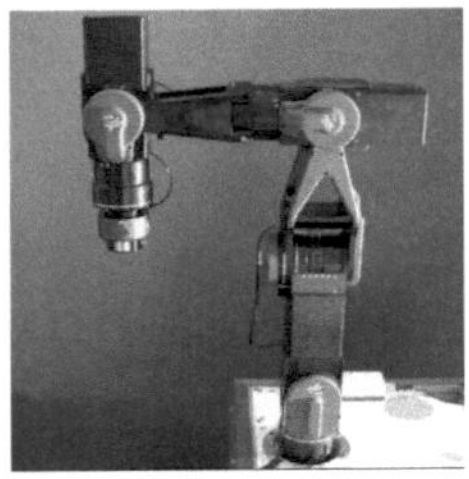

Fig. 8.1. Initial configuration for the simulations

All controllers contain a feedback of the joint torques $\boldsymbol{\tau}$. In order to get a meaningful comparison of the controllers, the proportional gain $\bar{\boldsymbol{K}}_\tau$ and the derivative gain $\bar{\boldsymbol{D}}_\tau$ of the torque feedback are chosen the same for all controllers. First, the performance of the controllers is compared with low torque control gains. Therefore, $\bar{\boldsymbol{K}}_\tau$ and $\bar{\boldsymbol{D}}_\tau$ are chosen as a diagonal matrices with the values from Table 8.2 as diagonal elements. In a second simulation the higher gain

Table 8.1. Stiffness and damping values for the simulations

Coord.	e_x	e_y	e_z	ϕ_x	ϕ_y	ϕ_z
Stiffness	2000 N/m	2000 N/m	2000 N/m	100 Nm/rad	100 Nm/rad	100 Nm/rad
Damping	313.05 Ns/m	313.05 Ns/m	313.05 Ns/m	70.0 Nms/rad	70.0 Nms/rad	70.0 Nms/rad

Table 8.2. Stiffness and damping values for the simulation with low torque controller gains (Case I)

i	1	2	3	4	5	6
$\bar{K}_{\tau,i}$	0.5	0.5	0.5	0.5	0.5	0.5
$\bar{D}_{\tau,i}$	0.041	0.047	0.041	0.033	0.034	0.041

Table 8.3. Stiffness and damping values for the simulation with high torque controller gains (Case II)

i	1	2	3	4	5	6
$\bar{K}_{\tau,i}$	3	3	3	3	3	3
$\bar{D}_{\tau,i}$	0.067	0.077	0.067	0.054	0.055	0.067

values from Table 8.3 were used. In both cases the damping values were chosen as $\bar{D}_{\tau,i} = 2\xi_\tau\sqrt{\bar{K}_{\tau,i}K_i/B_i}$ with a common *damping factor* of $\xi_\tau = 1.1$. This relative high value of ξ_τ was chosen in order to ensure that the resulting gain matrix $\bar{\boldsymbol{D}}_\tau$ can be also used for the backstepping controller. This will be explained in more detail in Section 8.1.3.

8.1.1 Singular Perturbation Based Controller

For the regulation case the singular perturbation based controller (5.32) from Chapter 5 can be written as

$$\begin{aligned}\boldsymbol{\tau}_m &= \boldsymbol{\tau}_d - \boldsymbol{K}_\tau(\boldsymbol{\tau} - \boldsymbol{\tau}_d) - \epsilon\boldsymbol{D}_\tau\dot{\boldsymbol{\tau}} \\ \boldsymbol{\tau}_d &= \boldsymbol{g}(\boldsymbol{q}) - \boldsymbol{J}(\boldsymbol{q})^T\left(\boldsymbol{K}_d\tilde{\boldsymbol{x}} + \boldsymbol{D}_d\dot{\tilde{\boldsymbol{x}}}\right) ,\end{aligned}$$

with positive definite controller gain matrices $\boldsymbol{K}_\tau$ and $\boldsymbol{D}_\tau$ and ϵ as the time scaling factor from the singular perturbation analysis. In order to simplify the comparison to the other controllers, the gain matrices are chosen as $\boldsymbol{K}_\tau = \bar{\boldsymbol{K}}_\tau$ and $\epsilon\boldsymbol{D}_\tau = \bar{\boldsymbol{D}}_\tau$ with diagonal matrices $\bar{\boldsymbol{K}}_\tau$ and $\bar{\boldsymbol{D}}_\tau$ according to the values from Table 8.2 and Table 8.3. The simulation results for the two cases of the step response are shown in Figure 8.2. Therein, the closed loop dynamics for the singular perturbation based controller is compared to the desired closed loop behavior according to (8.2) (dotted line in Figure 8.2). Notice that the desired closed loop behavior (8.2) does not use the modified inertia matrix $\boldsymbol{M}(\boldsymbol{q}) + (\boldsymbol{I} + \boldsymbol{K}_\tau)^{-1}\boldsymbol{B}$ of (5.28) from the singular perturbation analysis.

In Figure 8.2 one can see that all the coordinates deviate from the virtual equilibrium position in the transient stage due to the couplings in the inertia matrix but their steady state value is zero. The desired behavior is better approximated for the higher torque control gains (Case II, solid line). In the case of a low gain (Case I, dashed line), the difference to the desired step response is bigger.

8.1.2 Decoupling Based Controller

Next, the decoupling based controller from Chapter 6 is evaluated. For the purpose of readability the controller equation (6.30) is repeated here for the regulation case

$$\begin{aligned}\boldsymbol{\tau}_m &= \boldsymbol{\tau}_d + \boldsymbol{B}\boldsymbol{K}^{-1}\left(\ddot{\boldsymbol{\tau}}_d - \boldsymbol{D}_\tau(\dot{\boldsymbol{\tau}} - \dot{\boldsymbol{\tau}}_d) - \boldsymbol{K}_\tau(\boldsymbol{\tau} - \boldsymbol{\tau}_d)\right) \\ &\quad + \boldsymbol{B}\underbrace{\boldsymbol{M}(\boldsymbol{q})^{-1}\left(\boldsymbol{\tau} + \boldsymbol{\tau}_{ext} - \boldsymbol{C}(\boldsymbol{q},\dot{\boldsymbol{q}})\dot{\boldsymbol{q}} - \boldsymbol{g}(\boldsymbol{q})\right)}_{\ddot{\boldsymbol{q}}} , \\ \boldsymbol{\tau}_d &= \boldsymbol{g}(\boldsymbol{q}) - \boldsymbol{J}(\boldsymbol{q})^T\left(\boldsymbol{K}_d\tilde{\boldsymbol{x}} + \boldsymbol{D}_d\dot{\tilde{\boldsymbol{x}}}\right) .\end{aligned}$$

The matrices $\boldsymbol{K}_\tau$ and $\boldsymbol{D}_\tau$ are related to $\bar{\boldsymbol{K}}_\tau$ and $\bar{\boldsymbol{D}}_\tau$ from Section 8.1 via $\bar{\boldsymbol{K}}_\tau = \boldsymbol{B}\boldsymbol{K}^{-1}\boldsymbol{K}_\tau$ and $\bar{\boldsymbol{D}}_\tau = \boldsymbol{B}\boldsymbol{K}^{-1}\boldsymbol{D}_\tau$, respectively. The torque feedback matrices $\bar{\boldsymbol{K}}_\tau$ and $\bar{\boldsymbol{D}}_\tau$ were chosen in the same way as for the singular perturbation based controller according to Table 8.2 and Table 8.3.

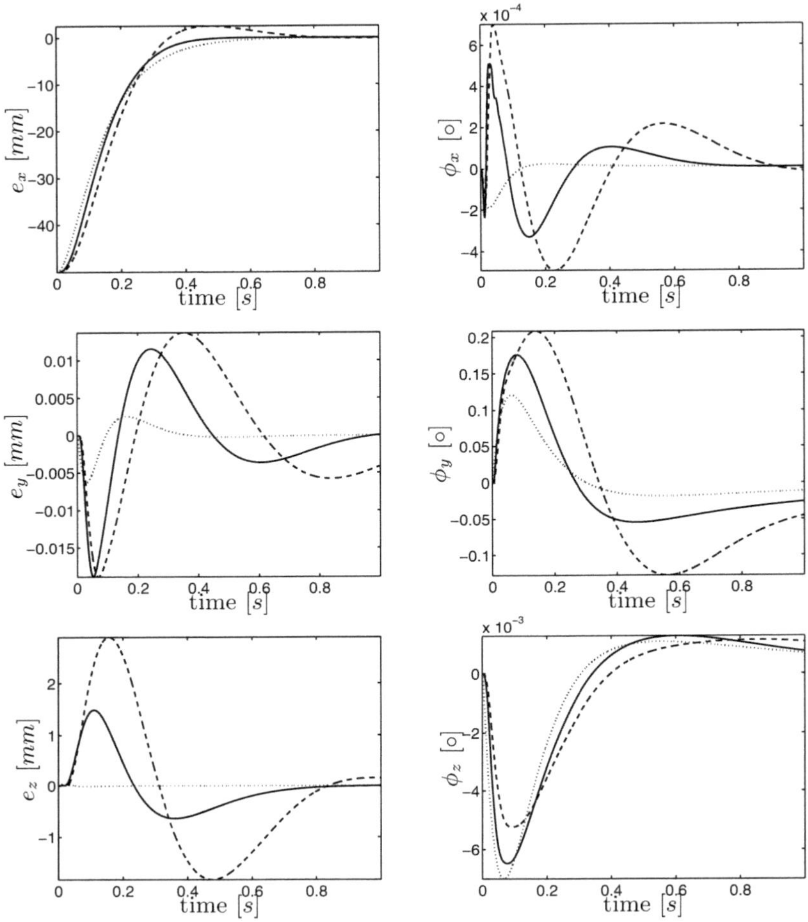

Fig. 8.2. Simulation results for the singular perturbation based controller for the two different torque control gain matrices. The dotted lines show the trajectories of the end-effector coordinates according to the desired dynamic behavior. The dashed lines show the results for the low torque control gains (Case I) and the solid lines show the results for the high torque control gains (Case II).

In Figure 8.3 the simulation results with the decoupling based controller are shown for the two different cases (low and high torque feedback gains). The dotted lines show the trajectories of the Cartesian coordinates for the desired dynamics (8.2). One can see that the deviations of the desired step responses is considerably smaller than for the singular perturbation based controller from Figure 8.2. Even for the low control gain (Case I) the simulation results corresponds quite well to the desired ones.

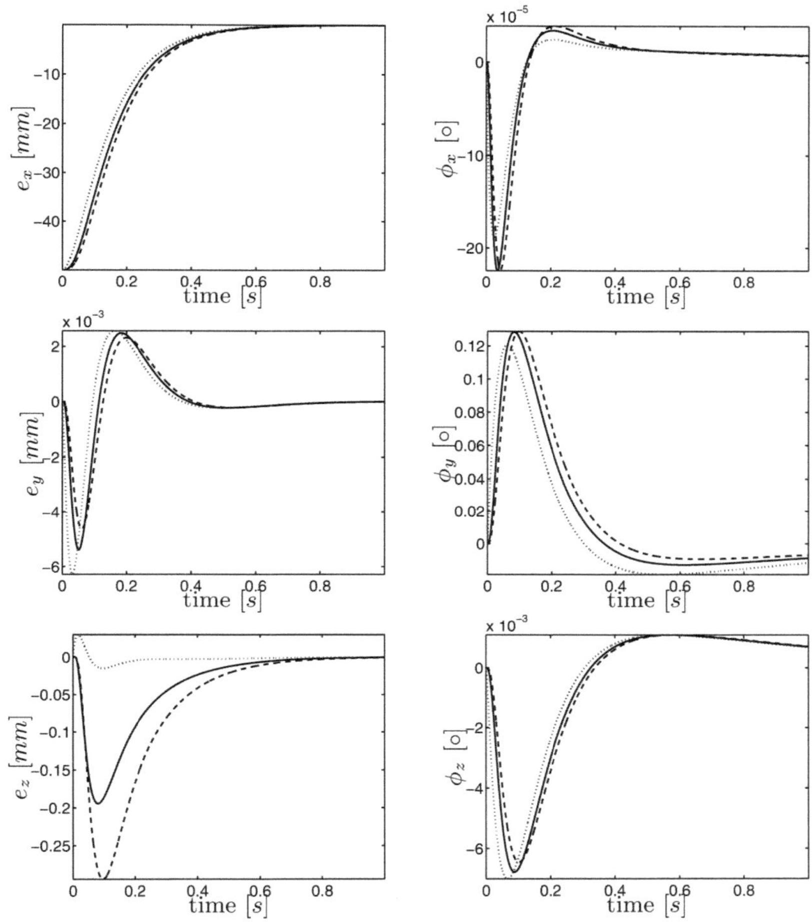

Fig. 8.3. Simulation results with the decoupling based controller for the two different torque control gain matrices. The dotted lines show the trajectories of the end-effector coordinates according to the desired dynamic behavior. The dashed lines show the results for the low torque control gains (Case I) and the solid lines show the results for the high torque control gains (Case II).

Notice that Figure 8.3 shows the results for a step response in the virtual equilibrium position. Regarding the closed loop dynamics (6.24)-(6.25) this corresponds to a stepwise excitation of the torque error dynamics (6.25). It should also be mentioned that the closed loop dynamics of the decoupling based controller fits exactly to the desired behavior for a step response in the external force (instead of a stepwise excitation of the torque error dynamics via the virtual equilibrium position). In this case the torque error would always be zero and the trajectory would fit exactly to the desired one. This holds of course only in the nominal case when all measurement signals are available exactly.

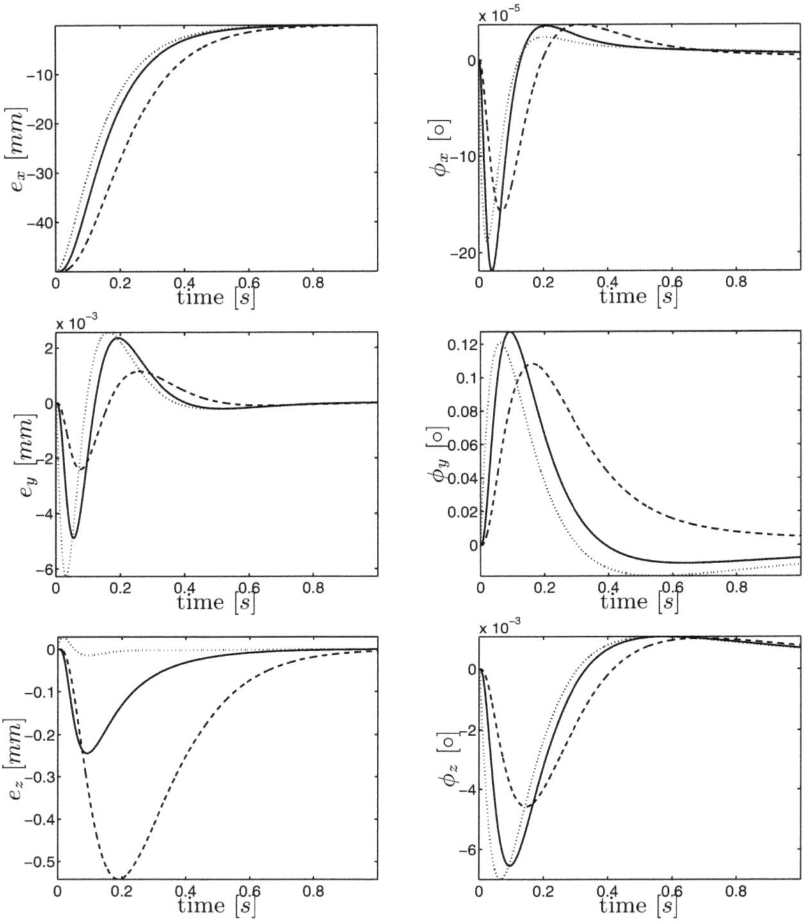

Fig. 8.4. Simulation results with the backstepping based controller for the two different torque control gain matrices. The dotted lines show the trajectories of the end-effector coordinates according to the desired dynamic behavior. The dashed lines show the results for the low torque control gains (Case I) and the solid lines show the results for the high torque control gains (Case II).

8.1.3 Backstepping Based Controller

The same simulation was then performed also for the backstepping based controller from (6.44) which is also repeated here

$$\begin{aligned}\boldsymbol{\tau}_m = &\boldsymbol{B}\ddot{\boldsymbol{q}} + \boldsymbol{\tau} + \boldsymbol{B}\boldsymbol{K}^{-1}\left(\ddot{\boldsymbol{\tau}}_d - (\bar{\boldsymbol{K}}_s\bar{\boldsymbol{K}}_t + \boldsymbol{C}_s^{-1}\boldsymbol{C}_t)\boldsymbol{e}_t + (\bar{\boldsymbol{K}}_s + \bar{\boldsymbol{K}}_t)\dot{\boldsymbol{e}}_t\right)\\ &-\boldsymbol{B}\boldsymbol{K}^{-1}(\boldsymbol{C}_t^{-1}\frac{\mathrm{d}}{\mathrm{d}t}(\boldsymbol{J}(\boldsymbol{q})^{-1}\dot{\tilde{\boldsymbol{x}}}) + \bar{\boldsymbol{K}}_s\boldsymbol{C}_t^{-1}\boldsymbol{J}(\boldsymbol{q})^{-1}\dot{\tilde{\boldsymbol{x}}})\ .\end{aligned}$$

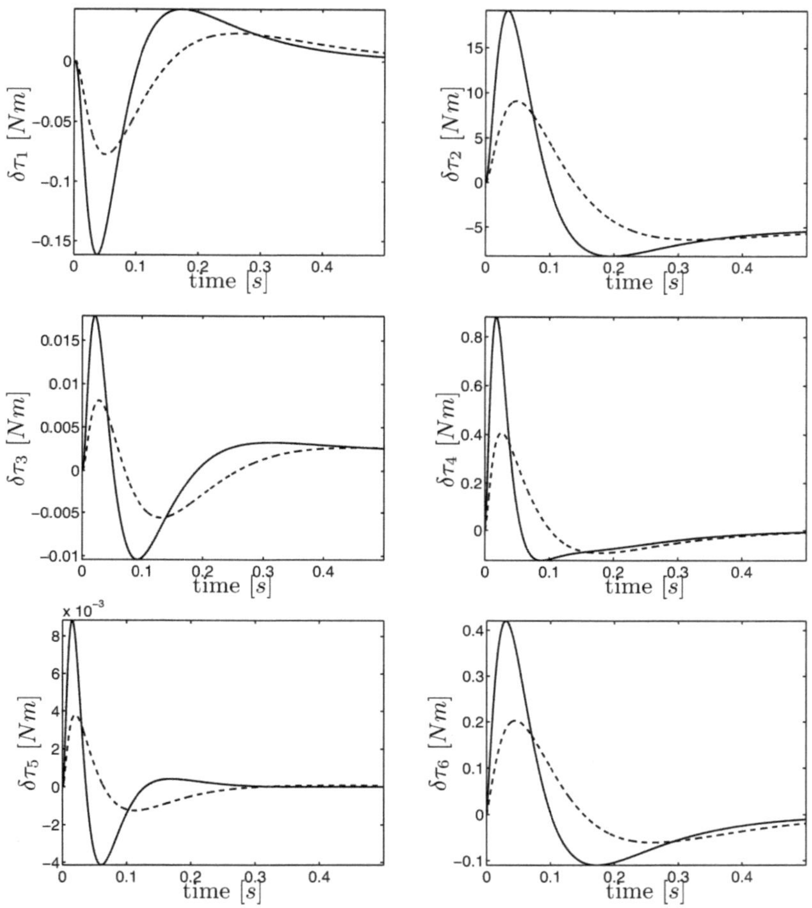

Fig. 8.5. Deviation of the joint torques from their initial values in the case of low torque control gains (Case I). The solid and dashed lines show the simulation results for the decoupling based controller and the backstepping based controller, respectively.

Notice that the controller parameter matrices $\boldsymbol{C}_s$, $\boldsymbol{C}_t$, $\bar{\boldsymbol{K}}_t$, and $\bar{\boldsymbol{K}}_s$ must all be positive definite. Herein, the parameters $\boldsymbol{C}_s$ and $\boldsymbol{C}_t$ are chosen as $\boldsymbol{C}_s = \boldsymbol{I}$ and $\boldsymbol{C}_t = \boldsymbol{I}$. The controller gain matrices $\bar{\boldsymbol{K}}_t$ and $\bar{\boldsymbol{K}}_s$ were designed such that the controller can be compared to the decoupling based and the singular perturbation based controller. Therefore, the gains can be related to $\bar{\boldsymbol{K}}_\tau$ and $\bar{\boldsymbol{D}}_\tau$ as follows

$$\begin{aligned} \boldsymbol{B}\boldsymbol{K}^{-1}\left(\bar{\boldsymbol{K}}_t\bar{\boldsymbol{K}}_s + \boldsymbol{C}_s^{-1}\boldsymbol{C}_t\right) &= \bar{\boldsymbol{K}}_\tau \\ \boldsymbol{B}\boldsymbol{K}^{-1}\left(\bar{\boldsymbol{K}}_t + \bar{\boldsymbol{K}}_s\right) &= \bar{\boldsymbol{D}}_\tau \ . \end{aligned}$$

The gain matrices $\bar{\boldsymbol{K}}_\tau$ and $\bar{\boldsymbol{D}}_\tau$ are chosen according to Table 8.2 and Table 8.3. In the design of these gains it has to be ensured that the above equations are satisfied for real-valued positive definite matrices $\bar{\boldsymbol{K}}_t$ and $\bar{\boldsymbol{K}}_s$. Therefore, the damping factor was chosen quite high as $\xi_\tau = 1.1$ in Section 8.1.

The results for the backstepping based controller are shown in Figure 8.4. Compared to the results of the decoupling based controller the Cartesian end-effector deviations are larger. The difference between the two controllers can also be seen by comparing the joint torques. Figure 8.5 shows the deviations $\delta\tau_i$, $i = 1, \cdots, 6$ of the torques τ_i from their starting values $\tau_i|_{t=0}$, i.e. $\delta\tau_i = \tau_i - \tau_i|_{t=0}$ for the decoupling based and the backstepping based controller in the case of low torque control gains (Case I). One can see that the backstepping based controller results in smaller torque differences than the decoupling based controller.

In the present comparison all the measurement signals are assumed to be available and are not corrupted by noise. Notice that the main difficulties in the implementation of the decoupling based and the backstepping based controllers for a *real* robot lie in the necessity of the measurement or estimation of the link side acceleration and the jerk as was explained in more detail in Chapter 6.

8.1.4 Passivity Based Controller

The same simulations were also performed for the passivity based controller from Chapter 7. The relation between the singular perturbation based controller and the passivity based controller was already described in Section 7.8.1. Therefore, the control law (7.76) can be written as

$$\begin{aligned}\boldsymbol{\tau}_m &= \boldsymbol{\tau}_d - \bar{\boldsymbol{K}}_\tau(\boldsymbol{\tau} - \boldsymbol{\tau}_d) - \bar{\boldsymbol{D}}_\tau\dot{\boldsymbol{\tau}} \\ \boldsymbol{\tau}_d &= \boldsymbol{g}(\bar{\boldsymbol{q}}(\boldsymbol{\theta})) - \boldsymbol{J}(\boldsymbol{\theta})^T \left(\boldsymbol{K}_d\tilde{\boldsymbol{x}}_\theta(\boldsymbol{\theta}) + \boldsymbol{D}_d\dot{\tilde{\boldsymbol{x}}}_\theta(\boldsymbol{\theta})\right) ,\end{aligned}$$

where the controller gain matrices $\bar{\boldsymbol{K}}_\tau$ and $\bar{\boldsymbol{D}}_\tau$ are related to the parameters $\boldsymbol{B}_\theta$ and $\boldsymbol{K}_s$ from (7.76) by $\bar{\boldsymbol{K}}_\tau = \boldsymbol{B}\boldsymbol{B}_\theta^{-1} - \boldsymbol{I}$ and $\bar{\boldsymbol{D}}_\tau = (\boldsymbol{I} - \boldsymbol{B}\boldsymbol{B}_\theta^{-1})\boldsymbol{K}_s\boldsymbol{K}^{-1}$, respectively.

For the implementation of the gravity compensation the function $\bar{\boldsymbol{q}}(\boldsymbol{\theta})$ must be computed. This was done by the iteration procedure described in Proposition 7.3. The iteration was started with $\hat{\boldsymbol{q}}_0 = \boldsymbol{\theta}$ and stopped after two iteration steps. In order to take account of the joint elasticity, the Cartesian controller stiffness matrix $\boldsymbol{K}_d$ was chosen according to the method described in Section 7.5.1.

Two different configurations of the controller were evaluated which differ in the choice of $\bar{\boldsymbol{D}}_\tau$. In both configurations the gain matrix $\bar{\boldsymbol{K}}_\tau$ is chosen in the same way as for the singular perturbation based controller according to Table 8.2 and Table 8.3 which leads to the motor scaling factors of $\boldsymbol{B}\boldsymbol{B}_\theta^{-1} = 1.5\boldsymbol{I}$ and $\boldsymbol{B}\boldsymbol{B}_\theta^{-1} = 4\boldsymbol{I}$, respectively. In the first configuration the feedback gain matrix $\bar{\boldsymbol{D}}_\tau$ was set to zero according to Section 7.2 while it was set to the values of Table 8.2 and Table 8.3 in the second configuration.

The controller performance for $\bar{\boldsymbol{D}}_\tau = \boldsymbol{0}$ is shown in Figure 8.6. Figure 8.7 shows the result for $\bar{\boldsymbol{D}}_\tau \neq \boldsymbol{0}$. One can see that the controller with non-zero $\bar{\boldsymbol{D}}_\tau$ has much better performance regarding the oscillation damping. The stability analysis in Section 7.5.2 was instead presented for the controller with $\bar{\boldsymbol{D}}_\tau = \boldsymbol{0}$.

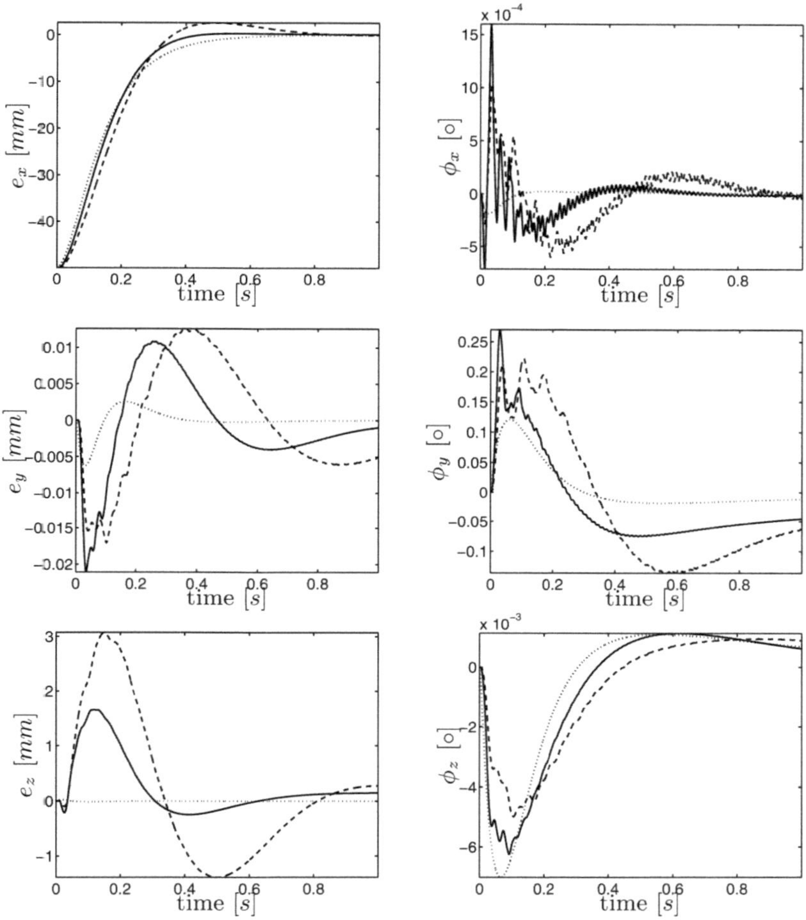

Fig. 8.6. Simulation results with the passivity based controller for the two different torque control gain matrices with $\boldsymbol{D}_\tau = \mathbf{0}$. The dotted lines show the trajectories of the end-effector coordinates according to the desired dynamic behavior. The dashed lines show the results for the low torque control gains (Case I) and the solid lines show the results for the high torque control gains (Case II).

The performance of the controller with $\bar{\boldsymbol{D}}_\tau \neq \mathbf{0}$ from Figure 8.7 is quite similar to the singular perturbation based controller. This is not surprising because the two controllers have quite a similar structure.

8.1.5 Robustness

Additionally to the simulation of the nominal performance of the controllers also the robustness of the controllers with respect to parameter uncertainties shall be considered.

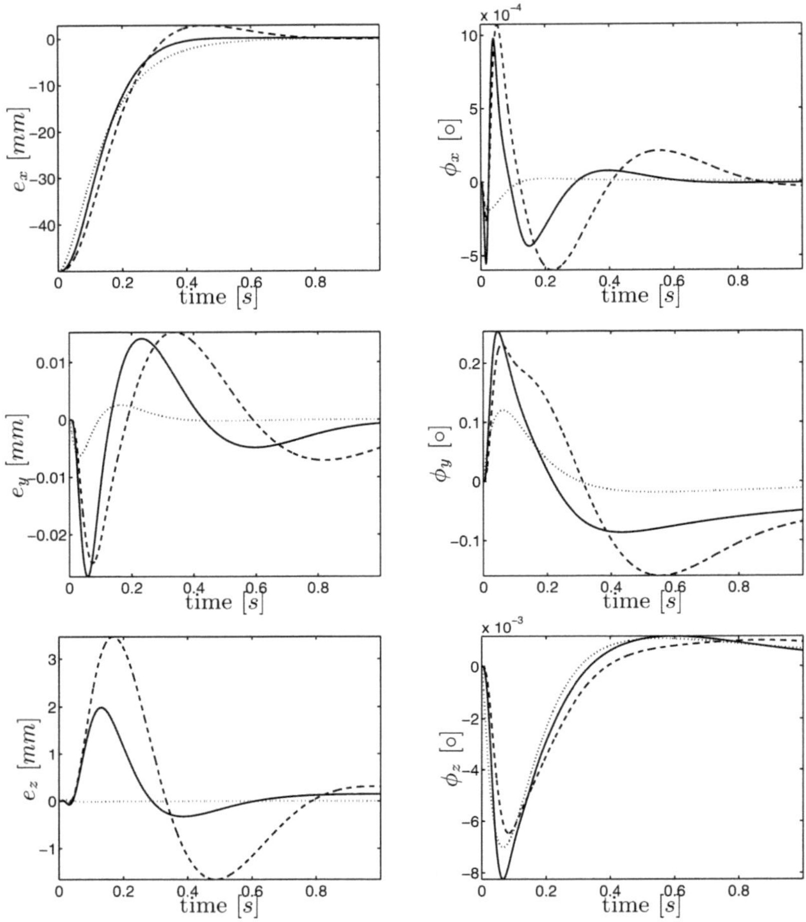

Fig. 8.7. Simulation results with the passivity based controller for the two different torque control gain matrices with $\boldsymbol{D}_\tau \neq \boldsymbol{0}$. The dotted lines show the trajectories of the end-effector coordinates according to the desired dynamic behavior. The dashed lines show the results for the low torque control gains (Case I) and the solid lines show the results for the high torque control gains (Case II).

The singular perturbation based controller does not contain model parameters explicitly. Similarly, also the stability statements of the passivity based controller are independent of the model parameters. If for instance the values of the motor inertia $\boldsymbol{B}$ are not known exactly, this only means that the scaled motor inertia $\boldsymbol{B}_\theta$ will be different than desired. But the stability is not affected by this.

The situation is different for the decoupling based controller and the backstepping controller. These controllers require the joint accelerations and the jerks. In the above simulations it was assumed that these signals can be obtained exactly.

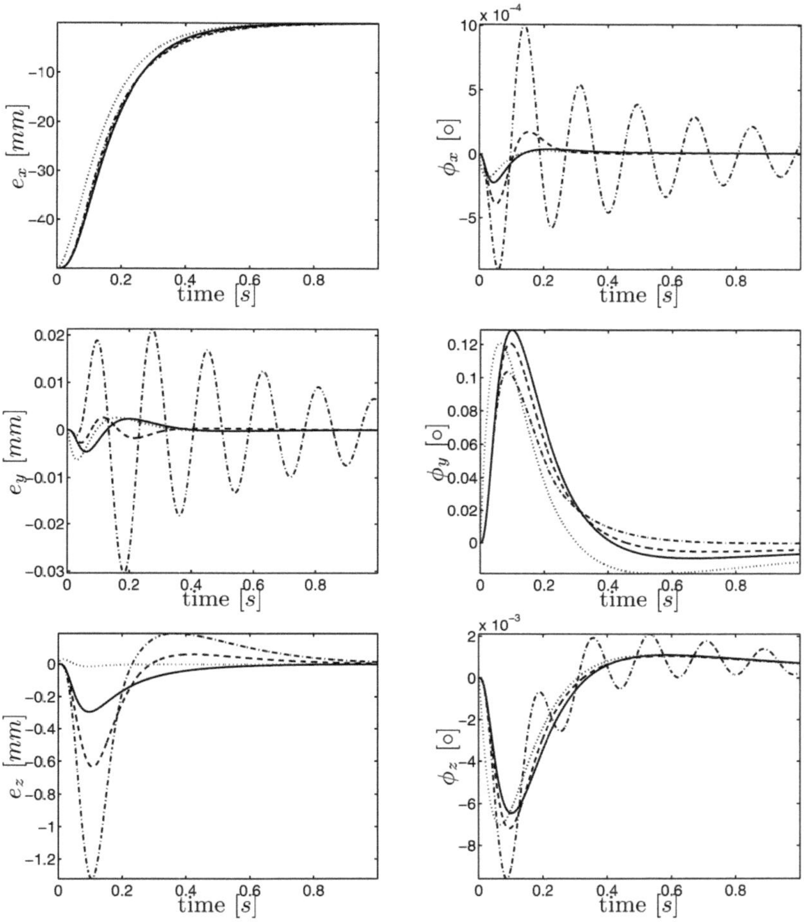

Fig. 8.8. Simulation results with the decoupling based controller for the robustness analysis. In all simulations the low torque control gains (Case I) were used. The dotted lines show the trajectories of the end-effector coordinates according to the desired dynamic behavior. The solid lines show the results without model uncertainties, i.e. for $f_m = 1$. The dashed and the dashed-dotted lines show the results for $f_m = 1.1$ and $f_m = 1.3$, respectively.

But in practice this will not be possible and one must compute these signals from other measurements using the model parameters via

$$\ddot{\boldsymbol{q}} = \boldsymbol{M}(\boldsymbol{q})^{-1} \left(\boldsymbol{\tau} + \boldsymbol{\tau}_{ext} - \boldsymbol{C}(\boldsymbol{q}, \dot{\boldsymbol{q}})\dot{\boldsymbol{q}} - \boldsymbol{g}(\boldsymbol{q})\right) \ , \tag{8.3}$$

$$\boldsymbol{q}^{(3)} = \boldsymbol{M}(\boldsymbol{q})^{-1} \left(\dot{\boldsymbol{\tau}} - \dot{\boldsymbol{M}}(\boldsymbol{q})\ddot{\boldsymbol{q}} + \frac{\mathrm{d}}{\mathrm{d}t}\left(\boldsymbol{\tau}_{ext} - \boldsymbol{C}(\boldsymbol{q}, \dot{\boldsymbol{q}})\dot{\boldsymbol{q}} - \boldsymbol{g}(\boldsymbol{q})\right)\right) \ . \tag{8.4}$$

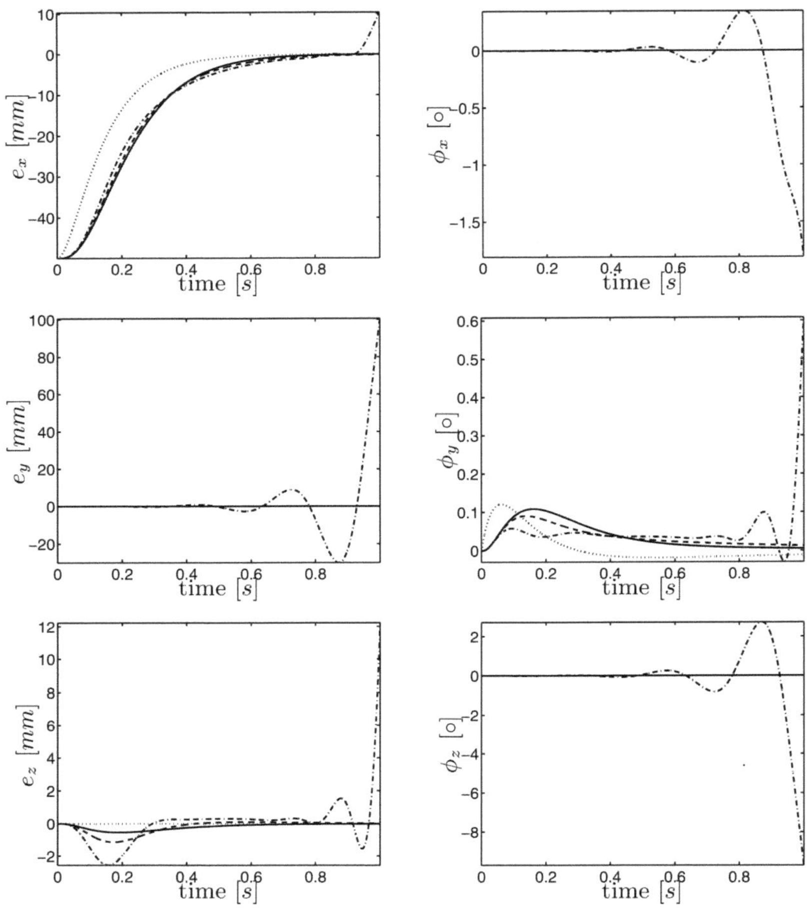

Fig. 8.9. Simulation results with the backstepping controller for the robustness analysis. In all simulations the low torque control gains (Case I) were used. The dotted lines show the trajectories of the end-effector coordinates according to the desired dynamic behavior. The solid lines show the results without model uncertainties, i.e. for $f_m = 1$. The dashed and the dashed-dotted lines show the results for $f_m = 1.1$ and $f_m = 1.3$, respectively.

Apart from this, the controller also contains the motor inertia. Even if the accelerations and the jerks can be exactly obtained, uncertainties in the motor inertia may affect the stability of the closed loop system.

In the following simulation the robustness of the decoupling based controller and the backstepping controller with respect to uncertainties of the motor inertia is analyzed. Similarly to the previous simulations, it is assumed that the joint accelerations and the jerks can be exactly obtained via (8.3) and (8.4). The controller uses the torque feedback according to Table 8.2. Figure 8.8 shows the

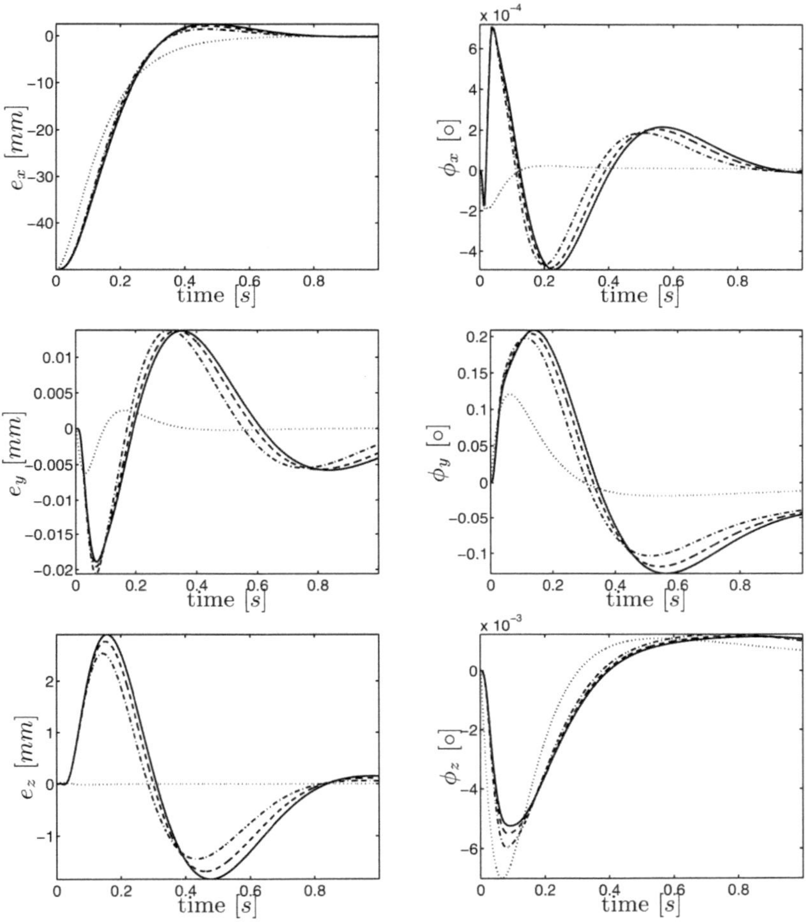

Fig. 8.10. Simulation results with the singular perturbation based controller for the robustness analysis. In all simulations the low torque control gains (Case I) were used. The dotted lines show the trajectories of the end-effector coordinates according to the desired dynamic behavior. The solid lines show the results without model uncertainties, i.e. for $f_m = 1$. The dashed and the dashed-dotted lines show the results for $f_m = 1.1$ and $f_m = 1.3$, respectively.

simulation results for different values of the motor inertia. The controller uses the nominal values $\boldsymbol{B}_{nom}$ of the motor inertia, while the model parameters were set to $\boldsymbol{B} = \boldsymbol{B}_{nom}/f_m$ for $f_m = 1$, $f_m = 1.1$, and $f_m = 1.3$.

The same simulation was also performed for the backstepping controller. The result is shown in Figure 8.9. One can see that the backstepping controller is more sensitive to the uncertainty in the motor inertia parameters than the decoupling

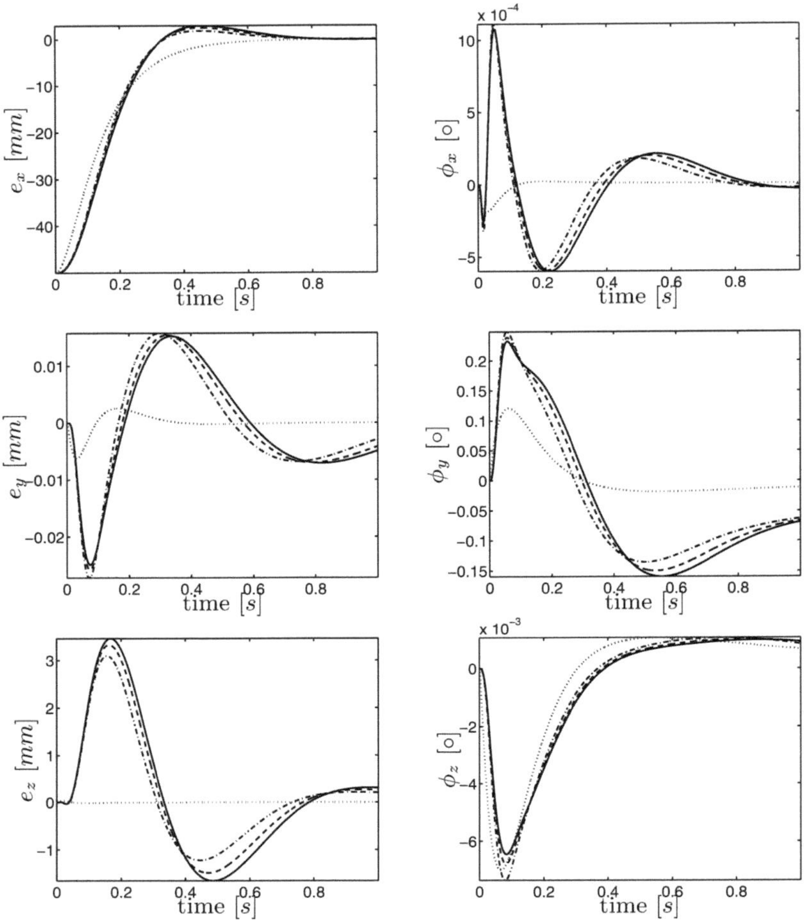

Fig. 8.11. Simulation results with the passivity based controller for the robustness analysis. In all simulations the low torque control gains (Case I) were used. The dotted lines show the trajectories of the end-effector coordinates according to the desired dynamic behavior. The solid lines show the results without model uncertainties, i.e. for $f_m = 1$. The dashed and the dashed-dotted lines show the results for $f_m = 1.1$ and $f_m = 1.3$, respectively.

based controller. This results from the fact that the backstepping controller contains additional feedback terms which depend on $\boldsymbol{B}$.

In order to evaluate the robustness of the singular perturbation based controller and the passivity based controller, the same experiment with the same controller gains is also performed with these controllers. The results are shown in Figure 8.10 and Figure 8.11. One can see that these controllers are robust with respect to the considered model uncertainty.

8.2 Experiments with the DLR-Lightweight-Robot-II

In this section some experiments with the seven-joint DLR-Lightweight-Robot-II concerning the implementation of a Cartesian impedance are shown. Figure 8.12 depicts the realtime controller structure of this robot. The robot is equipped with a signal processor in each joint. This allows the implementation of individual joint torque controllers with a high sampling rate (3 kHz). The measurement data (motor positions and joint torques) are transmitted via an optical bus[1] to a realtime computer system (500 MHz PowerPC using VxWorks as an Operating System). On this computer the Cartesian impedance controllers were implemented with a sampling rate of 1 kHz. The link side position $\boldsymbol{q}$ was computed based on the measurements and the joint stiffness via $\boldsymbol{q} = \boldsymbol{\theta} - \boldsymbol{K}^{-1}\boldsymbol{\tau}$. The time derivatives $\dot{\boldsymbol{\theta}}$, $\dot{\boldsymbol{q}}$, and $\dot{\boldsymbol{\tau}}$ were obtained via filtered numerical differentiation.

For the computation of more complex entities (i.e. the inertia matrix, the body Jacobian, the gravity term, and the Coriolis/centrifugal matrix) a supplementary task was used with a sampling time of 6 ms (see Figure 8.12). For the relevant Cartesian coordinates the error variable

$$\tilde{\boldsymbol{x}} = \begin{pmatrix} {}_t\boldsymbol{p}_{td}(\boldsymbol{q},t) \\ {}_t\boldsymbol{\phi}_{td}(\boldsymbol{q},t) \end{pmatrix} =: \begin{pmatrix} e_x \\ e_y \\ e_z \\ \phi_x \\ \phi_y \\ \phi_z \end{pmatrix} \tag{8.5}$$

was chosen (see also Section 3.5) with a roll-pitch-yaw representation for the Euler angles ${}_t\boldsymbol{\phi}_{td}(\boldsymbol{q},t)$.

Section 8.2.1 presents the experimental results with the singular perturbation based controller. This includes a verification of the achieved stiffness and damping behavior as well as a verification of the different nullspace projections from Chapter 4. The results with the passivity based controller are presented in Section 8.2.2. An experimental comparison of the decoupling based controller with a singular perturbation based controller similar to the one discussed herein is given in [OASKH03, OASKH05].

8.2.1 Singular Perturbation Based Controller

For the singular perturbation based controller two experiments are reported. In the first experiment the Cartesian impedance being obtained by the controller is evaluated. The second experiment deals with the nullspace stiffness.

According to Chapter 5 the controller can be written in the form

$$\begin{aligned} \boldsymbol{\tau}_m &= \boldsymbol{\tau}_d - \boldsymbol{K}_\tau(\boldsymbol{\tau} - \boldsymbol{\tau}_d) - \epsilon \boldsymbol{D}_\tau \dot{\boldsymbol{\tau}} \\ \boldsymbol{\tau}_d &= \boldsymbol{g}(\boldsymbol{q}) + \boldsymbol{J}(\boldsymbol{q})^T \left(\boldsymbol{\Lambda}(\boldsymbol{x})\ddot{\boldsymbol{x}}_d + \boldsymbol{\mu}(\boldsymbol{x},\dot{\boldsymbol{x}})\dot{\boldsymbol{x}}_d - \boldsymbol{K}_d\tilde{\boldsymbol{x}} - \boldsymbol{D}_d\dot{\tilde{\boldsymbol{x}}} \right) \end{aligned}$$

[1] The DLR lightweight robots use a SERCOS bus (see `http://www.sercos.org/`).

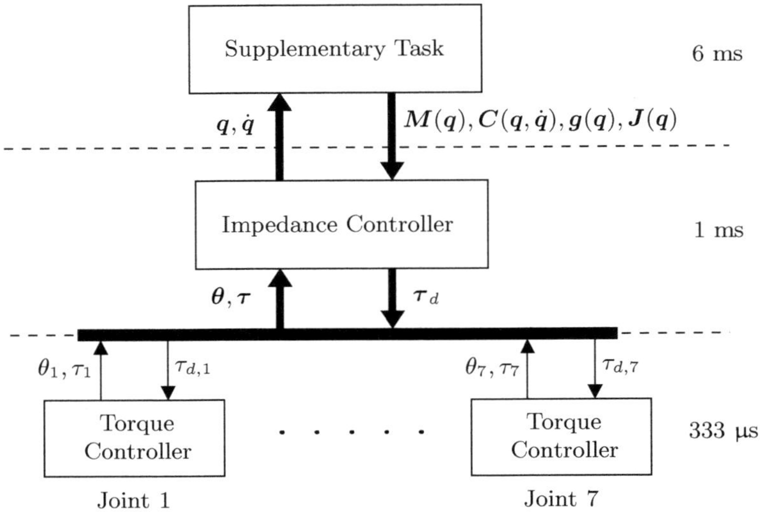

Fig. 8.12. Realtime control structure of the DLR-Lightweight-Robot-II

Table 8.4. Stiffness values for the impedance evaluation of the singular perturbation based controller

Coord.	e_x	e_y	e_z	ϕ_x	ϕ_y	ϕ_z
Stiffness	800 N/m	5000 N/m	5000 N/m	300 Nm/rad	300 Nm/rad	300 Nm/rad

with positive definite controller gain matrices $\boldsymbol{K}_\tau$, $\boldsymbol{D}_\tau$, $\boldsymbol{K}_d$, and $\boldsymbol{D}_d$. For the torque control gain matrices $\boldsymbol{K}_\tau$ and $\boldsymbol{D}_\tau$ diagonal matrices are chosen such that the inner torque controllers can be implemented on the local signal processors at a high sampling rate of 3 kHz (see Figure 8.12). The desired torque $\boldsymbol{\tau}_d$ is computed at a lower sampling rate of 1 kHz.

Cartesian Stiffness and Damping Evaluation

In the first experiment the Cartesian stiffness and damping are evaluated for the regulation case $\dot{\boldsymbol{x}}_d = \boldsymbol{0}$. As a desired stiffness matrix a diagonal matrix with the diagonal elements according to Table 8.4 was chosen. For the design of the damping matrix, the method from Section 3.3 was applied with a damping factor of $\xi = 0.7$. In the experiment a human user exerts (generalized) forces on the robot end-effector by pulling and pushing, mainly in the horizontal ($x-$ and $y-$coordinates) directions. Figure 8.1 shows the initial configuration. The interaction forces are measured by a six-degree-of-freedom force-torque-sensor[2] mounted on the end-effector. Notice that this sensor was not used in the implementation of the

[2] A JR3 sensor was used.

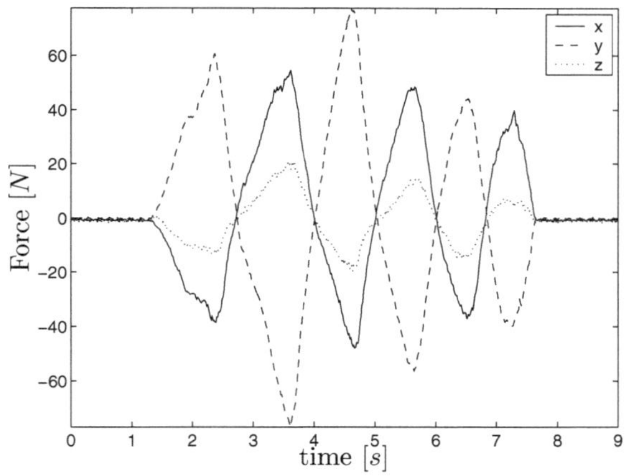

Fig. 8.13. Measured forces for the singular perturbation based controller. (solid: force in x-direction, dashed: force in y-direction, dotted: force in z-direction.)

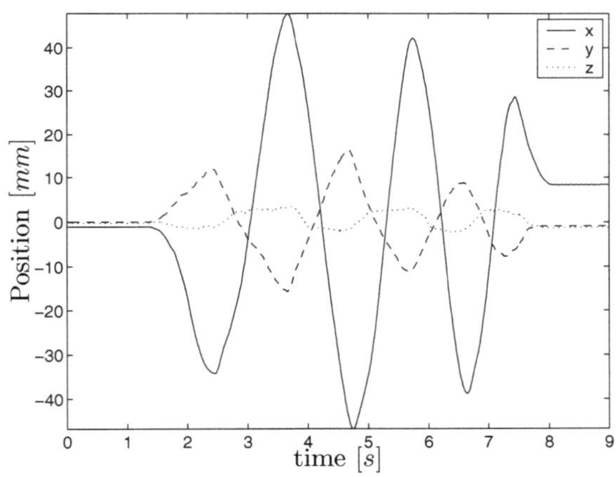

Fig. 8.14. End-effector deviation for the singular perturbation based controller. (solid: deviation in x-direction, dashed: deviation in y-direction, dotted: deviation in z-direction.)

impedance controller but is used only for the evaluation. Figure 8.13 shows the measured forces in $x-$, $y-$, and $z-$direction. The corresponding Cartesian end-effector deviations are shown in Figure 8.14. In order to evaluate the resulting stiffness and damping, the force and displacement in $x-$ and $y-$ direction are shown in Figure 8.15 and Figure 8.16, respectively. The corresponding static characteristic line according to the relevant stiffness value from Table 8.4 is shown by the dashed line. Notice that the *hysteresis*-like deviation from the static value is caused by the Cartesian damping. The dotted line shows additionally

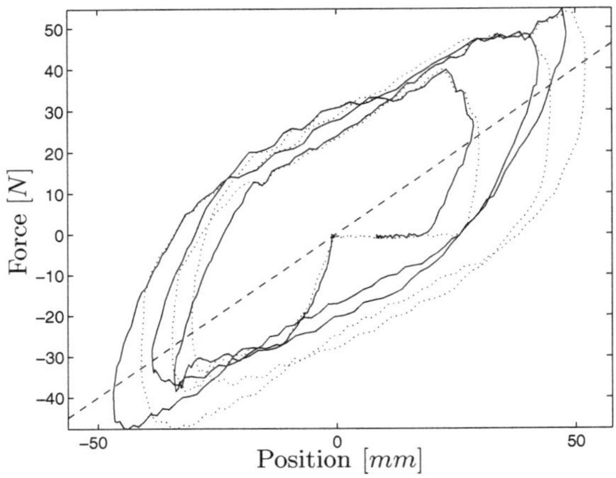

Fig. 8.15. Position and force in x-direction for the singular perturbation based controller. The dashed line represents the desired stiffness. The dotted line shows the simulation results.

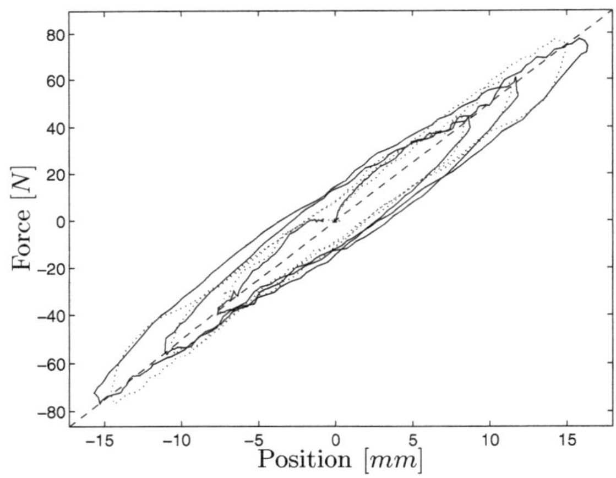

Fig. 8.16. Position and force in y-direction for the singular perturbation based controller. The dashed line represents the desired stiffness. The dotted line shows the simulation results.

the result of a simple simulation of the desired Cartesian impedance. In this simulation the measured contact force is used as an input and the Cartesian motion is the output. This simulation contains some further simplifications[3] and a simple friction model[4]. Notice that the simulation shows only the desired impedance and no joint elasticity is included. One can see that the experimental results fit

[3] The inertia matrix and the damping matrix were considered constant.
[4] The friction model for this robot was taken from [AS01].

quite well to the simulation of the desired impedance for low (Figure 8.15) and high (Figure 8.16) Cartesian stiffness.

Nullspace Stiffness

In the next experiment the nullspace projections from Chapter 4 are evaluated. The DLR-Lightweight-Robot-II has seven joints while for a Cartesian impedance six degrees-of-freedom are sufficient. Therefore, a one-dimensional nullspace remains. In this experiment a Cartesian impedance is used which is characterized by the stiffness values in Table 8.5 together with a damping factor of 0.7. In the experiment no external forces were present.

Table 8.5. Stiffness values for the nullspace experiment

Coord.	e_x	e_y	e_z	ϕ_x	ϕ_y	ϕ_z
Stiffness	1000 N/m	1000 N/m	1000 N/m	300 Nm/rad	300 Nm/rad	300 Nm/rad

In the following a comparison between the projection matrices $\boldsymbol{P}_1(\boldsymbol{q})$ from (4.20), $\boldsymbol{P}_2(\boldsymbol{q})$ from (4.22), and $\boldsymbol{P}_3(\boldsymbol{q})$ from (4.23) is given. As a considered metric for $\boldsymbol{P}_1(\boldsymbol{q})$ the Euclidean metric is used. Ideally, there should be no Cartesian end-effector deviation from the virtual equilibrium position $\boldsymbol{x}_d$. This is evaluated by the following procedure. For the joint space stiffness and the damping from (4.15) diagonal matrices $\boldsymbol{K}_n = k_n \boldsymbol{I}$ and $\boldsymbol{D}_n = d_n \boldsymbol{I}$ were chosen, and according to (4.16) the resulting joint torque is pre-multiplied by the different projection matrices. In order to generate an oscillating nullspace motion a positive value for the nullspace stiffness k_n was used together with a negative value for the damping factor d_n.

$$k_n = 50 \text{ Nm/rad} \tag{8.6}$$

$$d_n = -\frac{1}{2}\sqrt{k_n} \text{ Nm/rad/s} \tag{8.7}$$

The virtual equilibrium point for the nullspace motion was chosen as $\boldsymbol{q}_{d,0} = \boldsymbol{0}$. The choice of a negative damping value results in an unstable nullspace motion. When a joint reaches its mechanical end stop, the joint bounces off. This way an oscillating nullspace motion between the mechanical end stops is produced.

For an evaluation of the different nullspace projections the resulting end-effector motion during this oscillating nullspace motion is used. In this comparison only the translational motion of the end-effector is considered and the Cartesian error $||\tilde{\boldsymbol{x}}||_t$ denotes the Euclidean norm of the translational components of $\tilde{\boldsymbol{x}}$ only. In Figure 8.17 this Cartesian error is shown for the different nullspace projection matrices. Herein the projection matrices were switched online between $\boldsymbol{P}_1(\boldsymbol{q})$, $\boldsymbol{P}_3(\boldsymbol{q})$, and $\boldsymbol{P}_2(\boldsymbol{q})$ during the nullspace oscillation. One can see that the Cartesian errors are considerably larger in case of the static nullspace

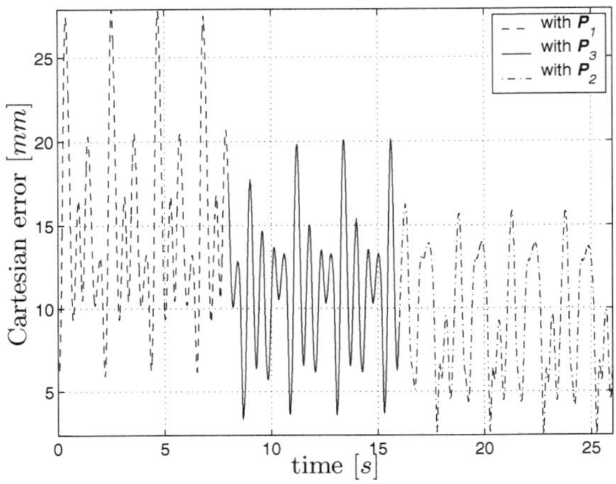

Fig. 8.17. Translational Cartesian error $||\tilde{\boldsymbol{x}}||_t$ for the different nullspace projections

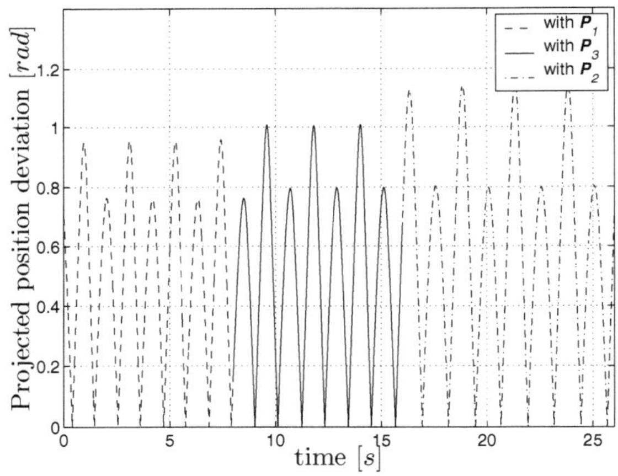

Fig. 8.18. Projected joint error $||\boldsymbol{Z}(\boldsymbol{q})^T(\boldsymbol{q} - \boldsymbol{q}_N)||_2$ for the different nullspace projections

projection via $\boldsymbol{P}_1(\boldsymbol{q})$, while the errors with the dynamically consistent projection matrices $\boldsymbol{P}_2(\boldsymbol{q})$ and $\boldsymbol{P}_3(\boldsymbol{q})$ are of similar magnitude.

In order to visualize also the nullspace motion, which was present during the generation of Figure 8.17, the projection of the joint position deviation $\boldsymbol{q} - \boldsymbol{q}_{d,0}$ into the nullspace is given in Figure 8.18. While the amplitudes of the oscillations in the nullspace for $\boldsymbol{P}_1(\boldsymbol{q})$ are the smallest (Figure 8.18), they result in high Cartesian errors (Figure 8.17). Notice that in the considered setting the nullspace motion was not symmetrical such that the amplitudes of the movement into the different directions are not identical. One can also see that the period of the oscillation is slightly different for the different nullspace projections. This results from the different weighting of the nullspace stiffness and the damping value by the projection matrices.

8.2.2 Passivity Based Controller

Finally, the passivity based impedance controller (7.36) from Chapter 7 is evaluated in two experiments. Figure 8.19 shows the DLR-Lightweight-Robot-II in the starting configuration for both of the two experiments. For the implementation the controller scheme of Figure 8.12 is used. The control law (7.3) is implemented with a sampling rate of 333 μs. The intermediate control input $\boldsymbol{u}$ from (7.36) is computed with a sampling time of 1 ms and is transmitted to the local signal processors.

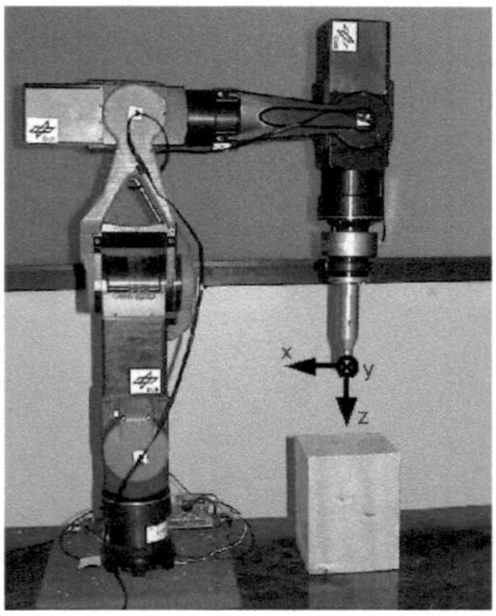

Fig. 8.19. DLR-Lightweight-Robot-II in the starting configuration

Stiffness and Damping

In this experiment the accuracy of the Cartesian impedance shall be evaluated when the robot interacts with a human user. The commanded values for the diagonal Cartesian stiffness matrix are shown in Table 8.6. For the damping matrix the design method from Section 3.3 is applied with an overall damping factor of 0.7. During the experiment a human user applies forces on the end-effector by pulling and pushing (mainly in horizontal directions). The forces are

Table 8.6. Stiffness values for the experiment with the passivity based controller

Coord.	e_x	e_y	e_z	ϕ_x	ϕ_y	ϕ_z
Stiffness	800 N/m	5000 N/m	5000 N/m	300 Nm/rad	300 Nm/rad	300 Nm/rad

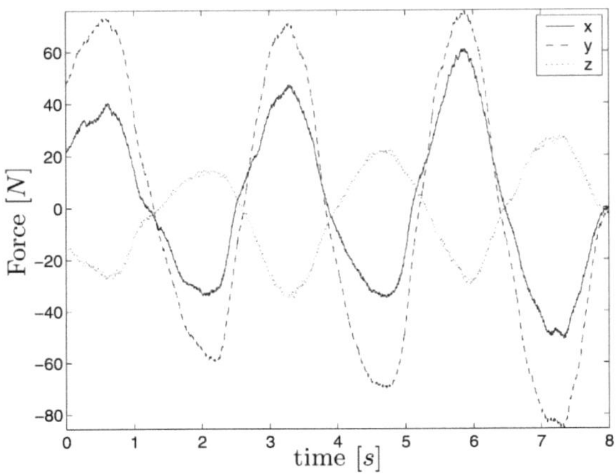

Fig. 8.20. Measured forces in the first experiment for the passivity based impedance controller. (solid: force in x-direction, dashed: force in y-direction, dotted: force in z-direction.)

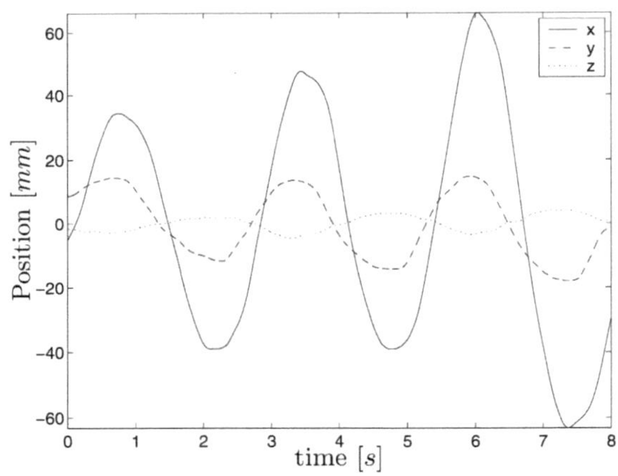

Fig. 8.21. End-effector deviation in the first experiment for the passivity based impedance controller.(solid: deviation in x-direction, dashed: deviation in y-direction, dotted: deviation in z-direction.)

measured by a six-degree-of-freedom force-torque-sensor mounted on the end-effector of the robot. This sensor is used for the evaluation only and is not used for the controller implementation.

The measured forces are shown in Figure 8.20. Figure 8.21 shows the Cartesian position of the end-effector. In Figure 8.22 and Figure 8.23 the force-position measurements are plotted for the x- and y-direction together with the desired static characteristic line. One can see that the mean slope fits well to the desired value. Notice that the measurement contains also the forces due to the desired

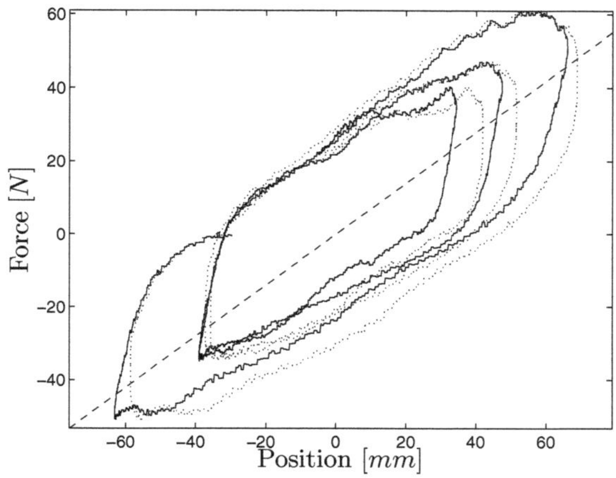

Fig. 8.22. Position and force in x-direction for the first experiment with the passivity based impedance controller. The dashed line represents the desired stiffness. The dotted line shows the simulation results.

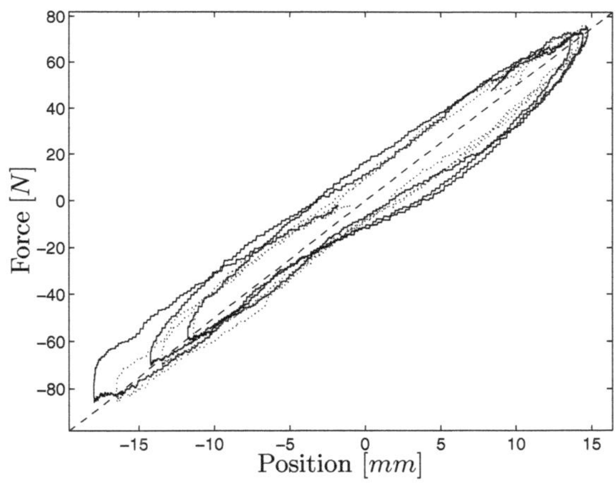

Fig. 8.23. Position and force in y-direction for the first experiment with the passivity based impedance controller. The dashed line represents the desired stiffness. The dotted line shows the simulation results.

damping which produces a *hysteresis*-like characteristics. The damping is evaluated by comparison with a simple simulation model of the desired dynamics. The dotted lines show the result of the simulation of the desired impedance in which the measured force is used as an input. This simulation contains also a simple Coulomb friction model and is further simplified by the use of a constant Cartesian inertia matrix. One can see that the simulation fits quite well to the measurement although the simulation results are quite sensitive to offset errors in the force measurement.

Table 8.7. Stiffness values for the impact experiment

Coord.	e_x	e_y	e_z	ϕ_x	ϕ_y	ϕ_z
Stiffness	5000 N/m	5000 N/m	700 N/m	300 Nm/rad	300 Nm/rad	300 Nm/rad

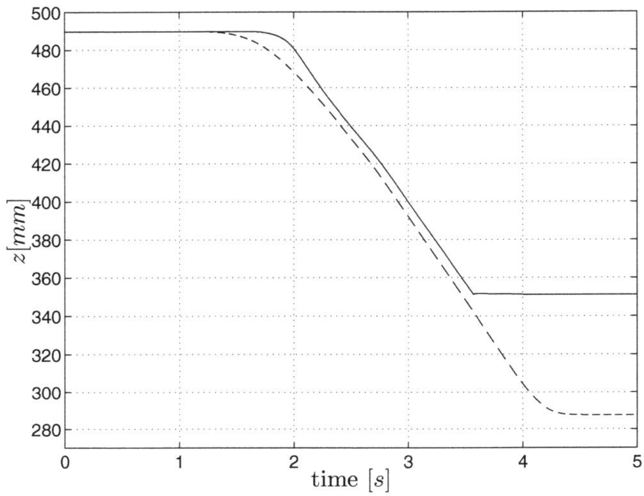

Fig. 8.24. Position in z-direction in the impact experiment for the passivity based impedance controller. The dashed line represents the virtual equilibrium position.

Impact Experiment

The second experiment shows the efficiency of the impedance controller during the impact on a stiff surface. The manipulator is commanded to move in the vertical $z-$direction[5]. After the impact with a wooden surface the force at the tip increases proportionally to the position error. The used stiffness values for this experiment are summarized in Table 8.7. The desired stiffness in vertical direction is chosen smaller such that the impact force is kept small. The desired damping is chosen according to a damping factor of 0.7. Figure 8.24 and 8.25 show the position and force in z-direction. At the time of impact the vertical velocity is about 85 mm/s. One can see that the initial peak in the contact force is reduced quite fast[6]. This is achieved by the lightweight construction of the arm and by the reduction of the effective motor inertia by a factor of about three via the torque feedback. Notice that impact experiments are commonly considered as

[5] It should be mentioned that in this experiment again the passivity based controller from (7.36) is used, but now the virtual equilibrium position is time-varying.

[6] It is reduced within 6 ms which is the used sampling interval of the force-torque-sensor.

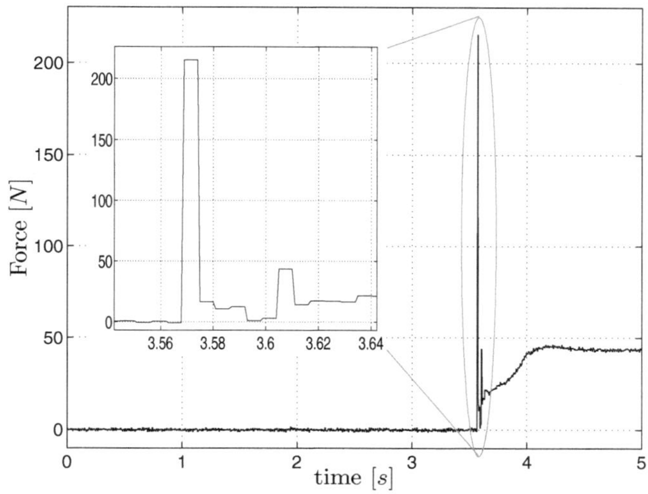

Fig. 8.25. Measured contact force in the vertical $z-$direction in the impact experiment for the passivity based impedance controller

challenging problems in robotics, in particular concerning the stability of control schemes based on the use of force-torque-sensors at the end-effector [Koe01]. With the presented controller no stability problems were observed in this impact experiment. This is ensured by the passivity properties of the used controller together with the high sampling time of the torque feedback.

9 Applications

In this short chapter some typical applications are presented for which the Cartesian impedance controllers of this monograph have been used. Impedance control indubitably is one of the core building blocks needed for the successful execution of advanced manipulation tasks [Hog85a, Hog85b, Hog85c, BS98]. Notice that in all the presented applications the Cartesian impedance control is only one part of a larger setup. In practice, therefore, the question how (for a given application) impedance control of a robot arm can best be combined with other robotics related subjects[1], such as path planning or computer vision, is an important issue, too. The purpose of this chapter is to give an overview of applications for which the discussed controllers were designed. In this presentation, consequently, only a rough description of the applications is given and the emphasis lies on the application of the Cartesian impedance controllers. Besides that also the generality of the impedance control concept for the use in manipulation tasks becomes visible.

The first four experiments in this chapter were performed with the DLR-Lightweight-Robot-II in different setups. Section 9.1 presents a typical service robotics task: wiping a table. Then, in Section 9.2, a medical robotics application is described, in which impedance control of the robot arm is utilized for spine surgery. Thereafter, the combination of the DLR-Lightweight-Robot-II with a mobile platform is described in Section 9.3. In Section 9.4, the DLR-Lightweight-Robot-II is used together with a mobile platform and an artificial four-finger hand for door opening. Finally, in section 9.5 the extension to more complex kinematic chains is exemplified by considering the two-arm manipulation with a humanoid upper-body system.

9.1 Table Wiping as a Service Robotics Task

Cleaning is a typical task in service robotics scenarios. In the following the wiping of a table is considered. Figure 9.1 shows the DLR-Lightweight-Robot-II with a simple rigid end-effector for a basic wiping application.

[1] Which are not part of this treatise.

C. Ott: Cartesian Imped. Cntrl. of Redund. & Flexible-Joint Robots, STAR 49, pp. 149–164, 2008.
springerlink.com

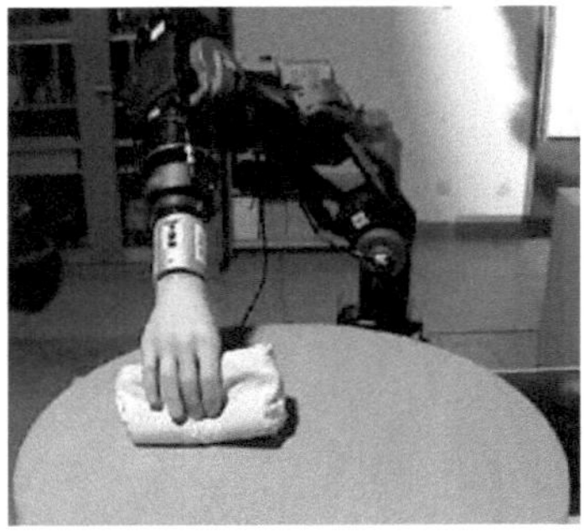

Fig. 9.1. DLR-Lightweight-Robot-II in a table wiping application

The whole application is split up into two steps. In the first step (*teaching step*) a human user can teach the robot a desired wiping movement. Therefore, a Cartesian impedance behavior is used for the robot in which the orientation stiffness is set high (100 Nm/rad), while the translational stiffness is set to zero such that the robot end-effector can be guided along a desired path. The end-effector path is stored and used in the second step (*replay step*) of the application as a virtual equilibrium point for the desired Cartesian impedance. In the replay step the desired stiffness is chosen such that the translational stiffness in a horizontal plane (along the table surface) is moderately high (2000 N/m). The translational stiffness corresponding to the vertical motion, instead, is set to zero. In order to ensure that the robot keeps in contact with the surface of the table a desired force of $f_d = -10$ N in vertical direction is added to the Cartesian force of the impedance controller.

For the implementation of the Cartesian impedance the controller from Chapter 7 was used. Since the orientational stiffness is set high in the teaching step and in the replay step, the Euler angle based stiffness from Section 3.5.2 is appropriate. If the robot comes in a singular configuration during the teaching step, the robot may get stuck at this point in the replay step. Therefore, the singularities must be avoided. This application indeed was the reason for implementing the singularity avoidance from Section 3.4. Furthermore, the projection based nullspace stiffness from Chapter 4 was used during the replay step. The nullspace stiffness was chosen quite low and was basically used in order to pull the joint configuration away from the joint limits.

In the basic setting of this application the robot simply wipes along the previously taught path. In more recent developments this application was then combined with a vision system and a path planning system in order to allow a

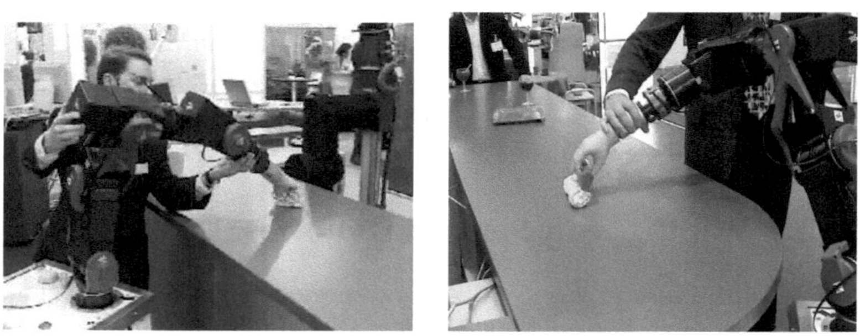

Fig. 9.2. Table wiping at the Hannover Fair 2002

more realistic setting. The vision system was used in order to detect a stain on the table surface. The taught path from the first step was then modified and generalized by the path planning system such that the commanded virtual equilibrium position for the impedance controller covers the detected stain [UASS04]. Furthermore, the DLR-Hand-II was used as an end-effector instead of the rigid end-effector from Figure 9.1.

The wiping application was presented at several faires[2] (Figure 9.2) and lab presentations with different settings. In these situations the robustness properties of the passivity based Cartesian impedance controller were very advantageous.

9.2 A Medical Scenario: Pedicle Screw Placement

In this scenario a medical robotics application is described for which the passivity based Cartesian impedance controller of Chapter 7 is used in combination with the quaternion based stiffness from Section 3.5.2. The presented application is topic of a joint project[3] between the Institute of Robotics and Mechatronics (German Aerospace Center, DLR e.V.), the company BrainLAB AG[4], the Technical University of Munich (TUM), and the BGU[5] Murnau.

The focus of this short description lies on the use of the Cartesian impedance controller rather than on the details of the application from a medical point of view. Notice also that this scenario actually is of interest for a new medical robot developed at the DLR. The whole application and work-flow was tested beforehand with the DLR-Lightweight-Robot-II. In this connection it is worth remarking that it is planned to use the same type of controllers as discussed herein also for the medical robot, since the mechanical construction is (from a control point of view) in some apects[6] quite similar to the DLR lightweight robots.

The purpose of the application is to support a surgeon during the placement of pedicle screws in spine surgery. In some cases, when a vertebra is broken or damaged, the structure can be stabilized by adjusting metal plates or pins. These plates are fixed to the spine by two screws which are drilled into the *pedicles* of adjacent vertebrae (see Figure 9.3). Figure 9.4 shows a typical placement of four pedicle screws on a spine model. The positions of the drilling points and the drilling axes usually are planned pre-operatively, based on CT[7] scans. During the operation the surgeon must inter-operatively find the pre-operatively planned poses on the spine of the patient. In a conventional surgery without navigation

[2] E.g. Hannover Fair 2002, Automatica Fair 2004.

[3] The BFS (*Bayrische Forschungsstiftung*) project *Naviped: Entwicklung eines navigationsgestützten halbautomatischen Robotersystems für die Platzierung von Schrauben an der menschlichen Wirbelsäule.*

[4] `http://www.brainlab.com`

[5] BGU ... Berufsgenossenschafliche Unfallklinik.

[6] The medical robot will for instance be equipped with Harmonic Drive® gears and joint torque sensors.

[7] CT ... Computer Tomography.

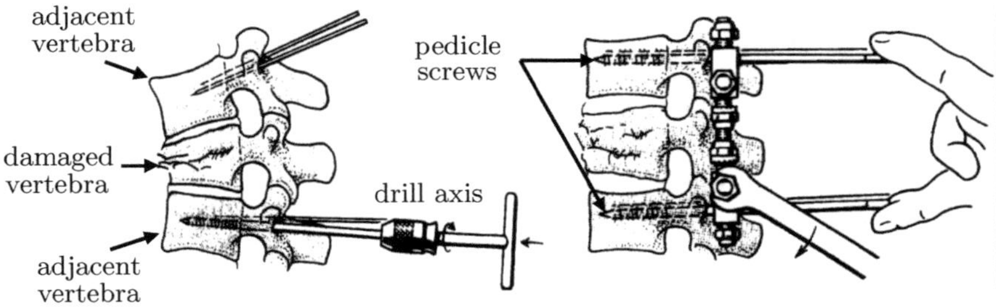

Fig. 9.3. Sketch of the pedicle screw placement (taken and modified from [MLSD99])

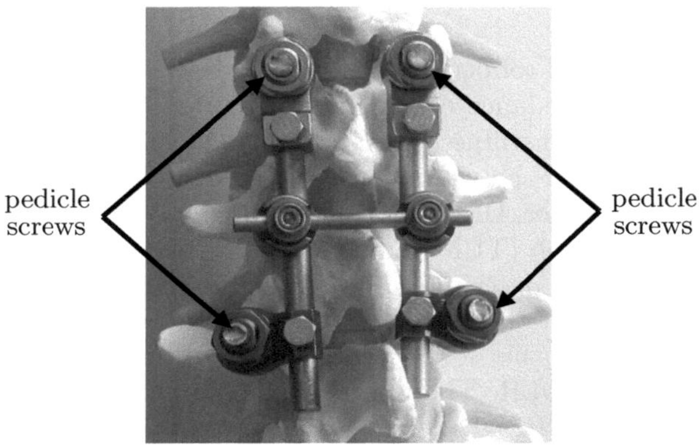

Fig. 9.4. Pedicle screw placement on a spine model. This picture was kindly provided by Ao. Univ.-Prof./USA Dr. med. Rudolf Beisse from the BGU Murnau.

system the surgeon iteratively adjusts the pose of the driller and detects its pose relative to the spine by taking several X-ray scans. In the presented navigation based robotics scenario, the robot should now be used to drill the holes for the screws at the previously planned poses. Markers are attached to a vertebra and to the robot (see Figure 9.5). Therefore, (after registration) the relative pose $\boldsymbol{h}_{tb}$ of the vertebra to the robot end-effector can be measured by a navigation system[8] which detects these markers. While the robot supports the placement of the driller at the planned pose on the pedicle, the drilling itself is done manually by the surgeon with a movable driller mounted on the robot end-effector. The drilling is performed manually because this allows the surgeon to feel the contact forces during the critical part of the operation. The use of a robot in this task has three main advantages. On the one hand the surgeon is less exposed to the X-ray radiation while verifying the planned position of the pedicle during the operation. On the other hand the robot prevents (at least to some extent) the driller from slipping off the pedicle. In combination with the navigation

[8] This navigation system is provided by the company BrainLAB AG.

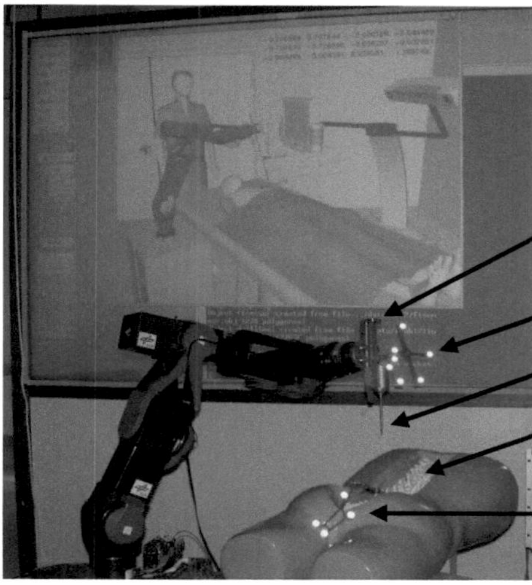

Fig. 9.5. Setup of the Naviped application

system it, furthermore, is possible to redeem displacements of the spine during the drilling task. Another important advantage of the navigation based robotic spine surgery is that the the hand-eye-coordination for the placement of the screw is easier than in a conventional surgery.

Given the difference $\boldsymbol{h}_{tb}$ between the vertebra and the tool frame, the pose of the vertebra can be computed by $\boldsymbol{h}_{sb}(t) = \boldsymbol{h}_{st}(\boldsymbol{q})\boldsymbol{h}_{tb}$. With regard to the control the vertebra position is considered as constant.

The work-flow of the operation is split up into several steps (see Figure 9.6). By the first five steps the robot end-effector (and hence the driller) is brought to a desired pose at the spine, where the actual drilling action can be performed. This desired pose is denoted as the *drill frame* in the following. The drilling action itself happens in step five. Afterwards the robot can be removed from the patient within step 6 and 7. In each of the steps a different Cartesian stiffness matrix is chosen.

- **Step 1:** In the first step the robot arm has zero Cartesian stiffness, and therefore can be moved freely by the user. In this mode the robot can be brought into an appropriate starting pose.
- **Step 2:** In this mode the desired stiffness matrix is defined such that the robot end-effector can be moved in a direction leading to the drill axis. The stiffness matrix has five positive eigenvalues and the remaining sixth eigenvalue is set to zero. Thus, the robot does not move autonomously to the drill frame, but can be moved towards the desired pose by the surgeon in an intuitive way. The orientation of the end-effector is adjusted automatically. When the distance of the end-effector to the drill axis gets smaller than a

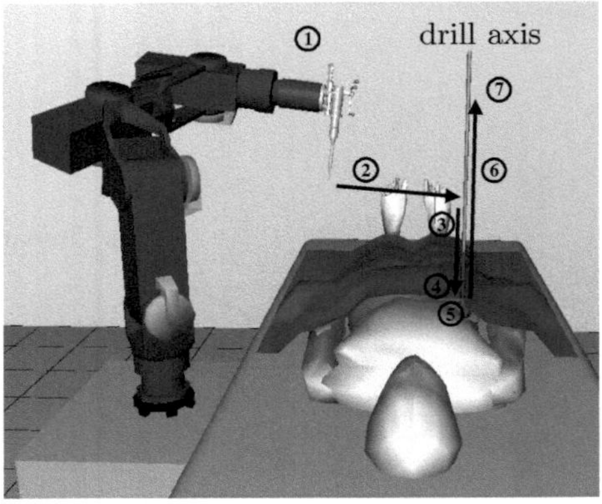

Fig. 9.6. Execution steps for the screw placement application

predefined threshold value of 10 mm, the sixth eigenvalue is set high and the virtual equilibrium pose is interpolated onto the drill axis.

- **Step 3:** Next the end-effector stiffness along the drilling axis is set to zero such that the robot can be guided along this axis. The virtual equilibrium pose is set onto the drill axis. The system changes into the next mode, when the distance of the end-effector to the drill point gets smaller than a predefined threshold value of 10 mm.
- **Step 4:** All Cartesian degrees-of-freedom have a large stiffness now and the virtual equilibrium point is interpolated to the origin of the drill frame. The robot then changes automatically into the next mode.
- **Step 5:** After reaching the drill frame, the actual drilling operation can be started. During the drilling operation a Cartesian position controller is used which is implemented similar to the Cartesian impedance controller but contains additionally an integral action in order to achieve a better position accuracy. The virtual desired pose of the robot is still adjusted according to the data from the navigation system such that displacements of the drill frame can be compensated (quasi-statically). In this mode the surgeon can perform the drilling action, with a driller mounted (movable along the drill axis) on the robot end-effector.
- **Step 6:** After finishing the drilling operation, in this mode the end-effector can be moved upward along the drilling axis by setting the stiffness along this axis to zero. The downward motion is prohibited by the use of a *virtual wall* which is moved upward together with the end-effector. When the distance of the end-effector from the drill point exceeds a predefined threshold value (of 150 mm) the next mode is activated.
- **Step 7:** In this mode the application has ended and the robot behavior is identical to step 1.

In this scenario one can see, how a complex task is split up into simple subtasks which use a different desired impedance behavior. The above described procedure presents the basic steps of the pedicle drilling application. Based on the presented impedance control concepts, this procedure can be adapted to the needs of the surgeon if desired.

9.3 Compliant Mobile Manipulation

In this book the Cartesian impedance control problem for robot arms is analyzed. In many applications it is desired to combine a robot arm with a mobile platform in order to extend the reachable workspace. In case that the mobile robot has only *holonomic* constraints[9] a straight-forward extension of the rigid body impedance control concepts from Chapter 3 to such systems is possible (see for instance [HK00, KYC+95]). In case that the mobile robot has also *non-holonomic* constraints the problem is more difficult.

In some applications, like for instance in the previously described table wiping, such a combination was also desired for the DLR lightweight robots. In this section a simple, i.e. rather pragmatic, solution is presented, how the Cartesian impedance controller can be combined with a non-holonomic mobile platform.

The considered system is shown in Figure 9.7. The mobile platform has four steerable wheels. Contrary to *caster-like wheels* the steering axes and the rotation axes of the wheels intersect. Therefore, the system is non-holonomic [Bla01]. In the following two different cases are considered. In the first scenario the Cartesian impedance is defined with respect to a world fixed coordinate system. In the second case it is, instead, defined with respect to a coordinate system fixed to the mobile platform.

Case 1: For the control of the combined platform-arm-system a rather pragmatic approach was followed. Notice that, for the ease of presentation, the following explanation is restricted to the translational motion, but it can be readily extended also to the end-effector orientation. Let the position of the mobile platform be denoted by $\boldsymbol{r}_{mob}$ and the end-effector position be given by $\boldsymbol{r}_{eef} = \boldsymbol{r}_{mob} + \boldsymbol{p}_{st}(\boldsymbol{q})$. The initial position of the arm with respect to the mobile platform is given by $\boldsymbol{p}_{st}(\boldsymbol{q}_0)$. For controlling the mobile platform a velocity controller is used, where the interaction between the arm and the mobile platform is considered as a disturbance. The controller of the robot arm, instead, is based on a Cartesian impedance controller which is designed for the end-effector position $\boldsymbol{r}_{eef}$ with respect to a world-fixed coordinate system. The set-point $\boldsymbol{v}_{mob,des}$ for the velocity controller of the mobile platform is then commanded by the simple feedback law $\boldsymbol{v}_{mob,des} = k_{mob}(\boldsymbol{p}_{st}(\boldsymbol{q}) - \boldsymbol{p}_{st}(\boldsymbol{q}_0))$, with a positive control gain k_{mob} such that the robot arm keeps away from its (platform-fixed) workspace boundaries. Thereby, it is assumed that in the initial configuration the end-effector is far from the workspace boundaries. The control gain k_{mob} of this simple feedback

[9] A set of *holonomic constraints* can be represented as a set of algebraic constraints on the configuration space. A set of constraints for the *generalized velocities* which is not equivalent to a set of holonomic constraints is said to be *non-holonomic*, see, e.g., [MLS94].

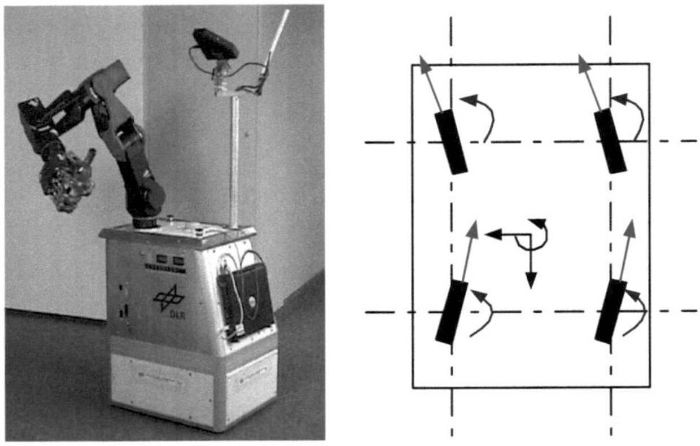

Fig. 9.7. Mobile manipulator system with the DLR-Lightweight-Robot-II and a non-holonomic mobile platform. The sketch on the right hand side shows the kinematic structure of the mobile platform.

law is set small such that the velocities of the platform are kept small and the motion of the platform can be neglected in the design of the Cartesian impedance controller.

The described control strategy of the platform-arm system is very simple and was motivated by the search for a pragmatic extension of the Cartesian impedance controller to a non-holonomic mobile manipulator system. In practice it turned out that this control strategy works very well in applications such as table wiping. From a theoretical point of view, the concept, however, is not sufficient, and indeed the extension of the Cartesian impedance controllers to systems with non-holonomic constraints can be considered as a challenging problem for future works.

Case 2: Figure 9.8 shows the mobile manipulator system in an application in which the Cartesian impedance controller is implemented with respect to a platform fixed coordinate system rather than to a world fixed coordinate system. The used Cartesian impedance controller was implemented according to the singular perturbation based approach (Chapter 5), with Cartesian coordinates chosen according to the modified Euler angle representation (see Section 3.5.2). The orientation stiffness values were chosen high (100 Nm/rad), while the translational stiffness values were chosen moderately stiff (500 N/m). The mobile platform was controlled as described above. This

Fig. 9.8. DLR-Lightweight-Robot-II mounted on a mobile platform in an application at the Hannover Fair 2002

results in a *joystick-like* behavior for the robot arm. By pulling or pushing the end-effector the mobile manipulator system can be moved around, since the velocity controller follows the deviations of the end-effector from the virtual equilibrium point which is now fixed to the mobile platform.

9.4 Opening a Door

In the next application the robot should open and move through a door. Therefore, the DLR-Lightweight-Robot-II is used in combination with a mobile platform and a dexterous four finger hand (DLR-Hand-II), see also [OBH+05]. The same system was previously used for several manipulation tasks with bottles and glasses in combination with a vision system for object recognition [HOB+04]. While those manipulation tasks were accomplished with a position controller for the arm, in the considered task of opening a door impedance control of the arm is much better. The use of an impedance controller (with moderate stiffness values) allows to keep the contact forces low, even if the environment is not exactly known.

Contrary to the modeling assumptions from Chapter 2 the robot arm is mounted on a mobile base in this application. It is, however, assumed that the motion of the mobile base is sufficiently slow and smooth such that the dynamical effects of the platform motion on the arm can be neglected. Moreover, the end-effector, i.e. the four finger hand, is included in the robot model in a simplified form as a rigid object. This means that the dynamical effects of the finger motion on the arm dynamics are also neglected. Therefore, the robot arm is controlled as a separate system and all dynamical couplings between the mobile base, the arm, and the artificial hand are neglected. Based on these modeling assumptions the presented Cartesian impedance controllers can be used without modifications. For the realization of this application the passivity based controller structure from Chapter 7 was used.

Before the door can be opened, the robot localizes[10] its relative position to the door hinge and the door handle is detected by computer vision. The arm approaches the handle based on coordinates from computer vision such that the hand is positioned at the right hand side of the handle as shown in Figure 9.9. Then the whole application is divided into two subtasks. The turning and opening of the door handle (subtask 1) and the moving through the door until it is sufficiently wide open (subtask 2). Furthermore, a joint level impedance controller (with moderate stiffness), with an equilibrium configuration as shown in Figure 9.9, is used for the four fingers of the artificial hand. The hand remains in this configuration for the whole application.

During the operation the contact force between the hand and the door handle shall be observed. Since the robot does not have a force-torque-sensor attached

[10] The localization of the mobile platform with respect to the door basically was achieved here by detecting some (known) markers on the door with a SICK laser scanner and fitting a model of the marker positions to those measurements.

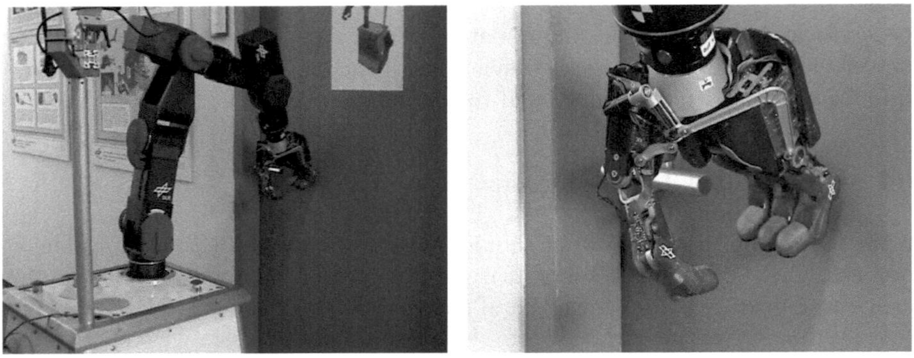

Fig. 9.9. Initial configuration for the door opening application

to the end-effector in this setup, the contact force must be estimated from other measurements. Therefore, a rough quasi-static estimation is done as follows. In a static equilibrium the external torques $\boldsymbol{\tau}_{ext}$ can be computed from the measurement of the joint torques $\boldsymbol{\tau}$ and the computation of the gravity torques $\boldsymbol{g}(\boldsymbol{q})$ via $\boldsymbol{\tau}_{ext} = \boldsymbol{\tau} - \boldsymbol{g}(\boldsymbol{q})$. If the external torques are assumed to be exerted via the generalized force $\boldsymbol{F}_{ext}$, then the relationship $\boldsymbol{J}(\boldsymbol{q})^T \boldsymbol{F}_{ext} = \boldsymbol{\tau} - \boldsymbol{g}(\boldsymbol{q})$ holds. This equation can be used for a *rough*, i.e. a quasi-static, estimation of $\boldsymbol{F}_{ext}$. The estimation of the generalized contact force allows to implement simple stopping conditions for establishing the contact. For simplicity the Euclidean norm of the translational components from $\boldsymbol{F}_{ext}$ is used here for these stopping conditions.

The turning and opening of the door handle is split up into several steps as depicted in Figure 9.10. For the actual impedance implementation the quaternion based stiffness from Section 3.5.2 was applied. With the first two steps the contact of the hand with the handle is established (step 1 and 2 in Figure 9.10). Each of these two steps finishes when a predefined threshold value of 10 N for the estimated contact force is exceeded. Then the actual handle turning motion follows (step 3), which ends when the contact force exceeds another threshold value of 30 N signalizing that the handle is turned to its maximum angle. After that the door is opened by a horizontal movement into the normal direction of the door surface (step 4) and the handle is turned back to its rest configuration (step 5 and 6). Now the first part of the door opening application is finished.

In the second part, the door shall be moved until it is fully open such that the mobile platform can drive through. For the robot arm a Cartesian impedance controller is used which allows to keep the door at a distance while the mobile platform drives through the door hinge. The desired Cartesian impedance behavior is defined via a potential function as follows. Let $\boldsymbol{r}(\boldsymbol{q}) = (r_x(\boldsymbol{q}), r_y(\boldsymbol{q}), r_z(\boldsymbol{q}))^T$ denote the position of the end-effector with respect to a platform fixed coordinate system. The desired impedance shall be designed such that the end-effector holds its initial height and is free to move along a horizontal circularly shaped

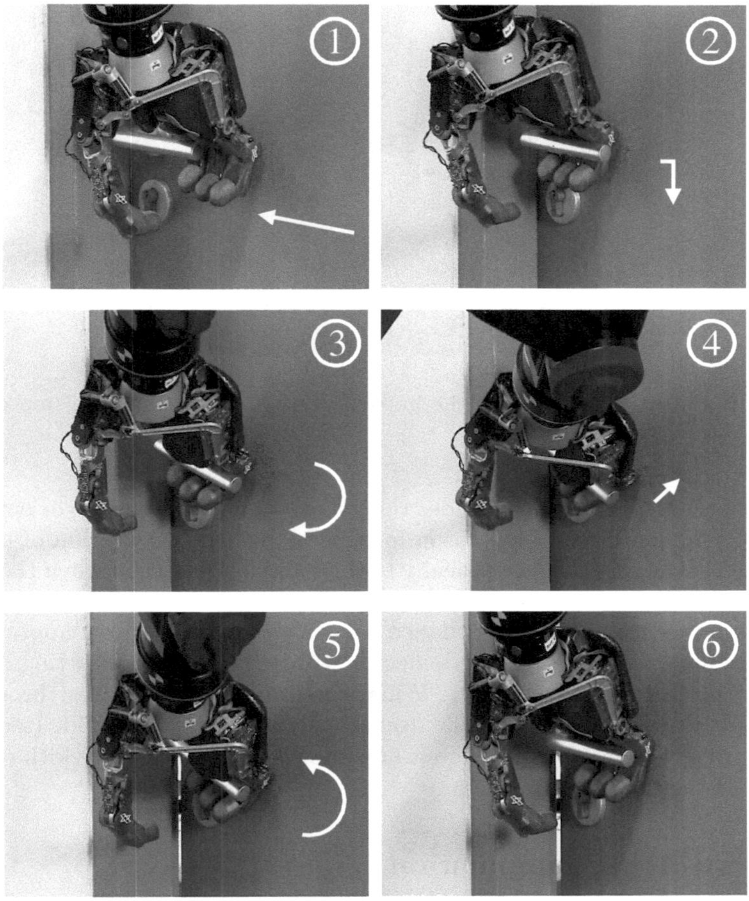

Fig. 9.10. Impedance controlled movements for the handle opening

path[11] fixed to the mobile platform. By choosing positive stiffness factors k_r and k_z this can easily be formulated via the potential function

$$V_d(\boldsymbol{q}) = k_r \left(\sqrt{r_x(\boldsymbol{q})^2 + r_y(\boldsymbol{q})^2} - d_0 \right)^2 + k_z (r_z(\boldsymbol{q}) - r_z(\boldsymbol{q}_0))^2 \ ,$$

$$d_0 = \sqrt{r_x(\boldsymbol{q}_0)^2 + r_y(\boldsymbol{q}_0)^2} \ ,$$

which replaces the potential function of the stiffness from Section 3.5. This potential function does not have an isolated minimum but has its minimum value along a circular path (see Figure 9.11). A translational stiffness can be implemented according to this potential function simply by computing the differential

[11] The center of this circular path lies in the initial height at a point above $\boldsymbol{r}_c$.

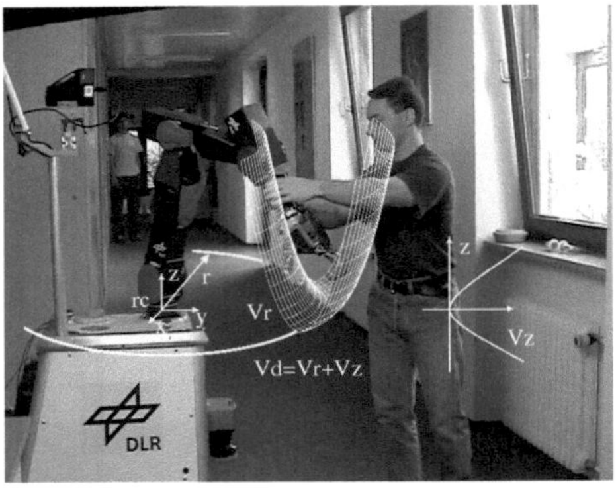

Fig. 9.11. Potential function for the implementation of the Cartesian impedance in the door opening application

of $V_d(\boldsymbol{q})$ with respect to[12] $\boldsymbol{q}$. Besides this translational stiffness, no orientational stiffness is implemented. The reason for this is that the orientation of the end-effector will then adjust automatically to the handle orientation when the handle is grasped by the artificial hand. By using the above potential function for the implementation of the desired impedance behavior together with an appropriately designed damping term[13] the end-effector will be kept at the starting distance from the platform fixed point $\boldsymbol{r}_c$. When the mobile robot moves through the door hinge, the arm therefore will automatically move the door aside (see Figure 9.12). Notice that this works without any knowledge of the door width or of the position of the door rotation axis.

9.5 Two-Arm Manipulation

Finally, as a last application the two-armed manipulation for DLR's humanoid manipulator "Justin" [OEF+06] is presented. This system consists of two DLR-Lightweight-Robots-III mounted on a movable torso, two artificial hands, and a sensor head mounted on a two degrees-of-freedom pan-tilt-unit (see Fig. 9.13). Overall, Justin has 43 degrees-of-freedom. Also, all the joints of the arms, the hands, and the torso are equipped with joint torque sensors in addition to the common motor position sensors. The following discussion will be restricted to the control of the arms and the torso. Further extensions which take full advantage of the articulated hands can be found in [WmOH06, WmOH07].

The considered task is the coordinated grasping and manipulation of a large object by both arms. As an underlying control approach, the controller from

[12] But notice that in the implementation of this impedance behavior within the framework of the passivity based controller from Chapter 7 the motor side position $\boldsymbol{\theta}$ is used instead of the link side position $\boldsymbol{q}$. See Chapter 7 for more details on this.

[13] The design of the damping in this application is not crucial and therefore is omitted here for the ease of presentation.

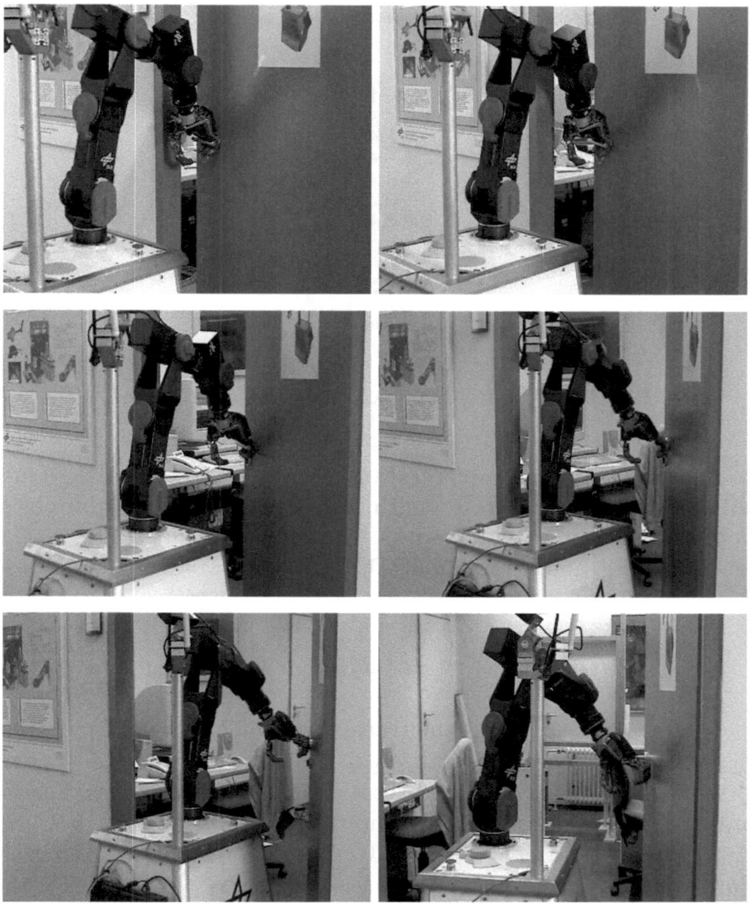

Fig. 9.12. Opening of the door

section 7.2.1 is applied. However, for the two-arm system it is useful to consider a more general impedance structure than a single spatial spring connected to the end-effector. Therefore, a somewhat more abstract viewpoint is adopted. In (7.10) the stiffness term is now replaced by the differential of a more general potential function $V_{imp}(\boldsymbol{\theta})$ and the damping is realized via an appropriate damping matrix $\boldsymbol{D}_j(\boldsymbol{\theta})$ in joint space, i.e. instead of (7.10) the control law

$$\boldsymbol{u}_{imp} = -\frac{\partial V_{imp}(\boldsymbol{\theta})}{\partial \boldsymbol{\theta}} - \boldsymbol{D}_j(\boldsymbol{\theta})\dot{\boldsymbol{\theta}} \tag{9.1}$$

is used. For the design of the potential function $V_{imp}(\boldsymbol{\theta})$ the potential function $V_S(\cdot)$ according to one spatial spring shall be used as a basic building block. This potential function $V_S(\boldsymbol{h}_{s1}, \boldsymbol{h}_{s2}, \mathcal{K})$ depends on two frames $\boldsymbol{h}_{s1} := [\boldsymbol{R}_{s1}, \boldsymbol{p}_{s1}] \in SE(3)$ and $\boldsymbol{h}_{s2} := [\boldsymbol{R}_{s2}, \boldsymbol{p}_{s2}] \in SE(3)$ between which the spring

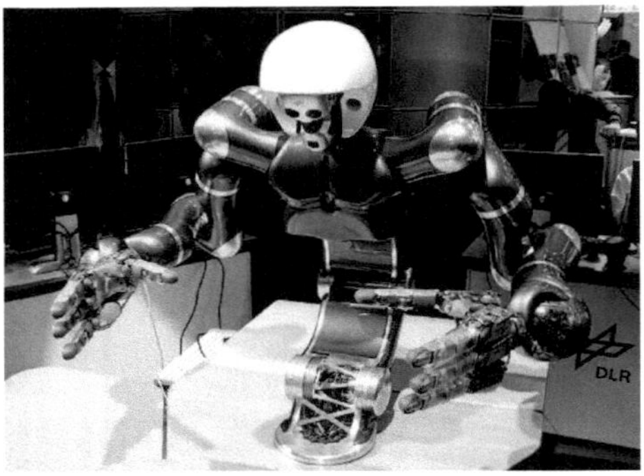

Fig. 9.13. DLR's humanoid manipulator "Justin"

is acting, and also on some configuration-independent internal parameters $\mathcal{K}$, like the stiffness values or the rest length. Appropriate implementations for such spatial springs can found in the works by Fasse and Zhang[FB97], Caccavale and Natale[CNSV99], and Stramigioli[SD01]. For the two-arm manipulation with Justin this potential function was chosen as

$$V_S(\boldsymbol{h}_{s1}, \boldsymbol{h}_{s2}, \mathcal{K}) := \frac{1}{2}\boldsymbol{e}_{12}^T \boldsymbol{K}_t \boldsymbol{e}_{12} + 2\boldsymbol{\epsilon}_{12}^T \boldsymbol{K}_r \boldsymbol{\epsilon}_{12} \ , \tag{9.2}$$

where $\boldsymbol{e}_{12} := \boldsymbol{p}_{s1} - \boldsymbol{p}_{s2} - \boldsymbol{p}_{12,0}$ is the translational deviation of the displacement $\boldsymbol{p}_{s1} - \boldsymbol{p}_{s2}$ from its desired rest length $\boldsymbol{p}_{12,0}$. The matrices $\boldsymbol{K}_t \in \mathbb{R}^{3\times3}$ and $\boldsymbol{K}_r \in \mathbb{R}^{3\times3}$ represent the translational and rotational stiffness matrices. Furthermore, $\boldsymbol{\epsilon}_{12}$ denotes the unit quaternion representation of $\boldsymbol{R}_{v1,v2} := \boldsymbol{R}_{1,v1}^T \boldsymbol{R}_{s1}^T \boldsymbol{R}_{s2} \boldsymbol{R}_{2,v2}$ with $\boldsymbol{R}_{1,v1} \in SO(3)$ and $\boldsymbol{R}_{2,v2} \in SO(3)$ as the "rotational rest length" of the spring. Consequently, $\mathcal{K}$ contains the stiffness matrices $\boldsymbol{K}_t, \boldsymbol{K}_r$ as well as the translational and rotational rest length $\boldsymbol{p}_{12,0}, \boldsymbol{R}_{1,v1}, \boldsymbol{R}_{2,v2}$. For the design of the damping matrix $\boldsymbol{D}_j(\boldsymbol{\theta})$ one can again utilize the method discussed in Section 3.3.

The forward kinematics mappings for the right and the left arm with respect to the base of the torso can be computed by (2.3) and will be denoted by $\boldsymbol{h}_{sr}(\cdot)$ and $\boldsymbol{h}_{sl}(\cdot)$.

Based on the above definition of $V_S(\cdot)$ an impedance behavior for a two-arm manipulator could simply be designed by using two spatial springs $\mathcal{K}_r$ and $\mathcal{K}_l$ for the right and the left arm, which connect the end-effectors to the virtual equilibrium frames $\boldsymbol{h}_{r,d}$ and $\boldsymbol{h}_{l,d}$, respectively. If the springs are implemented as complete 6D-springs (i.e. with full rank translational and rotational stiffness matrices), then the complete Cartesian motion of the arms can be influenced already via only these two springs. For some applications it is useful if some part of the motion is instead defined via an additional coupling spring $\mathcal{K}_c$ between the arms. Clearly, in such a configuration (as shown in Fig. 9.14) both the rest lengths as well as the stiffness values of the individual springs and the coupling spring should be chosen in a *compatible* way such that the springs do not interfere with each other. The complete potential function for implementing such a two-arm impedance via (9.1) is given by

$$V_{imp}(\boldsymbol{\theta}) = V_S(\boldsymbol{h}_{sr}(\boldsymbol{\theta}), \boldsymbol{h}_{r,d}, \mathcal{K}_r) + V_S(\boldsymbol{h}_{sl}(\boldsymbol{\theta}), \boldsymbol{h}_{l,d}, \mathcal{K}_l) + V_S(\boldsymbol{h}_{sr}(\boldsymbol{\theta}), \boldsymbol{h}_{sl}(\boldsymbol{\theta}), \mathcal{K}_c) \ . \tag{9.3}$$

With an impedance structure as depicted in Fig. 9.14 one can thus implement different behaviors, ranging from an independent control of the arms via $\mathcal{K}_r$ and $\mathcal{K}_l$ to a pure coupling based on $\mathcal{K}_c$. One disadvantage of this structure, however, is the fact that the springs should not be designed independently, but in a compatible way as explained above.

Instead of the two individual springs attached to the end-effectors one can define a virtual object frame $\boldsymbol{h}_{so}(\boldsymbol{h}_{sr}(\boldsymbol{\theta}), \boldsymbol{h}_{sl}(\boldsymbol{\theta}))$ depending on the two end-effector frames as sketched in Fig. 9.15. Such a virtual object was used by Natale in [Nat03]. The object frame was chosen as $\boldsymbol{h}_{so} = [\boldsymbol{R}_{so}, \boldsymbol{p}_{so}]$ with $\boldsymbol{p}_{so} = \frac{1}{2}(\boldsymbol{p}_{sr} + \boldsymbol{p}_{sl})$

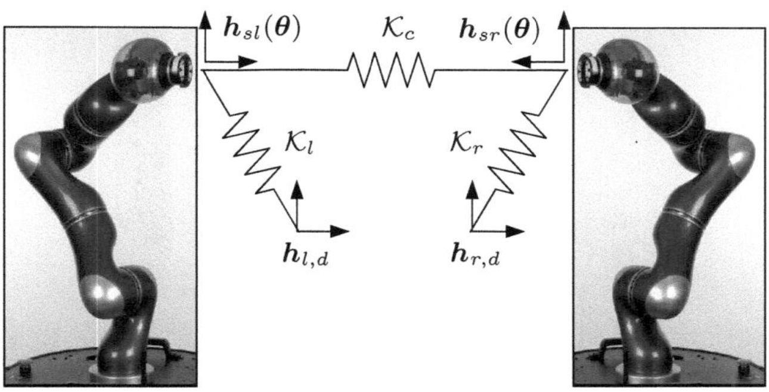

Fig. 9.14. Two-arm impedance behavior by adding a coupling spring to the individual arm springs

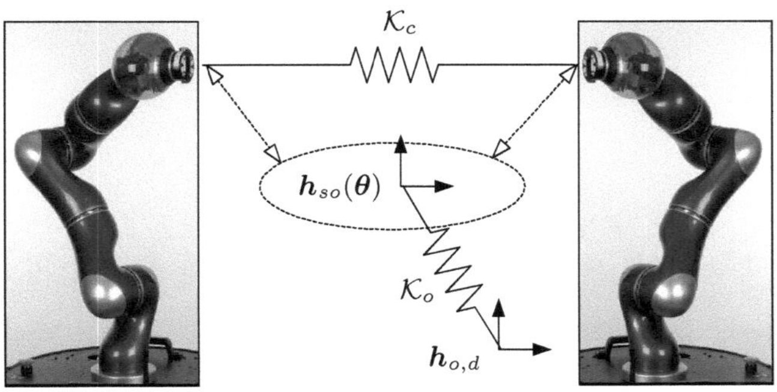

Fig. 9.15. Two-arm impedance behavior by combining a coupling spring with an object level spring

and $\boldsymbol{R}_{so} = \boldsymbol{R}_{sr}\bar{\boldsymbol{R}}_{rl}$, where $\bar{\boldsymbol{R}}_{rl}$ denotes *a half*[14] of the rotation matrix $\boldsymbol{R}_{sr}^{-1}\boldsymbol{R}_{sl}$. This virtual object is connected via a spatial spring $\mathcal{K}_o$ to a virtual equilibrium pose $\boldsymbol{h}_{o,d}$. In combination with the coupling stiffness, one can intuitively define an impedance behavior which is useful for grasping large objects with two arms. The relevant potential function is given by

$$V_{imp}(\boldsymbol{\theta}) = V_S(\boldsymbol{h}_{so}(\boldsymbol{h}_{sr}(\boldsymbol{\theta}), \boldsymbol{h}_{sl}(\boldsymbol{\theta})), \boldsymbol{h}_{o,d}, \mathcal{K}_o) + V_S(\boldsymbol{h}_{sr}(\boldsymbol{\theta}), \boldsymbol{h}_{sl}(\boldsymbol{\theta}), \mathcal{K}_c) \ . \tag{9.4}$$

As an example task for these impedance behaviors the grasping and manipulation of a cylindrical trash bin has been implemented. In the experiments the coupling spring $\mathcal{K}_c$ realized a translational stiffness acting only in a horizontal plane. The springs $\mathcal{K}_l$ and $\mathcal{K}_r$ on the other hand implemented a translational stiffness according to the vertical movement of the arms, as well as an orientation stiffness. For the hands simple joint impedance controllers were used. That way the *grasping* was performed basically by the arms.

Fig. 9.16. The humanoid manipulator Justin manipulating a cylindrical trash bin

[14] This *halving* of the rotation matrix can easily be done using an angle-axis representation.

10 Controller Comparison and Conclusions

In Chapter 5 to 7 several control approaches for implementing a Cartesian impedance behavior with a flexible joint robot were discussed in detail. An experimental evaluation of these controllers was already presented in Chapter 8. Finally, an overall comparison highlighting the respective advantages and disadvantages of the presented control approaches shall be given in this concluding Chapter. Table 10.1 summarizes the properties of the different controllers. Such a comparison clearly is in some aspects subjective and therefore the following few comments on the ratings are necessary.

Table 10.1. Comparison of the presented controllers

	Singular Perturbation	Decoupling & Backstepping	Passivity Based
Model	Sing. Pert.	Reduced	Reduced (+D)
Theor. Justification	-	+	+
Robustness	0	-	+
Tracking & IS[1]	+	+	-
Low Complexity	+	-	+
State Variables	$\boldsymbol{\tau}, \dot{\boldsymbol{\tau}}, \boldsymbol{q}, \dot{\boldsymbol{q}}$	$\boldsymbol{\tau}, \dot{\boldsymbol{\tau}}, \boldsymbol{q}, \dot{\boldsymbol{q}}$	$\boldsymbol{\theta}, \dot{\boldsymbol{\theta}}, \boldsymbol{q}, \dot{\boldsymbol{q}}$
Feedback Variables	$\boldsymbol{\tau}, \dot{\boldsymbol{\tau}}, \boldsymbol{q}, \dot{\boldsymbol{q}}$	$\boldsymbol{\tau}, \dot{\boldsymbol{\tau}}, \boldsymbol{q}, \dot{\boldsymbol{q}}, \ddot{\boldsymbol{q}}, \boldsymbol{q}^{(3)}$	$\boldsymbol{\tau}, (\dot{\boldsymbol{\tau}}), \boldsymbol{\theta}, \dot{\boldsymbol{\theta}}$

Model: In the singular perturbation approach the controller design is only based on an approximative model in which the inner torque control loop is neglected for the design of the impedance controller. The other methods, instead, refer to the *reduced model* of a flexible joint robot. For the passivity based controller even an extension to a model including joint damping is straightforward.

Theoretical Justification: This rating is a consequence of the considered model assumptions. Here, the singular perturbation based controller clearly has its main disadvantage.

[1] IS ... Inertia Shaping.

C. Ott: Cartesian Imped. Cntrl. of Redund. & Flexible-Joint Robots, STAR 49, pp. 165–167, 2008.
springerlink.com

Robustness: The structural properties of the passivity based controller imply robustness to modeling uncertainties with respect to the link side dynamics. For the decoupling and the backstepping based controller no robustness analysis can be done, if the joint acceleration and the jerk cannot be measured but must be calculated from other measurements. In the regulation case the singular perturbation based controller has similar robustness properties as the passivity based controller. But this holds only for the approximative singular perturbation model of the system.

Tracking and Inertia Shaping: The extension to the tracking case and inertia shaping can be done in a similar way for the singular perturbation based controller and for the decoupling and backstepping based controllers, but leads to more difficult controllers. The passivity based controller originally was designed for the regulation case. An extension for the tracking case was given, but here the controller needs additionally the measurements of the link accelerations and the jerks.

Low Complexity: The complexity rating refers to the implementation costs of the controller. Here, the decoupling and the backstepping based controllers have their main disadvantage due to the additional feedback of the link accelerations and the jerks. While these signals can in principle be computed based on the other measurement data and on the model parameters, the computation is critical from a practical point of view.

State Variables and Feedback Variables: In the analysis of the singular perturbation based, the decoupling based, and the backstepping controllers the torque and the link side position are used. These controllers are (implicitly) designed in form of a cascaded control structure with an inner torque control loop and an outer impedance control loop. The analysis of the passivity based controller on the other hand uses the state variables of the motors and the links. The term *feedback variables*, instead, refers to the variables which are actually needed for an implementation of the controllers. Since the link side accelerations and jerks usually cannot be measured, they must be computed based on the model and the state variables.

To sum it up, one can say that the decoupling based and the backstepping controller are in practice difficult to be implemented, but they represent a complete solution to the impedance control problem for the (reduced) flexible joint robot model including the tracking case and inertia shaping. From a practical point of view, instead, the singular perturbation approach and the passivity based controller are to be favored due to their low complexity and their robustness properties. While the singular perturbation based controller allows a straightforward extension to the tracking case and also inertia shaping, its theoretical justification on the basis of *Tychonov's theorem* is weak. The passivity based controller, on the other hand, clearly has its main advantages in the theoretical justification and the low complexity.

With the presented controllers several solutions to the Cartesian impedance control problem for flexible joint robots are available and a wide range of applications can be handled with these controllers. Some typical applications were discussed in Chapter 9.

If the desired impedance is defined differently than considered in this work, it will often be possible to modify the controllers accordingly by only minor changes, similar as it was exemplified in Section 9.1. The problem of how to define the desired impedance for a given task, however, is an interesting problem for itself.

References

[Arn89] Arnold, V.I.: Mathematical Methods of Classical Mechanics, 2nd edn. Springer, Heidelberg (1989)

[AS01] Albu-Schäffer, A.: Regelung von Robotern mit elastischen Gelenken am Beispiel der DLR-Leichtbauarme. PhD thesis, Technische Universität München (2001)

[ASFS+04] Albu-Schäffer, A., Fischer, M., Schreiber, G., Schoeppe, F., Hirzinger, G.: Soft robotics: What cartesian stiffness can we obtain with passively compliant, uncoupled joints? In: IEEE/RSJ International Conference on Intelligent Robots and Systems, pp. 3295–3301 (2004)

[ASH00] Albu-Schäffer, A., Hirzinger, G.: State feedback controller for flexible joint robots: A globally stable approach implemented on dlr's light-weight robots. In: IEEE/RSJ International Conference on Intelligent Robots and Systems, pp. 1087–1094 (2000)

[ASH01a] Albu-Schäffer, A., Hirzinger, G.: A globally stable state-feedback controller for flexible joint robots. Journal of Advanced Robotics, Special Issue: Selected Papers from IROS 2000 15(8), 799–814 (2001)

[ASH01b] Albu-Schäffer, A., Hirzinger, G.: Parameter identification and passivity based joint control for a 7dof torque controlled light weight robot. In: IEEE International Conference on Robotics and Automation, pp. 1087–1093 (2001)

[ASmOH07] Albu-Schäffer, A., Ott, C., Hirzinger, G.: A unified passivity-based control framework for position, torque and impedance control of flexible joint robots. The International Journal of Robotics Research 26(1), 23–39 (2007)

[ASOFH03] Albu-Schäffer, A., Ott, C., Frese, U., Hirzinger, G.: Cartesian impedance control of redundant robots: Recent results with the dlr-light-weight-arms. In: IEEE International Conference on Robotics and Automation, pp. 3704–3709 (2003)

[ASOH04a] Albu-Schäffer, A., Ott, C., Hirzinger, G.: Passivity based cartesian impedance control for flexible joint manipulators. In: IFAC Symposium on Nonlinear Control Systems (NOLCOS) (2004)

[ASOH04b] Albu-Schäffer, A., Ott, C., Hirzinger, G.: A passivity based cartesian impedance controller - part II: Full state feedback, impedance design and experiments. In: IEEE International Conference on Robotics and Automation, pp. 2666–2672 (2004)

[Bai85] Baillieul, J.: Kinematic programming alternatives for redundant manipulators. In: IEEE International Conference on Robotics and Automation, pp. 722–728 (1985)

[BCLM01] Bloch, A.M., Chang, D.E., Leonard, N.E., Marsden, J.E.: Controlled lagrangians and the stabilization of mechanical systems ii: Potential shaping. IEEE Transactions on Automatic Control 46, 1556–1571 (2001)

[BK00] Bruyninckx, H., Khatib, O.: Gauss' principle and the dynamics of redundant and constrained manipulators. In: IEEE International Conference on Robotics and Automation, pp. 2563–2568 (2000)

[Bla01] Blaimschein, P.: Modellierung und Regelung mobiler Robotersysteme. PhD thesis, Johannes Kepler Universität Linz (2001)

[BLM00] Bloch, A.M., Leonard, N.E., Marsden, J.E.: Controlled lagrangians and the stabilization of mechanical systems i: The first matching theorem. IEEE Transactions on Automatic Control 45(12), 2253–2270 (2000)

[BM92] Basile, G., Marro, G.: Controlled and Conditioned Invariants in Linear Systems Theory. Prentice Hall, Englewood Cliffs (1992)

[BOL95] Brogliato, B., Ortega, R., Lozano, R.: Global tracking controllers for flexible-joint manipulators: a comparative study. Automatica 31(7), 941–956 (1995)

[Boo03] Boothby, W.M.: An Introduction to Differentiable Manifolds and Riemannian Geometry, 2nd edn. Academic Press, Elsevier Science (2003)

[BOvdS02] Blankenstein, G., Ortega, R., van der Schaft, A.J.: The matching conditions of controlled lagrangians and interconnection and damping assignment passivity based control. International Journal of Control 75(9), 645–665 (2002)

[BRT01] Bicchi, A., Lodi Rizzini, S., Tonietti, G.: Compliant design for intrinsic safety: General issues and preliminary design. In: IEEE/RSJ International Conference on Intelligent Robots and Systems, pp. 1864–1869 (2001)

[BS98] Beccari, G., Stramigioli, S.: Impedance control as merging mechanism for a behavior-based architecture. In: IEEE International Conference on Robotics and Automation, pp. 1429–1434 (1998)

[BTBP03] Bicchi, A., Tonietti, G., Bavaro, M., Piccigallo, M.: Variable stiffness actuators for fast and safe motion control. In: Siciliano, B., Khatib, O., Groen, F.C.A. (eds.) Proceedings of ISRR 2003, Springer Tracts in Advanced Robotics (STAR). Springer, Heidelberg (2003)

[CH89] Colgate, E., Hogan, N.: An analysis of contact instability in terms of passive physical equivalents. In: IEEE International Conference on Robotics and Automation, pp. 404–409 (1989)

[CK95] Chang, K.-S., Khatib, O.: Manipulator control at kinematic singularities: A dynamically consistent strategy. In: IEEE/RSJ International Conference on Intelligent Robots and Systems, pp. 84–88 (1995)

[CK02] Chen, S.F., Kao, I.: Geometric approach to the conservative congruence transformation (cct) for robotic stiffness control. In: IEEE International Conference on Robotics and Automation, pp. 544–549 (2002)

[CNSV99] Caccavale, F., Natale, C., Siciliano, B., Villani, L.: Six-dof impedance control based on angle/axis representations. IEEE Transactions on Robotics and Automation 15(2), 289–299 (1999)

[CW93] Chen, Y.-C., Walker, I.D.: A consistent null-space based approach to inverse kinematics of redundant robots. In: IEEE International Conference on Robotics and Automation, pp. 374–381 (1993)

[DL00] De Luca, A.: Feedforward/feedback laws for the control of flexible robots. In: IEEE International Conference on Robotics and Automation, pp. 233–240 (2000)

[DLL98] De Luca, A., Lucibello, P.: A general algorithm for dynamic feedback linearization of robots with elastic joints. In: IEEE International Conference on Robotics and Automation, pp. 504–510 (1998)

[DLSZ05] De Luca, A., Siciliano, B., Zollo, L.: Pd control with on-line gravity compensation for robots with elastic joints: Theory and experiments. Automatica 41(10), 1809–1819 (2005)

[DLT96] De Luca, A., Tomei, P.: Theory of Robot Control. In: Elastic Joints, pp. 179–218. Springer, London (1996)

[Duf90] Duffy, J.: The fallacy of modern hybrid control theory that is based on orthogonal complements of twist and wrench spaces. Journal of Robotic Systems 7(2), 139–144 (1990)

[dWSB96] de Wit, C., Siciliano, B., Bastin, G. (eds.): Theory of Robot Control. Communications and Control Engineering. Springer, London (1996)

[Fas97] Fasse, E.D.: On the spatial compliance of robotic manipulators. ASME Journal of Dynamic Systems, Measurement, and Control 119, 839–844 (1997)

[FB97] Fasse, E.D., Broenink, J.F.: A spatial impedance controller for robotic manipulation. IEEE Transactions on Robotics and Automation 13(4), 546–556 (1997)

[Fea04] Featherstone, R.: An empirical study of the joint space inertia matrix. The International Journal of Robotics Research 23(9), 856–871 (2004)

[FH95] Fasse, E.D., Hogan, N.: Control of physical contact and dynamic interaction. In: 7th International Symposium of Robotics Research, pp. 28–38 (1995)

[FLMR95] Fliess, M., Lévine, J., Martin, P., Rouchon, P.: Flatness and defect of nonlinear systems: introductory theory and examples. International Journal of Control 61(6), 1327–1361 (1995)

[Fra97] Frankel, T.: The Geometry of Physics. Cambridge University Press, Cambridge (1997)

[Ge96] Ge, S.S.: Adaptive controller design for flexible joint manipulators. Automatica 32(2), 273–278 (1996)

[Gei03] Geist, C.: Untersuchung des Einsatzes von Gelenk-Beschleunigungssensoren zur Regelung eines elastischen Robotergelenkes. Master's thesis, Fachhochschule Niederrhein (2003)

[GFG99] Goldsmith, P.B., Francis, B.A., Goldenberg, A.A.: Stability of hybrid position/force control applied to manipulators with flexible joints. International Journal of Robotics and Automation 14(4), 146–160 (1999)

[GHS89] Ghorbel, F., Hung, J.Y., Spong, M.W.: Adaptive control of flexible joint manipulators. In: IEEE International Conference on Robotics and Automation, pp. 1188–1193 (1989)

[Har97] Harville, D.: Matrix Algebra from a Statistician's Perspective. Springer, Heidelberg (1997)

[HG93] Hsu, L., Guenther, R.: Variable structure adaptive cascade control of multi-link robot manipulators with flexible joints: the case of arbitrary uncertain flexibilities. In: IEEE International Conference on Robotics and Automation, pp. 340–345 (1993)

[HHS89] Hsu, P., Hauser, J., Sastry, S.: Dynamic control of redundant manipulators. Journal of Robotic Systems 6(2), 133–148 (1989)

[HJ90] Horn, R.A., Johnson, C.R.: Matrix Analysis. Cambridge University Press, Cambridge (1990)

[HK00] Holmberg, R., Khatib, O.: Development and control of a holonomic mobile robot for mobile manipulation tasks. The International Journal of Robotic Research 19(11), 1066–1074 (2000)

[HOB+04] Hillenbrand, U., Ott, C., Brunner, B., Borst, C., Hirzinger, G.: Towards service robots for the human environment: the robutler. In: Mechatronics and Robotics (MECHROB), pp. 1497–1502 (2004)

[Hog85a] Hogan, N.: Impedance control: An approach to manipulation, part I - theory. ASME Journal of Dynamic Systems, Measurement, and Control 107, 1–7 (1985)

[Hog85b] Hogan, N.: Impedance control: An approach to manipulation, part II - implementation. ASME Journal of Dynamic Systems, Measurement, and Control 107, 8–16 (1985)

[Hog85c] Hogan, N.: Impedance control: An approach to manipulation, part III - applications. ASME Journal of Dynamic Systems, Measurement, and Control 107, 17–24 (1985)

[HV91] Huang, M.Z., Varma, H.: Optimal rate allocation in kinematically redundant manipulators - the dual projection method. In: IEEE International Conference on Robotics and Automation, pp. 702–707 (1991)

[IKO96] Iggidr, A., Kalitine, B., Outbib, R.: Semidefinite lyapunov functions: Stability and stabilization. Mathematics of Control, Signals, and Systems 9, 95–106 (1996)

[KG97] Kircanski, N.M., Goldenberg, A.A.: An experimental study of nonlinear stiffness, hysteresis, and friction effects in robotic joints with harmonic drives and torque sensors. The International Journal of Robotics Research 16(2), 214–239 (1997)

[Kha87] Khatib, O.: A unified approach for motion and force control of robot manipulators: The operational space formulation. IEEE Journal of Robotics and Automation 3(1), 1115–1120 (1987)

[Kha02] Khalil, H.K.: Nonlinear Systems, 3rd edn. Prentice Hall, Englewood Cliffs (2002)

[KKO86] Kokotović, P.V., Khalil, H.K., O'Reilly, J.: Singular Perturbation Methods in Control: Analysis and Design. Academic Press, London (1986)

[Koe01] Koeppe, R.: Robot Compliant Motion based on Human Skill. PhD thesis, ETH Zürich (2001)

[KS98] Kelly, R., Santibanez, V.: Global regulation of elastic joint robots based on energy shaping. IEEE Transactions on Automatic Control 43(10), 1451–1456 (1998)

[KS02a] Kugi, A., Schlacher, K.: Analyse und synthese nichtlinearer dissipativer systeme: Ein überblick (teil 1). at - Automatisierungstechnik 2, 63–69 (2002)

[KS02b] Kugi, A., Schlacher, K.: Analyse und synthese nichtlinearer dissipativer systeme: Ein überblick (teil 2). at - Automatisierungstechnik 3, 103–111 (2002)

[Kug01] Kugi, A.: Non-linear Control Based on Physical Models. Springer, Heidelberg (2001)

[KYC+95] Khatib, O., Yokoi, K., Chang, K., Ruspini, D., Holmberg, R., Casal, A., Baader, A.: Force strategies for cooperative tasks in multiple mobile manipulation systems. In: Int. Symp. of Robotics Research (1995)

[LB92] Lozano, R., Brogliato, B.: Adaptive control of robot manipulators with flexible joints. IEEE Transactions on Automatic Control 37(2), 174–181 (1992)

[LD88] Lipkin, H., Duffy, J.: Hybrid twist and wrench control for a robotic manipulator. ASME Journal of Mechanisms, Transmissions, and Automation in Design 110, 138–144 (1988)

[LG95] Lin, T., Goldenberg, A.A.: Robust adaptive control of flexible joint robots with joint torque feedback. In: IEEE International Conference on Robotics and Automation, pp. 1229–1234 (1995)

[LG96] Lin, T., Goldenberg, A.A.: A unified approach to motion and force control of flexible joint robots. In: IEEE International Conference on Robotics and Automation, pp. 1115–1120 (1996)

[LKCC91] Laurin-Kovitz, K.F., Colgate, J.E., Carnes, S.D.R.: Design of programmable passive impedance. In: IEEE International Conference on Robotics and Automation, pp. 1476–1481 (1991)

[LO95] Loria, A., Ortega, R.: On tracking control of rigid and flexible joint robots. Applied Mathematics and Computer Science, special issue on Mathematical Methods in Robotics 5(2), 101–113 (1995)

[Lor01] Loria, A.: Cascaded nonlinear time-varying systems: analysis and design. Minicourse at the Congreso Internacional de Computacion, Cd. Mexico (2001)

[Mas81] Mason, M.T.: Compliance and force control for computer controlled manipulators. IEEE Transactions on Systems, Man, and Cybernetics 11(6), 418–432 (1981)

[Mil63] Milnor, J.: Morse Theory. Princton University Press (1963)

[MLS94] Murray, R.M., Li, Z., Sastry, S.S.: A Mathematical Introduction to Robotic Manipulation. CRC Press, Boca Raton (1994)

[MLSD99] Middelanis, I., Liehn, M., Steinmüller, L., Döhler, R.: OP-Handbuch - Grundlagen, Instrumentarium, OP-Ablauf, 2nd edn. Springer, Heidelberg (1999)

[Nat03] Natale, C.: Interaction Control of Robot Manipulators: Six-Degrees-of-Freedom Tasks. Springer Tracts in Advanced Robotics (STAR), vol. 3. Springer, Heidelberg (2003)

[NSV99] Natale, C., Siciliano, B., Villani, L.: Spatial impedance control of redundant manipulators. In: IEEE International Conference on Robotics and Automation, pp. 1788–1793 (1999)

[NT92] Nicosia, S., Tomei, P.: A method to design adaptive controllers for flexible joint robots. In: IEEE International Conference on Robotics and Automation, pp. 701–706 (1992)

[NT95] Nicosia, S., Tomei, P.: A global output feedback controller for flexible joint robots. Automatica 31(10), 1465–1469 (1995)

[OASH02] Ott, C., Albu-Schäffer, A., Hirzinger, G.: Comparison of adaptive and nonadaptive tracking control laws for a flexible joint manipulator. In: IEEE/RSJ International Conference on Intelligent Robots and Systems, pp. 2018–2024 (2002)

[OASK+04a] Ott, C., Albu-Schäffer, A., Kugi, A., Stramigioli, S., Hirzinger, G.: Ein passivitätsbasierter ansatz zur kartesischen impedanzregelung von robotern mit elastischen gelenken. In: Robotik 2004, München, June 2004. VDI Berichte Nr. 1841, pp. 71–79. VDI/VDE-Gesellschaft Mess- und Automatisierungstechnik, VDI Verlag (2004)

[OASK+04b] Ott, C., Albu-Schäffer, A., Kugi, A., Stramigioli, S., Hirzinger, G.: A passivity based cartesian impedance controller - part I: Torque feedback and gravity compensation. In: IEEE International Conference on Robotics and Automation, pp. 2659–2665 (2004)

[OASK+05] Ott, C., Albu-Schäffer, A., Kugi, A., Stramigioli, S., Hirzinger, G.: Cartesian impedance control of flexible joint robots: A passivity based approach. at-Automatisierungstechnik, 378–388 (2005)

[OASKH03] Ott, C., Albu-Schäffer, A., Kugi, A., Hirzinger, G.: Decoupling based cartesian impedance control of flexible joint robots. In: IEEE International Conference on Robotics and Automation, pp. 3101–3107 (2003)

[OASKH05] Ott, C., Albu-Schäffer, A., Kugi, A., Hirzinger, G.: Cutting Edge Robotics. In: Cartesian Impedance Control of Flexible Joint Robots: A Decoupling Approach, Advanced Robotic Systems International, pp. 671–682. Pro literatur Verlag (2005)

[OASKH08] Ott, C., Albu-Schäffer, A., Kugi, A., Hirzinger, G.: On the passivity based impedance control of flexible joint robots. IEEE Transactions on Robotics 24(2), 416–429 (2008)

[OBH+05] Ott, C., Borst, C., Hillenbrand, U., Brunner, B., Bäuml, B., Hirzinger, G.: The robutler: Towards service robots for the human environment. In: Video, IEEE International Conference on Robotics and Automation Video Proceedings (2005)

[OCY97] Oh, Y., Chung, W.K., Youm, Y.: Extended impedance control of redundant manipulators using joint space decomposition. In: IEEE International Conference on Robotics and Automation, pp. 1080–1087 (1997)

[OCY98] Oh, Y., Chung, W., Youm, Y.: Extended impedance control of redundant manipulators based on weighted decomposition of joint space. Journal of Robotic Systems 15(5), 231–258 (1998)

[OCYS98] Oh, Y., Chung, W.K., Youm, Y., Suh, I.H.: Motion/force decomposition of redundant manipulator and its application to hybrid impedance control. In: IEEE International Conference on Robotics and Automation, pp. 1441–1446 (1998)

[OEF+06] Ott, C., Eiberger, O., Friedl, W., Bäuml, B., Hillenbrand, U., Borst, C., Albu-Schäffer, A., Brunner, B., Hirschmüller, H., Kielhöfer, S., Konietschke, R., Suppa, M., Wimböck, T., Zacharias, F., Hirzinger, G.: A humanoid two-arm system for dexterous manipulation. In: IEEE-RAS International Conference on Humanoid Robots, pp. 276–283 (2006)

[OKL95] Ortega, R., Kelly, R., Loria, A.: A class of output feedback globally stabilizing controllers for flexible joints robots. IEEE Transactions on Robotics and Automatic 11(5), 766–770 (1995)

[OL97] Oh, J.H., Lee, J.S.: Control of flexible joint robot system by backstepping design approach. In: IEEE International Conference on Robotics and Automation, pp. 3435–3440 (1997)

[OSGEB02] Ortega, R., Spong, M.W., Gomez-Estern, F., Blankenstein, G.: Stabilization of a class of underactuated mechanical systems via interconnection and damping assignment. IEEE Transactions on Automatic Control 47(8), 1218–1233 (2002)

[OvdSME02] Ortega, R., van der Schaft, A.J., Maschke, B., Escobar, G.: Interconnection and damping assignment passivity-based control of port-controlled hamiltonian systems. Automatica 38(4), 585–596 (2002)

[Par99] Park, J.: Analysis and Control of Kinematically Redundant Manipulators: An Approach based on Kinematically Decoupled Joint Space Decomposition. PhD thesis, Pohang University of Science and Technology (POSTECH) (1999)

[PCY99] Park, J., Chung, W.K., Youm, Y.: On dynamical decoupling of kinematically redundant manipulators. In: IEEE/RSJ International Conference on Intelligent Robots and Systems, pp. 1495–1500 (1999)

[PCY00] Park, J., Chung, W.K., Youm, Y.: Unified motion specification and control of kinematically redundant manipulators. In: IEEE International Conference on Robotics and Automation, pp. 3946–3952 (2000)

[PP88] Paden, B., Panja, R.: Globally asymptotically stable 'pd+' controller for robot manipulators. International Journal of Control 47(6), 1697–1712 (1988)

[RC81] Raibert, M.H., Craig, J.J.: Hybrid position/force control of manipulators. ASME Journal of Dynamical Systems, Measurement and Control 105, 126–133 (1981)

[SD01] Stramigioli, S., Duindam, V.: Variable spatial springs for robot control applications. In: IEEE/RSJ International Conference on Intelligent Robots and Systems, pp. 1906–1911 (2001)

[Sel96] Selig, J.M.: Geometrical Methods in Robotics. Monographs in Computer Science. Springer, Heidelberg (1996)

[Sic90] Siciliano, B.: Kinematic control of redundant robot manipulators: A tutorial. Journal of Intelligent and Robotic Systems 3, 201–212 (1990)

[SJK97] Sepulchre, R., Jankovic, M., Kokotovic, P.V.: Constructive Nonlinear Control. Springer, Heidelberg (1997)

[SK97a] Santibanez, V., Kelly, R.: Energy shaping based controllers for rigid and elastic joint robots: Analysis via passivity theorems. In: IEEE International Conference on Robotics and Automation, pp. 2225–2231 (1997)

[SK97b] Santibanez, V., Kelly, R.: Strict lyapunov functions for control of robot manipulators. Automatica 33(4), 675–682 (1997)

[SL87] Slotine, J.J.E., Li, W.: On the adaptive control of robot manipulators. International Journal of Robotics Research 6(3), 49–59 (1987)

[Spo87a] Spong, M.W.: Modeling and control of elastic joint robots. IEEE Journal of Robotics and Automation 3, 291–300 (1987)

[Spo87b] Spong, M.W.: Modeling and control of elastic joint robots. Transactions of the ASME: Journal of Dynamic Systems, Measurement, and Control 109, 310–319 (1987)

[Spo89] Spong, M.W.: Adaptive control of flexible joint manipulators. Systems and Control Letters 13(1), 15–21 (1989)

[Spo95] Spong, M.W.: Adaptive control of flexible joint manipulators: Comments on two papers. Automatica 31(4), 585–590 (1995)

[SRS88] Sira-Ramirez, H., Spong, M.W.: Variable structure control of flexible joint manipulators. International Journal of Robotics and Automation 3(2), 57–64 (1988)

[SS90] Seibert, P., Suarez, R.: Global stabilization of nonlinear cascade systems. Systems and Control Letters 14(4), 347–352 (1990)

[SS96] Sciavicco, L., Siciliano, B.: Modeling and Control of Robot Manipulators. Hill Companies. McGraw-Hill Companies, New York (1996)

[Str01] Stramigioli, S.: Modeling and IPC Control of Interactive Mechanical Systems: A Coordinate-free Approach. Lecture Notes in Control and Information Sciences, vol. 266. Springer, Heidelberg (2001)

[Tom91] Tomei, P.: A simple pd controller for robots with elastic joints. IEEE Transactions on Automatic Control 35, 1208–1213 (1991)

[TS96] Tuttle, T.D., Seering, W.P.: A nonlinear model of a harmonic drive gear transmission. IEEE Transactions on Robotics and Automation 12(3), 368–374 (1996)

[UASS04] Urbanek, H., Albu-Schäffer, A., Smagt, P.v.d.: Learning from demonstration repetitive movements for autonomous service robotics. In: IEEE/RSJ International Conference on Intelligent Robots and Systems, pp. 3495–3500 (2004)

[vdS00] van der Schaft, A.: L_2-Gain and Passivity Techniques in Nonlinear Control, 2nd edn. Springer, Heidelberg (2000)

[Vid93] Vidyasagar, M.: Nonlinear Systems Analysis, 2nd edn. Prentice Hall, Englewood Cliffs (1993)

[WmOH06] Wimböck, T., Ott, C., Hirzinger, G.: Passivity-based object-level impedance control for a multifingered hand. In: IEEE/RSJ International Conference on Intelligent Robots and Systems, pp. 4621–4627 (2006)

[WmOH07] Wimböck, T., Ott, C., Hirzinger, G.: Impedance behaviors for two-handed manipulation: Design and experiments. In: IEEE International Conference on Robotics and Automation, pp. 4182–4189 (2007)

[Yos90] Yoshikawa, T.: Foundations of Robotics. The MIT Press, Cambridge (1990)

[ZF00] Zhang, S., Fasse, E.D.: Spatial compliance modeling using a quaternion-based potential function method. Multibody System Dynamics 4, 75–101 (2000)

[ZKRS02] Zinn, M., Khatib, O., Roth, B., Salisbury, J.K.: Towards a human-centered intrinsically-safe robotic manipulator. In: Second IARP - IEEE/RAS Joint Workshop on Technical Challenges for Dependable Robots in Human Environments, Toulouse (2002)

[ZLS04] Zollo, L., De Luca, A., Siciliano, B.: Regulation with on-line gravity compensation for robots with elastic joints. In: IEEE International Conference on Robotics and Automation, pp. 2687–2692 (2004)

[ZSL+03] Zollo, L., Siciliano, B., De Luca, A., Guglielmelli, E., Dario, P.: Compliance control for a robot with elastic joints. In: IEEE International Conference of Advanced Robotics, pp. 1411–1416 (2003)

[ZSL+05] Zollo, L., Siciliano, B., De Luca, A., Guglielmelli, E., Dario, P.: Compliance control for an anthropomorphic robot with elastic joints: Theory and experiments. ASME Journal of Dynamic Systems, Measurements, and Control 127(3), 321–328 (2005)

A Appendix

A.1 Time-Varying Systems

In this section some definitions and lemmata are reported which are relevant for the stability analysis of time-varying systems. This appendix is intended to clarify some terms used in this book without going into the details. More comprehensive treatments of the stability theory for time-varying systems can be found in one of the many books on nonlinear control, like for instance [Kha02] or [Vid93].

In the stability theory of time-varying systems positive definite functions play an important role. Therefore, the notion of positive definiteness shall be defined. This is done first for the time invariant case in Definition A.1 and subsequently for the time-varying case in Definition A.2.

Definition A.1. *Given an open set $\mathcal{D} \subseteq \mathbb{R}^n$ which contains the origin $\mathbf{0} \in \mathcal{D}$. A continuous function $V : \mathcal{D} \rightarrow \mathbb{R}$ is said to be locally positive semi-definite, if it fulfills the conditions $V(\mathbf{0}) = 0$ and $V(\boldsymbol{x}) \geq 0 \;\; \forall \boldsymbol{x} \in \mathcal{D}$. It is said to be locally positive definite if the stronger condition $V(\boldsymbol{x}) > 0 \;\; \forall \boldsymbol{x} \in \mathcal{D} \backslash \{0\}$ holds.*

It is positive definite, if $\mathcal{D} = \mathbb{R}^n$ and there exists a constant $r > 0$ such that

$$\inf_{||\boldsymbol{x}|| \geq r} V(\boldsymbol{x}) > 0$$

holds. Furthermore, a positive definite function $V(\boldsymbol{x})$ is said to be radially unbounded, if the condition

$$\lim_{||\boldsymbol{x}|| \rightarrow \infty} V(\boldsymbol{x}) = \infty$$

holds.

A function $V(\boldsymbol{x})$ is said to be negative (semi-)definite, if the function $-V(\boldsymbol{x})$ is positive (semi-)definite.

The above definition can be extended to the time-varying case as follows. Only the global definitions shall be given here for brevity.

C. Ott: Cartesian Imped. Cntrl. of Redund. & Flexible-Joint Robots, STAR 49, pp. 177–190, 2008.
springerlink.com

Definition A.2. *A (time-varying) function* $V : \mathbb{R}\times\mathbb{R}^n \rightarrow \mathbb{R}$ *is said to be positive semi-definite, if it fulfills the conditions* $V(t,\mathbf{0}) = 0\ \forall t \in \mathbb{R}$ *and* $V(t,\boldsymbol{x}) \geq 0\ \ \forall(t,\boldsymbol{x}) \in \mathbb{R}\times\mathbb{R}^n$*.*

It is said to be positive definite, if it additionally fulfills the condition

$$V(t,\boldsymbol{x}) \geq W_1(\boldsymbol{x})\ \ \forall(t,\boldsymbol{x}) \in \mathbb{R}\times\mathbb{R}^n$$

for some positive definite function $W_1 : \mathbb{R}^n \rightarrow \mathbb{R}$*. Furthermore, it is radially unbounded, if* $W_1(\boldsymbol{x})$ *is so.*

Finally, $V(t,\boldsymbol{x})$ *is said to be decrescent, if it fulfills the condition*

$$V(t,\boldsymbol{x}) \leq W_2(\boldsymbol{x})\ \ \forall(t,\boldsymbol{x}) \in \mathbb{R}\times\mathbb{R}^n$$

for some positive definite function $W_2 : \mathbb{R}^n \rightarrow \mathbb{R}$*.*

In the time-varying case the solution of an ordinary differential equation depends on the initial state and also on the initial time t_0. Usually, it is then desired that all convergence and stability properties hold *uniformly* in the initial time. In the following the definition of special comparison functions, known as class $\mathcal{K}$ (resp. $\mathcal{K}_\infty$) functions, is given. These functions will allow transparent definitions of uniform boundedness and uniform (asymptotic) stability.

Definition A.3. *A continuous function* $\alpha : \mathcal{A} \rightarrow \mathbb{R}$*, defined on an interval* $\mathcal{A} = [0,a)$ *with* $a > 0$*, is said to belong to class* $\mathcal{K}$*, if it is strictly increasing and* $\alpha(0) = 0$*. It is said to belong to class* $\mathcal{K}_\infty$*, if* $\mathcal{A} = [0,\infty)$ *and* $\alpha(a) \rightarrow \infty$ *for* $a \rightarrow \infty$*.*

Class $\mathcal{K}$ (resp. $\mathcal{K}_\infty$) functions have the following important properties [Kha02].

Lemma A.4. *Let* $\alpha_1 : \mathcal{A} \rightarrow \mathbb{R}$ *and* $\alpha_2 : \mathcal{A} \rightarrow \mathbb{R}$ *be class* $\mathcal{K}$ *(resp.* $\mathcal{K}_\infty$*) functions. Then the inverse* α_1^{-1} *and the concatenation* $\alpha_1 \circ \alpha_2$ *are also class* $\mathcal{K}$ *(resp.* $\mathcal{K}_\infty$*) functions.*

The following Lemma, taken from [Kha02], gives an important relation between positive definite functions and class $\mathcal{K}$ functions.

Lemma A.5. *Let* $V : \mathcal{D} \rightarrow \mathbb{R}$ *be a continuous and positive definite function, which is defined on a set* $\mathcal{D} \subset \mathbb{R}^n$ *that contains the origin, i.e.* $\mathbf{0} \in \mathcal{D}$*. Let the ball* $\mathcal{B}_r = \{\boldsymbol{x} \in \mathcal{D}|\ ||\boldsymbol{x}|| < r\} \subset \mathcal{D}$ *be contained in this domain for some* $r > 0$*. Then there exist class* $\mathcal{K}$ *functions* α_1 *and* α_2*, which are defined on the interval* $[0,r)$*, such that the inequalities*

$$\alpha_1(||\boldsymbol{x}||) \leq V(\boldsymbol{x}) \leq \alpha_2(||\boldsymbol{x}||) \quad \forall \boldsymbol{x} \in \mathcal{B}_r$$

hold. In case $\mathcal{D} = \mathbb{R}^n$*, the functions* α_1 *and* α_2 *will be defined on* $[0,\infty)$ *and the above mentioned inequalities will hold for all* $\boldsymbol{x} \in \mathbb{R}^n$*. Furthermore, if the function* $V(\boldsymbol{x})$ *is radially unbounded, then* α_1 *and* α_2 *can be found such that they belong to class* $\mathcal{K}_\infty$*.*

A similar statement can also be given for time-varying functions, as shown in the next lemma, which treats only the global case.

Corollary A.6. *Let $V : \mathbb{R} \times \mathbb{R}^n \to \mathbb{R}$ be a continuous, positive definite, radially unbounded, and decrescent function. Then there exist class $\mathcal{K}_\infty$ functions α_1 and α_2 such that the inequalities*

$$\alpha_1(||\boldsymbol{x}||) \leq V(t, \boldsymbol{x}) \leq \alpha_2(||\boldsymbol{x}||) \quad \forall (t, \boldsymbol{x}) \in \mathbb{R} \times \mathbb{R}^n$$

hold.

Proof. From the fact that the function $V(t, \boldsymbol{x})$ is positive definite and radially unbounded it follows that there exists a positive definite and radially unbounded function $W_1(\boldsymbol{x})$ such that $V(t, \boldsymbol{x}) \geq W_1(\boldsymbol{x})$ $\forall (t, \boldsymbol{x}) \in \mathbb{R} \times \mathbb{R}^n$. Due to Lemma A.5 there exists a class $\mathcal{K}_\infty$ function α_1, such that $V(t, \boldsymbol{x}) \geq W_1(\boldsymbol{x}) \geq \alpha_1(||\boldsymbol{x}||)$ $\forall (t, \boldsymbol{x}) \in \mathbb{R} \times \mathbb{R}^n$ holds. Therefore, the existence of the lower bound is shown. Due to the decrescence (see Definition A.2) of $V(t, \boldsymbol{x})$ there exists also a positive definite function $W_2(\boldsymbol{x})$ which fulfills the condition $V(t, \boldsymbol{x}) \leq W_2(\boldsymbol{x})$ $\forall (t, \boldsymbol{x}) \in \mathbb{R} \times \mathbb{R}^n$. Again due to Lemma A.5 one can find a class $\mathcal{K}_\infty$ function α_2, such that $V(t, \boldsymbol{x}) \leq W_2(\boldsymbol{x}) \leq \alpha_2(||\boldsymbol{x}||)$ $\forall (t, \boldsymbol{x}) \in \mathbb{R} \times \mathbb{R}^n$ which completes the proof.

Using the notion of class $\mathcal{K}$ (resp. $\mathcal{K}_\infty$) functions, one can define the property of uniform global boundedness of a trajectory as follows. For this, a system of the form

$$\dot{\boldsymbol{x}} = \boldsymbol{f}(t, \boldsymbol{x}) \tag{A.1}$$

is considered for which it is assumed that the solution $\boldsymbol{x}(t, t_0, \boldsymbol{x}_0)$ exists for $t \geq t_0$ and for any initial time t_0 and any initial state $\boldsymbol{x}(t_0) = \boldsymbol{x}_0$.

Definition A.7. *The solution $\boldsymbol{x}(t, t_0, \boldsymbol{x}_0)$ of (A.1) with initial state $\boldsymbol{x}_0$ and initial time t_0 is said to be uniformly globally bounded if there exists a class $\mathcal{K}_\infty$ function α and a number $c > 0$ such that*

$$||\boldsymbol{x}(t, t_0, \boldsymbol{x}_0)|| \leq \alpha(||\boldsymbol{x}_0||) + c \quad \forall t \geq t_0 \tag{A.2}$$

holds.

In order to show that the solution of a dynamical system is uniformly globally bounded, the following lemma is useful.

Lemma A.8. *If there exists a continuously differentiable, positive definite, radially unbounded, and decrescent function $V(t, \boldsymbol{x})$, for which the time derivative of $V(t, \boldsymbol{x})$ along the solutions of (A.1)*

$$\dot{V}(t, \boldsymbol{x}) = \frac{\partial V(t, \boldsymbol{x})}{\partial \boldsymbol{x}} \boldsymbol{f}(t, \boldsymbol{x}) + \frac{\partial V(t, \boldsymbol{x})}{\partial t} \tag{A.3}$$

is negative semi-definite, i.e. $\dot{V}(t, \boldsymbol{x}) \leq 0$, then the solutions of (A.1) will be uniformly globally bounded.

Proof. Due to Corollary A.6 there exists class $\mathcal{K}_\infty$ functions α_1 and α_2, such that

$$\alpha_1(||\boldsymbol{x}||) \le V(t,\boldsymbol{x}) \le \alpha_2(||\boldsymbol{x}||) \quad \forall (t,\boldsymbol{x}) \in \mathbb{R} \times \mathbb{R}^n$$

holds. Let $\boldsymbol{x}(t, t_0, \boldsymbol{x}_0)$ be the solution of (A.1) with initial state $\boldsymbol{x}_0$ and initial time t_0 at time $t \ge t_0$. Then due to $\dot{V}(t,\boldsymbol{x}) \le 0$, the inequality $V(t, \boldsymbol{x}(t, t_0, \boldsymbol{x}_0)) \le V(t_0, \boldsymbol{x}_0)$ must hold for $t \ge t_0$. With $\alpha_1(||\boldsymbol{x}(t, t_0, \boldsymbol{x}_0)||) \le V(t, \boldsymbol{x}(t, t_0, \boldsymbol{x}_0))$ and $V(t_0, \boldsymbol{x}_0) \le \alpha_2(||\boldsymbol{x}_0||)$ it follows that also the inequality

$$\alpha_1(||\boldsymbol{x}(t, t_0, \boldsymbol{x}_0)||) \le \alpha_2(||\boldsymbol{x}_0||) \quad \forall t \ge t_0$$

must hold. Due to Lemma A.4 and the fact that class $\mathcal{K}_\infty$ functions are strictly increasing (see Definition A.3) one can follow that the inequality

$$||\boldsymbol{x}(t, t_0, \boldsymbol{x}_0)|| \le \alpha(||\boldsymbol{x}_0||) \quad \forall t \ge t_0 \tag{A.4}$$

holds for the class $\mathcal{K}_\infty$ function $\alpha = \alpha_1^{-1} \circ \alpha_2$. Therefore, the requirement of Definition A.7 is fulfilled and the uniform global boundedness of $\boldsymbol{x}(t, t_0, \boldsymbol{x}_0)$ follows from (A.4).

Lemma A.8 is used in the stability proof of the decoupling based controller from Section 6.1.

As already mentioned above, in the time-varying case one is also interested in a stability property, which holds uniformly in the initial time t_0. For this it is assumed in the following that the origin $\boldsymbol{x} = \boldsymbol{0}$ is an equilibrium point of (A.1).

Definition A.9. *The equilibrium point $\boldsymbol{x} = \boldsymbol{0}$ of (A.1) is*

- *stable if, for each $\epsilon > 0$, there is $\delta = \delta(\epsilon, t_0) > 0$ such that*

$$||\boldsymbol{x}(t_0)|| < \delta \;\Rightarrow\; ||\boldsymbol{x}(t)|| < \epsilon \;, \quad \forall\, t > t_0 \ge 0 \;. \tag{A.5}$$

- *uniformly stable if, for each $\epsilon > 0$, there is $\delta = \delta(\epsilon) > 0$, independent of t_0, such that (A.5) is satisfied.*
- *unstable if it is not stable.*
- *asymptotically stable if it is stable and there is a positive constant $c = c(t_0)$ such that $\boldsymbol{x}(t) \to \boldsymbol{0}$ as $t \to \infty$, for all $||\boldsymbol{x}(t_0)|| < c$.*
- *uniformly asymptotically stable if it is uniformly stable and there is a positive constant c, independent of t_0, such that for all $||\boldsymbol{x}(t_0)|| < c$, $\boldsymbol{x}(t) \to \boldsymbol{0}$ as $t \to \infty$, uniformly in t_0; i.e., for each $\eta > 0$, there is $T = T(\eta) > 0$ such that*

$$||\boldsymbol{x}(t)|| < \eta, \quad \forall\, t > t_0 + T(\eta),\ \forall\ ||\boldsymbol{x}(t_0)|| < c \;.$$

- *uniformly globally asymptotically stable if it is uniformly stable, $\delta(\epsilon)$ can be chosen to satisfy $\lim_{\epsilon\to\infty} \delta(\epsilon) = \infty$, and, for each pair of positive numbers η and c, there is $T = T(\eta, c) > 0$ such that*

$$||\boldsymbol{x}(t)|| < \eta, \quad \forall\, t > t_0 + T(\eta, c),\ \forall\ ||\boldsymbol{x}(t_0)|| < c \;.$$

Equivalent definitions of uniform stability and uniform asymptotic stability based on class $\mathcal{K}$ functions can be found in [Kha02]. The following two theorems, taken from [Kha02], show how the Lyapunov stability theory for autonomous systems can be extended to nonautonomous systems.

Theorem A.10 *Let $\boldsymbol{x} = \mathbf{0}$ be an equilibrium point for (A.1) and $D \subset \mathbb{R}^n$ be a domain containing $\boldsymbol{x} = \mathbf{0}$. Let $V : [0, \infty) \times D \to \mathbb{R}$ be a continuously differentiable function such that*

$$W_1(\boldsymbol{x}) \leq V(t, \boldsymbol{x}) \leq W_2(\boldsymbol{x})$$
$$\frac{\partial V(t, \boldsymbol{x})}{\partial t} + \frac{\partial V(t, \boldsymbol{x})}{\partial \boldsymbol{x}} \boldsymbol{f}(t, \boldsymbol{x}) \leq 0$$

$\forall\, t \geq 0$ and $\forall\, \boldsymbol{x} \in D$, where $W_1(\boldsymbol{x})$ and $W_2(\boldsymbol{x})$ are continuously positive definite functions on D. Then, $\boldsymbol{x} = \mathbf{0}$ is uniformly stable.

Theorem A.11 *Suppose the assumptions of Theorem A.10 are satisfied and that additionally the condition*

$$\frac{\partial V(t, \boldsymbol{x})}{\partial t} + \frac{\partial V(t, \boldsymbol{x})}{\partial \boldsymbol{x}} \boldsymbol{f}(t, \boldsymbol{x}) \leq -W_3(\boldsymbol{x}) \quad \forall\, t \geq 0,\ \forall\, \boldsymbol{x} \in D \tag{A.6}$$

holds for a continuous positive definite function $W_3 : \mathbb{R}^n \to \mathbb{R}$. Then, $\boldsymbol{x} = \mathbf{0}$ is uniformly asymptotically stable. Moreover, if $D = \mathbb{R}^n$ and $W_1(\boldsymbol{x})$ is radially unbounded, then $\boldsymbol{x} = \mathbf{0}$ is uniformly globally asymptotically stable.

A function $V(t, \boldsymbol{x})$, which fulfills the conditions of Theorem A.10 is referred to as a *(time-varying) Lyapunov function.* If it fulfills also the condition (A.6) of Theorem A.11 it will be called a *strict (time-varying) Lyapunov function.*

A.2 Conditional Stability

In Chapter 4 *semi-definite Lyapunov functions* are used for the stability analysis. Therefore, the notion of *conditional stability* is needed, as it is defined in [SJK97]. The following definitions are reported in order to clarify the stability properties being used.

Consider therefore a time-invariant system of the form

$$\dot{\boldsymbol{x}} = \boldsymbol{f}(\boldsymbol{x}) \tag{A.7}$$

with state $\boldsymbol{x} \in \mathbb{R}^n$. Assume that the point $\boldsymbol{x}_s$ is a stationary point of (A.7), i.e. $\boldsymbol{f}(\boldsymbol{x}_s) = \mathbf{0}$. Suppose that the solution $\boldsymbol{x}(t)$ to (A.7) with initial state $\boldsymbol{x}(0) = \boldsymbol{x}_0$ exists for all $t > 0$. For the conditional stability all requirements of the stability definitions must hold only for those initial conditions which lie in a particular set $\mathcal{A} \subset \mathbb{R}^n$. The notion of conditional stability is therefore weaker than the usual (Lyapunov) stability.

Definition A.12. *A stationary point* $\boldsymbol{x}_s$ *of (A.7) is said to be stable conditionally to* $\mathcal{A} \subset \mathbb{R}^n$*, if* $\boldsymbol{x}_s \in \mathcal{A}$ *and for each* $\epsilon > 0$ *there exists* $\delta(\epsilon) > 0$ *such that for any initial condition* $\boldsymbol{x}_0 \in \mathcal{A}$ *the following implication holds:*

$$||\boldsymbol{x}_0 - \boldsymbol{x}_s|| < \delta(\epsilon) \Rightarrow ||\boldsymbol{x}(t) - \boldsymbol{x}_s|| < \epsilon \ , \quad \forall t \geq 0 \ . \tag{A.8}$$

Definition A.13. *A stationary point* $\boldsymbol{x}_s$ *of (A.7) is said to be attractive conditionally to* $\mathcal{A} \subset \mathbb{R}^n$*, if* $\boldsymbol{x}_s \in \mathcal{A}$ *and there exists an* $\eta(\boldsymbol{x}_s) > 0$ *such that for any initial condition* $\boldsymbol{x}_0 \in \mathcal{A}$ *the following implication holds:*

$$||\boldsymbol{x}_0 - \boldsymbol{x}_s|| < \eta(\boldsymbol{x}_s) \Rightarrow \lim_{t \to \infty} \boldsymbol{x}(t) = \boldsymbol{x}_s \ . \tag{A.9}$$

Definition A.14. *A stationary point* $\boldsymbol{x}_s$ *of (A.7) is said to be asymptotically stable conditionally to* $\mathcal{A} \subset \mathbb{R}^n$ *if it is both stable and attractive conditionally to* $\mathcal{A}$.

Definition A.15. *The stationary point* $\boldsymbol{x}_s$ *of (A.7) is said to be globally asymptotically stable conditionally to* $\mathcal{A} \subset \mathbb{R}^n$ *if it is asymptotically stable conditionally to* $\mathcal{A}$ *and* $\eta(\boldsymbol{x}_s) = +\infty$.

A.3 Passivity Definitions

In several parts of this book the passivity properties of some systems are used. Therefore, the definitions of *dissipativity* and *passivity*, as given in [Kug01, vdS00, KS02a, KS02b, SJK97], are reported here for further reference. Consider a time-invariant system of the form

$$\dot{\boldsymbol{x}} = \boldsymbol{f}(\boldsymbol{x}, \boldsymbol{u}) \tag{A.10}$$

$$\boldsymbol{y} = \boldsymbol{h}(\boldsymbol{x}, \boldsymbol{u}) \tag{A.11}$$

with state $\boldsymbol{x} \in \mathbb{R}^n$, input $\boldsymbol{u} \in \mathbb{R}^m$, and output $\boldsymbol{y} \in \mathbb{R}^m$.

Definition A.16. *The system (A.10)-(A.11) is said to be dissipative with supply rate* $s(\boldsymbol{u}, \boldsymbol{y})$ *if there exists a non-negative function* $S(\boldsymbol{x})$*, called* storage function, *such that the inequality*

$$S(\boldsymbol{x}(t)) - S(\boldsymbol{x}(t_0)) \leq \int_{t_0}^{t} s(\boldsymbol{u}(\tau), \boldsymbol{y}(\tau)) \mathrm{d}\tau$$

holds for all admissible $\boldsymbol{u}$*, all initial conditions* $\boldsymbol{x}(t_0)$ *and all* $t > t_0$.

Definition A.17. *The system (A.10)-(A.11) is said to be*

- *passive if it is dissipative with supply rate* $s(\boldsymbol{u}, \boldsymbol{y}) = \boldsymbol{y}^T \boldsymbol{u}$.
- *strict input passive if there exists* $\alpha > 0$ *such that the system is dissipative with supply rate* $s(\boldsymbol{u}, \boldsymbol{y}) = \boldsymbol{y}^T \boldsymbol{u} - \alpha ||\boldsymbol{u}||^2$.

- *strict output passive if there exists* $\beta > 0$ *such that the system is dissipative with supply rate* $s(\boldsymbol{u}, \boldsymbol{y}) = \boldsymbol{y}^T \boldsymbol{u} - \beta ||\boldsymbol{y}||^2$.

A.4 Some Matrix Lemmata and Definitions

A matrix $\boldsymbol{A} \in \mathbb{R}^{n\times m}$ can be seen as a linear transformation from $\mathbb{R}^m$ to $\mathbb{R}^n$. The following definition clarifies some relevant spaces.

Definition A.18. *Consider a matrix* $\boldsymbol{A} \in \mathbb{R}^{n\times m}$ *which represents a linear transformation from* $\mathbb{R}^m$ *to* $\mathbb{R}^n$*. The* domain *of this transformation is* $\mathbb{R}^m$*; its* co-domain *is* $\mathbb{R}^n$*. The* kernel *(or nullspace) of* $\boldsymbol{A}$ *is defined as*

$$\ker(\boldsymbol{A}) := \{\boldsymbol{v} \in \mathbb{R}^m | \boldsymbol{A}\boldsymbol{v} = \boldsymbol{0}\} ,$$

and such is a subspace of the domain. Finally, the range *of* $\boldsymbol{A}$ *is defined as*

$$\text{range}(\boldsymbol{A}) := \{\boldsymbol{y} \in \mathbb{R}^n | \exists \boldsymbol{v} \in \mathbb{R}^n : \boldsymbol{y} = \boldsymbol{A}\boldsymbol{v}\} ,$$

and such is a subspace of the co-domain.

Definition A.19. *Let* $\mathcal{A}$ *be a subset of* $\mathbb{R}^n$*. Given a positive definite matrix* $\boldsymbol{G} \in \mathbb{R}^{n\times n}$ *as a metric of* $\mathbb{R}^n$*, the* orthogonal complement *of* $\mathcal{A}$ *is defined as*

$$\mathcal{A}^{\perp} := \{\boldsymbol{y} \in \mathbb{R}^n | \ \boldsymbol{y}^T \boldsymbol{G}\boldsymbol{x} = 0 \ \ \forall \boldsymbol{x} \in \mathcal{A}\} .$$

The properties of positive definite matrices are of interest for the stability analysis of linear as well as of nonlinear systems. The following two lemmata, which are used in this book, are well known and are reported here without proof.

Lemma A.20. *Suppose that a symmetric matrix* $\boldsymbol{M} \in \mathbb{R}^{n\times n}$ *is partitioned as*

$$\boldsymbol{M} = \begin{bmatrix} \boldsymbol{A} & \boldsymbol{C} \\ \boldsymbol{C}^T & \boldsymbol{B} \end{bmatrix} \tag{A.12}$$

where $\boldsymbol{A}$ *and* $\boldsymbol{B}$ *are symmetric square matrices. Then the matrix* $\boldsymbol{M}$ *is positive definite* ($\boldsymbol{M} > 0$) *if and only if* $\boldsymbol{A}$ *is positive definite and* $\boldsymbol{B} - \boldsymbol{C}^T \boldsymbol{A}^{-1} \boldsymbol{C}$ *is positive definite.*

Proof. See [HJ90].

Lemma A.21. *Given an arbitrary positive definite matrix* $\boldsymbol{Q}$*, one can find a unique positive definite solution* $\boldsymbol{P}$ *of the Lyapunov equation* $\boldsymbol{A}^T\boldsymbol{P} + \boldsymbol{P}\boldsymbol{A} = -\boldsymbol{Q}$ *if and only if the matrix* $\boldsymbol{A}$ *is Hurwitz.*

Proof. See [Vid93].

Finally, a useful lemma from [vdS00] concerning skew symmetric matrices is given without proof. Notice that this lemma is of interest for Property 2.6 and Lemma 3.2.

Lemma A.22. *Given a symmetric matrix* $\boldsymbol{A} \in \mathbb{R}^{n\times n}$ *and another square matrix* $\boldsymbol{B} \in \mathbb{R}^{n\times n}$ *of the same dimension. Then the matrix* $\boldsymbol{C} = \boldsymbol{A} - 2\boldsymbol{B}$ *is skew symmetric if and only if the equality* $\boldsymbol{A} = \boldsymbol{B} + \boldsymbol{B}^T$ *holds.*

Lemma A.23. *Given a function* $\boldsymbol{A}(\boldsymbol{x}) : \mathbb{R}^n \to \mathbb{R}^{n\times n}$. *Suppose that the integrals*

$$\boldsymbol{i}_1(\boldsymbol{x}_1, \boldsymbol{x}_2) := \int_{x_1}^{x_2} \boldsymbol{A}(\boldsymbol{x})\mathrm{d}\boldsymbol{x}$$
$$i_2(\boldsymbol{x}_1, \boldsymbol{x}_2) := \int_{x_1}^{x_2} \left(\int_{x_1}^{x} \boldsymbol{A}(\boldsymbol{\xi})\mathrm{d}\boldsymbol{\xi}\right)^T \mathrm{d}\boldsymbol{x}$$

are path independent for all $\boldsymbol{x}_1, \boldsymbol{x}_2 \in \mathbb{R}^n$. *Then the inequalities*

$$||\boldsymbol{i}_1(\boldsymbol{x}_1, \boldsymbol{x}_2)||_2 \leq \sup_{\forall \boldsymbol{x}\in\mathbb{R}^n} ||\boldsymbol{A}(\boldsymbol{x})||_{i2} ||\boldsymbol{x}_2 - \boldsymbol{x}_1||_2$$
$$|i_2(\boldsymbol{x}_1, \boldsymbol{x}_2)| \leq \frac{1}{2} \sup_{\forall \boldsymbol{x}\in\mathbb{R}^n} ||\boldsymbol{A}(\boldsymbol{x})||_{i2} ||\boldsymbol{x}_2 - \boldsymbol{x}_1||_2^2$$

hold, where $||\cdot||_2$ *and* $||\cdot||_{i2}$ *are the Euclidean norm for vectors and the spectral norm for matrices, respectively.*

Proof. The proof can be done by choosing particular paths of integration. Consider the paths $\boldsymbol{x}(s) : [0,1] \to \mathbb{R}^n$ and $\boldsymbol{\xi}(t) : [0,1] \to \mathbb{R}^n$

$$\boldsymbol{x}(s) := \boldsymbol{x}_1 + (\boldsymbol{x}_2 - \boldsymbol{x}_1)s$$
$$\boldsymbol{\xi}(t) := \boldsymbol{x}_1 + (\boldsymbol{x} - \boldsymbol{x}_1)t$$

By evaluating the integration for $\boldsymbol{i}_1$, one obtains

$$\boldsymbol{i}_1(\boldsymbol{x}_1, \boldsymbol{x}_2) = \int_0^1 \boldsymbol{A}(\boldsymbol{x}_1 + (\boldsymbol{x}_2 - \boldsymbol{x}_1)s)(\boldsymbol{x}_2 - \boldsymbol{x}_1)\mathrm{d}s$$
$$||\boldsymbol{i}_1(\boldsymbol{x}_1, \boldsymbol{x}_2)||_2 \leq \int_0^1 ||\boldsymbol{A}(\boldsymbol{x}_1 + (\boldsymbol{x}_2 - \boldsymbol{x}_1)s)(\boldsymbol{x}_2 - \boldsymbol{x}_1)||_2\mathrm{d}s$$
$$||\boldsymbol{i}_1(\boldsymbol{x}_1, \boldsymbol{x}_2)||_2 \leq \sup_{\forall \boldsymbol{x}\in\mathbb{R}^n} ||\boldsymbol{A}(\boldsymbol{x})||_{i2} ||\boldsymbol{x}_2 - \boldsymbol{x}_1||_2 \ .$$

Similarly, for i_2 one can compute

$$i_2(\boldsymbol{x}_1, \boldsymbol{x}_2) = \int_0^1 \left(\int_0^1 \boldsymbol{A}(\boldsymbol{x}_1 + (\boldsymbol{x}_2 - \boldsymbol{x}_1)st)(\boldsymbol{x}_2 - \boldsymbol{x}_1)s\ \mathrm{d}t\right)^T (\boldsymbol{x}_2 - \boldsymbol{x}_1)\mathrm{d}s$$
$$|i_2(\boldsymbol{x}_1, \boldsymbol{x}_2)| \leq \int_0^1 \int_0^1 \left|(\boldsymbol{x}_2 - \boldsymbol{x}_1)^T \boldsymbol{A}(\boldsymbol{x}_1 + (\boldsymbol{x}_2 - \boldsymbol{x}_1)st)(\boldsymbol{x}_2 - \boldsymbol{x}_1)\right| s\ \mathrm{d}t\mathrm{d}s$$
$$|i_2(\boldsymbol{x}_1, \boldsymbol{x}_2)| \leq \int_0^1 \int_0^1 \sup_{\forall \boldsymbol{x}\in\mathbb{R}^n} ||\boldsymbol{A}(\boldsymbol{x})||_{i2} ||\boldsymbol{x}_2 - \boldsymbol{x}_1||_2^2 s\ \mathrm{d}t\mathrm{d}s$$
$$|i_2(\boldsymbol{x}_1, \boldsymbol{x}_2)| \leq \frac{1}{2} \sup_{\forall \boldsymbol{x}\in\mathbb{R}^n} ||\boldsymbol{A}(\boldsymbol{x})||_{i2} ||\boldsymbol{x}_2 - \boldsymbol{x}_1||_2^2 \ .$$

A.5 Inversion of the Extended Jacobian $\boldsymbol{J}_N(\boldsymbol{q})$

In this section an explicit form for the inverse of the matrix

$$\boldsymbol{J}_N(\boldsymbol{q}) = \begin{pmatrix} \boldsymbol{J}(\boldsymbol{q}) \\ \left(\boldsymbol{Z}(\boldsymbol{q})\boldsymbol{M}(\boldsymbol{q})\boldsymbol{Z}(\boldsymbol{q})^T\right)^{-1} \boldsymbol{Z}(\boldsymbol{q})\boldsymbol{M}(\boldsymbol{q}) \end{pmatrix} = \begin{pmatrix} \boldsymbol{J}(\boldsymbol{q}) \\ \boldsymbol{N}(\boldsymbol{q}) \end{pmatrix} \tag{A.13}$$

from Section 4.4 is derived. The following properties of the matrices $\boldsymbol{J}(\boldsymbol{q}) \in \mathbb{R}^{m\times n}$, $\boldsymbol{Z}(\boldsymbol{q}) \in \mathbb{R}^{(n-m)\times n}$ and $\boldsymbol{M}(\boldsymbol{q}) \in \mathbb{R}^{n\times n}$ are assumed:

- The matrix $\boldsymbol{J}(\boldsymbol{q})$ has full row rank.
- The row vectors of the matrix $\boldsymbol{Z}(\boldsymbol{q})$ span the (right) nullspace of $\boldsymbol{J}(\boldsymbol{q})$, such that $\boldsymbol{J}(\boldsymbol{q})\boldsymbol{Z}(\boldsymbol{q})^T = \boldsymbol{0}$ holds and $\boldsymbol{Z}(\boldsymbol{q})$ has full row rank.
- The matrix $\boldsymbol{M}(\boldsymbol{q})$ is positive definite.

First the invertibility of $\boldsymbol{J}_N(\boldsymbol{q})$ is shown by the following lemma.

Lemma A.24. *Let $\boldsymbol{J} \in \mathbb{R}^{m\times n}$, with $n > m$, be a rectangular matrix which has full row rank. Let further be $\boldsymbol{Z} \in \mathbb{R}^{(n-m)\times n}$ a full rank right annihilator of $\boldsymbol{J}$, such that $\boldsymbol{J}\boldsymbol{Z}^T = \boldsymbol{0}$ holds. Then the matrix*

$$\boldsymbol{J}_N = \begin{pmatrix} \boldsymbol{J} \\ \left(\boldsymbol{Z}\boldsymbol{W}\boldsymbol{Z}^T\right)^{-1} \boldsymbol{Z}\boldsymbol{W} \end{pmatrix}$$

is non-singular for any positive definite matrix $\boldsymbol{W} \in \mathbb{R}^{n\times n}$.

Proof.[1] Notice first that, since the row vectors of $\boldsymbol{Z}$ span the right nullspace of $\boldsymbol{J}$, the same is also true for the matrix $(\boldsymbol{Z}\boldsymbol{W}\boldsymbol{Z}^T)^{-1}\boldsymbol{Z}$ for any positive definite matrix $\boldsymbol{W}$. The proof of the Lemma is now done by contradition. Suppose that $\boldsymbol{J}_N$ were singular. Then there would exist a vector $\boldsymbol{v} \neq \boldsymbol{0}$ such that $\boldsymbol{J}_N\boldsymbol{v} = \boldsymbol{0}$. From this, one can follow[2]

$$\boldsymbol{J}\boldsymbol{v} = \boldsymbol{0} \Rightarrow \boldsymbol{v} \in \ker(\boldsymbol{J}) \ ,$$
$$\left(\boldsymbol{Z}\boldsymbol{W}\boldsymbol{Z}^T\right)^{-1} \boldsymbol{Z}\boldsymbol{W}\boldsymbol{v} = \boldsymbol{0} \Rightarrow \boldsymbol{W}\boldsymbol{v} \perp \ker(\boldsymbol{J}) \ .$$

But this implies that

$$\boldsymbol{v} \perp \boldsymbol{W}\boldsymbol{v} \Rightarrow \boldsymbol{v}^T\boldsymbol{W}\boldsymbol{v} = \boldsymbol{0}$$

must hold, which is clearly in contradiction to the assumption that the matrix $\boldsymbol{W}$ is positive definite. Hence one can conclude by contradiction that the matrix $\boldsymbol{J}_N$ cannot be singular.

[1] The author would like to thank Udo Frese for pointing out this proof.

[2] Notice that in the following the symbol $\perp$ refers to the orthogonality in the Euclidean sense.

The existence of the inverse of the square matrix $\boldsymbol{J}_N(\boldsymbol{q})$ follows immediately from the above Lemma. Let the inverse $\boldsymbol{J}_N(\boldsymbol{q})^{-1}$ now be partitioned as

$$\boldsymbol{J}_N(\boldsymbol{q})^{-1} = \begin{bmatrix}\boldsymbol{A}(\boldsymbol{q}) & \boldsymbol{B}(\boldsymbol{q})\end{bmatrix} , \tag{A.14}$$

with non-square matrices $\boldsymbol{A}(\boldsymbol{q}) \in \mathbb{R}^{n\times m}$ and $\boldsymbol{B}(\boldsymbol{q}) \in \mathbb{R}^{n\times(n-m)}$. From

$$\begin{aligned}\boldsymbol{J}_N(\boldsymbol{q})\boldsymbol{J}_N(\boldsymbol{q})^{-1} &= \begin{bmatrix}\boldsymbol{I} & \boldsymbol{0}\\ \boldsymbol{0} & \boldsymbol{I}\end{bmatrix}\\ &= \begin{bmatrix}\boldsymbol{J}(\boldsymbol{q})\boldsymbol{A}(\boldsymbol{q}) & \boldsymbol{J}(\boldsymbol{q})\boldsymbol{B}(\boldsymbol{q})\\ \boldsymbol{N}(\boldsymbol{q})\boldsymbol{A}(\boldsymbol{q}) & \boldsymbol{N}(\boldsymbol{q})\boldsymbol{B}(\boldsymbol{q})\end{bmatrix}\end{aligned} \tag{A.15}$$

one gets the four identities

$$\boldsymbol{J}(\boldsymbol{q})\boldsymbol{A}(\boldsymbol{q}) = \boldsymbol{I} , \tag{A.16}$$

$$\left(\boldsymbol{Z}(\boldsymbol{q})\boldsymbol{M}(\boldsymbol{q})\boldsymbol{Z}(\boldsymbol{q})^T\right)^{-1}\boldsymbol{Z}(\boldsymbol{q})\boldsymbol{M}(\boldsymbol{q})\boldsymbol{A}(\boldsymbol{q}) = \boldsymbol{0} , \tag{A.17}$$

$$\left(\boldsymbol{Z}(\boldsymbol{q})\boldsymbol{M}(\boldsymbol{q})\boldsymbol{Z}(\boldsymbol{q})^T\right)^{-1}\boldsymbol{Z}(\boldsymbol{q})\boldsymbol{M}(\boldsymbol{q})\boldsymbol{B}(\boldsymbol{q}) = \boldsymbol{I} , \tag{A.18}$$

$$\boldsymbol{J}(\boldsymbol{q})\boldsymbol{B}(\boldsymbol{q}) = \boldsymbol{0} . \tag{A.19}$$

One can now determine the matrix $\boldsymbol{A}(\boldsymbol{q})$ from the first two identities and the matrix $\boldsymbol{B}(\boldsymbol{q})$ from the last two as follows:

Equation A.17 is obviously satisfied for any matrix $\boldsymbol{A}(\boldsymbol{q})$ of the form

$$\boldsymbol{A}(\boldsymbol{q}) = \boldsymbol{M}(\boldsymbol{q})^{-1}\boldsymbol{J}(\boldsymbol{q})^T\boldsymbol{Q}_A(\boldsymbol{q}) \tag{A.20}$$

with an arbitrary matrix $\boldsymbol{Q}_A(\boldsymbol{q}) \in \mathbb{R}^{m\times m}$. The matrix $\boldsymbol{Q}_A(\boldsymbol{q})$ can then be determined from equation A.16 easily as $\boldsymbol{Q}_A(\boldsymbol{q}) = (\boldsymbol{J}(\boldsymbol{q})\boldsymbol{M}(\boldsymbol{q})^{-1}\boldsymbol{J}(\boldsymbol{q})^T)^{-1}$ and hence

$$\boldsymbol{A}(\boldsymbol{q}) = \boldsymbol{M}(\boldsymbol{q})^{-1}\boldsymbol{J}(\boldsymbol{q})^T(\boldsymbol{J}(\boldsymbol{q})\boldsymbol{M}(\boldsymbol{q})^{-1}\boldsymbol{J}(\boldsymbol{q})^T)^{-1} . \tag{A.21}$$

The same procedure can be applied to get $\boldsymbol{B}(\boldsymbol{q})$. Equation A.19 is satisfied for any matrix $\boldsymbol{B}(\boldsymbol{q})$ of the form

$$\boldsymbol{B}(\boldsymbol{q}) = \boldsymbol{Z}(\boldsymbol{q})^T\boldsymbol{Q}_B(\boldsymbol{q}) \tag{A.22}$$

with an arbitrary matrix $\boldsymbol{Q}_B(\boldsymbol{q}) \in \mathbb{R}^{(n-m)\times(n-m)}$. The matrix $\boldsymbol{Q}_B(\boldsymbol{q})$ can then be determined from equation A.18 as $\boldsymbol{Q}_B(\boldsymbol{q}) = \boldsymbol{I}$ and hence

$$\boldsymbol{B}(\boldsymbol{q}) = \boldsymbol{Z}(\boldsymbol{q})^T \tag{A.23}$$

Therefore, the solution is given by

$$\boldsymbol{J}_N(\boldsymbol{q})^{-1} = \begin{bmatrix}\boldsymbol{M}(\boldsymbol{q})^{-1}\boldsymbol{J}(\boldsymbol{q})^T(\boldsymbol{J}(\boldsymbol{q})\boldsymbol{M}(\boldsymbol{q})^{-1}\boldsymbol{J}(\boldsymbol{q})^T)^{-1} & \boldsymbol{Z}(\boldsymbol{q})^T\end{bmatrix} . \tag{A.24}$$

A.6 Model Parameters for the DLR-Lightweight-Robot-II

Figure A.1 shows the kinematic structure of the DLR-Lightweight-Robot-II. The parameters needed for the computation of the robot kinematics from Chapter 2.1 are given in Table A.1.

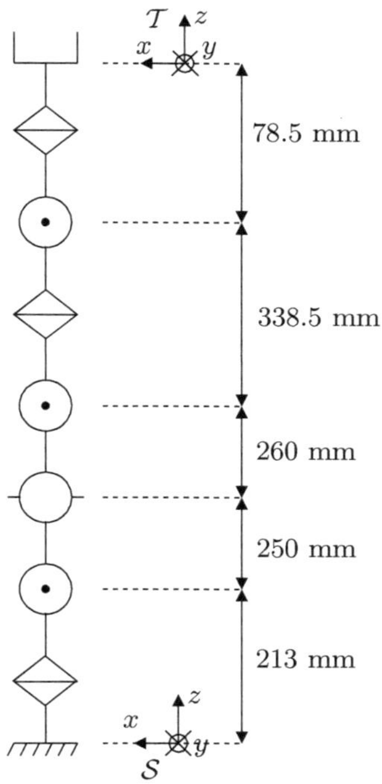

Fig. A.1. Kinematic structure of the DLR-Lightweight-Robot-II

Table A.1. Kinematics data of the DLR-Lightweight-Robot-II: $d_1 = 0.213$ m, $d_2 = 0.463$ m, $d_3 = d_4 = 0.723$ m, $d_5 = 1.0615$ m, $d_6 = 1.14$ m

i	1	2	3	4	5	6	7
$\boldsymbol{\omega}_i$	$\begin{pmatrix}0\\0\\1\end{pmatrix}$	$\begin{pmatrix}0\\1\\0\end{pmatrix}$	$\begin{pmatrix}-1\\0\\0\end{pmatrix}$	$\begin{pmatrix}0\\1\\0\end{pmatrix}$	$\begin{pmatrix}0\\0\\1\end{pmatrix}$	$\begin{pmatrix}0\\1\\0\end{pmatrix}$	$\begin{pmatrix}0\\0\\1\end{pmatrix}$
$\boldsymbol{l}_i$[m]	$\begin{pmatrix}0\\0\\0\end{pmatrix}$	$\begin{pmatrix}0\\0\\d_1\end{pmatrix}$	$\begin{pmatrix}0\\0\\d_2\end{pmatrix}$	$\begin{pmatrix}0\\0\\d_3\end{pmatrix}$	$\begin{pmatrix}0\\0\\d_4\end{pmatrix}$	$\begin{pmatrix}0\\0\\d_5\end{pmatrix}$	$\begin{pmatrix}0\\0\\d_6\end{pmatrix}$

Table A.2. Motor inertia and joint stiffness values of the DLR-Lightweight-Robot-II. The gear ratios have a value of 160 (for all the joints) and are already taken into account for the motor inertia values.

Joint i	Motor Inertia B_i [kgm^2]	Joint Stiffness K_i [Nm/rad]
1	1.836	8000.0
2	4.447	15000.0
3	3.242	14000.0
4	1.817	12000.0
5	1.392	8800.0
6	1.402	6000.0
7	1.392	8800.0

A.7 List of Used Symbols

Throughout this book scalar quantities are denoted by plain letters (e.g. m, n, δ, ϵ). For vector and matrix quantities bold letters are used instead. Dots denote derivatives with respect to the time t up to the second order (i.e. $\dot{\boldsymbol{q}} := \frac{\mathrm{d}}{\mathrm{d}t}\boldsymbol{q}, \ddot{\boldsymbol{q}} := \frac{\mathrm{d}^2}{\mathrm{d}t^2}\boldsymbol{q}$). Higher time derivatives are denoted in the form $\boldsymbol{q}^{(k)} := \frac{\mathrm{d}^k}{\mathrm{d}t^k}\boldsymbol{q}\ ,\ \ k > 2$.

The following list contains the symbols which are used in several chapters.

$n \in \mathbb{Z}$ ··· Number of joints.
$m \in \mathbb{Z}$ ··· Number of Cartesian coordinates (usually $m = 6$).
$r \in \mathbb{Z}$ ··· Degree of redundancy $r = n - m$.
$t \in \mathbb{R}$ ··· Time variable.
$\mathcal{S}$ ··· Base (or world) frame.
$\mathcal{T}$ ··· Tool frame at the robot end-effector.
$\mathcal{D}$ ··· *Desired* frame, i.e. virtual equilibrium frame.
$\lambda_i(\boldsymbol{A})$ ··· The ith eigenvalue of a symmetric matrix $\boldsymbol{A}$.
$\lambda_{min}(\boldsymbol{A})$ ··· The minimum eigenvalue of a symmetric matrix $\boldsymbol{A}$.
$\lambda_{max}(\boldsymbol{A})$ ··· The maximum eigenvalue of a symmetric matrix $\boldsymbol{A}$.
$\lambda_{m,M}$ ··· The minimum eigenvalue of the inertia matrix, i.e. $\lambda_{m,M} := \inf_{\boldsymbol{q}\in\mathcal{Q}^p} \min_{i=1..n} \lambda_i(\boldsymbol{M}(\boldsymbol{q}))$.
$\lambda_{M,M}$ ··· The maximum eigenvalue of the inertia matrix, i.e. $\lambda_{M,M} := \sup_{\boldsymbol{q}\in\mathcal{Q}^p} \max_{i=1..n} \lambda_i(\boldsymbol{M}(\boldsymbol{q}))$.
$\sigma_i(\boldsymbol{A})$ ··· The ith singular value of a matrix $\boldsymbol{A}$.
$\sigma_{min}(\boldsymbol{A})$ ··· The minimum singular value of a matrix $\boldsymbol{A}$.
$\sigma_{max}(\boldsymbol{A})$ ··· The maximum singular value of a matrix $\boldsymbol{A}$.

$\sigma_{m,J}$	$\cdots$ The minimum singular value of the Jacobian matrix, i.e. $\sigma_{m,J} := \inf_{\boldsymbol{q}\in\bar{\mathcal{Q}}^p} \min_{i=1..n} \sigma_i(\boldsymbol{J}(\boldsymbol{q}))$.
$\sigma_{M,J}$	$\cdots$ The maximum singular value of the Jacobian matrix, i.e. $\sigma_{M,J} := \sup_{\boldsymbol{q}\in\mathcal{Q}^p} \max_{i=1..n} \sigma_i(\boldsymbol{J}(\boldsymbol{q}))$.
$\mathcal{Q}$	$\cdots$ Configuration space
$\mathcal{Q}^p$	$\cdots$ Subsection of the configuration space, where all prismatic joints keep bounded, cf. (2.36).
$\bar{\mathcal{Q}}^p$	$\cdots$ Subsection of the configuration space, where the Jacobian is non-singular, cf. (3.6).
$\bar{\mathcal{Q}}^p_c$	$\cdots$ $\bar{\mathcal{Q}}^p_c := \boldsymbol{f}(\bar{\mathcal{Q}}^p)$.
$\boldsymbol{q} \in \mathbb{R}^n$	$\cdots$ Vector of link side joint angles.
$\boldsymbol{\theta} \in \mathbb{R}^n$	$\cdots$ Vector of motor side joint angles.
$\boldsymbol{\tau} \in \mathbb{R}^n$	$\cdots$ Vector of joint torques, $\boldsymbol{\tau} = \boldsymbol{K}(\boldsymbol{\theta} - \boldsymbol{q})$.
$\boldsymbol{\tau}_m \in \mathbb{R}^n$	$\cdots$ Vector of motor torques.
$\boldsymbol{\tau}_{ext} \in \mathbb{R}^n$	$\cdots$ Vector of external torques.
$\boldsymbol{f}(\boldsymbol{q}) \in \mathbb{R}^m$	$\cdots$ Forward Kinematics for a set of Cartesian Coordinates.
$\boldsymbol{J}(\boldsymbol{q}) \in \mathbb{R}^{m\times,n}$	$\cdots$ Analytical Jacobian, $\boldsymbol{J}(\boldsymbol{q}) = \partial\boldsymbol{f}(\boldsymbol{q})/\partial\boldsymbol{q}$.
$\boldsymbol{J}^b(\boldsymbol{q}) \in \mathbb{R}^{m\times,n}$	$\cdots$ Body (or geometrical) Jacobian.
$\boldsymbol{V}^b$	$\cdots$ Body twist.
$\boldsymbol{F}_{ext} \in \mathbb{R}^m$	$\cdots$ Vector of generalized external forces.
$\boldsymbol{x} \in \mathbb{R}^m$	$\cdots$ Vector of Cartesian coordinates, $\boldsymbol{x} = \boldsymbol{f}(\boldsymbol{q})$.
$\boldsymbol{x} \in \mathbb{R}^n$	$\cdots$ Generic state vector (used only in the appendix).
$\boldsymbol{x}_d \in \mathbb{R}^m$	$\cdots$ *Desired* position, i.e. virtual equilibrium position.
$\tilde{\boldsymbol{x}} \in \mathbb{R}^m$	$\cdots$ Cartesian error, $\tilde{\boldsymbol{x}} = \boldsymbol{x} - \boldsymbol{x}_d$.
$\boldsymbol{\phi} \in \mathbb{R}^3$	$\cdots$ Roll-pitch-yaw Euler angles.
$\boldsymbol{h}_{ab} \in SE(3)$	$\cdots$ Homogeneous transformation from frame $\mathcal{B}$ to frame $\mathcal{A}$.
$\boldsymbol{R}_{ab} \in SO(3)$	$\cdots$ Rotation matrix according go $\boldsymbol{h}_{ab}$.
${}_a\boldsymbol{p}_{ab} \in \mathbb{R}^3$	$\cdots$ Translation according go $\boldsymbol{h}_{ab}$.
$\boldsymbol{p}_{ab} \in \mathbb{R}^3$	$\cdots$ Vector from the origin of frame $\mathcal{A}$ to frame $\mathcal{B}$, represented in the base frame $\mathcal{S}$, $\boldsymbol{p}_{ab} = \boldsymbol{R}_{sa\,a}\boldsymbol{p}_{ab}$.
${}_b\boldsymbol{v}_{sb} \in \mathbb{R}^3$	$\cdots$ Velocity of frame $\mathcal{B}$ with respect to frame $\mathcal{B}$, represented in $\mathcal{B}$.
${}_b\boldsymbol{\omega}_{sb} \in \mathbb{R}^3$	$\cdots$ Angular velocity of frame $\mathcal{B}$ with respect to frame $\mathcal{B}$, represented in $\mathcal{B}$.
$\boldsymbol{I}$	$\cdots$ Identity matrix.
B_i	$\cdots$ Rotor inertia of motor i, for $i = 1, \cdots, n$.
K_i	$\cdots$ Stiffness of joint i, for $i = 1, \cdots, n$.
$\boldsymbol{B} \in \mathbb{R}^{n\times n}$	$\cdots$ Motor inertia matrix of the manipulator, $\boldsymbol{B} = \mathrm{diag}(B_i)$.

$\boldsymbol{K} \in \mathbb{R}^{n \times n}$	... Joint stiffness matrix, $\boldsymbol{K} = \mathrm{diag}(K_i)$.
$\boldsymbol{K}_d \in \mathbb{R}^{m \times m}$	... Matrix of desired stiffness.
$\boldsymbol{K}_n \in \mathbb{R}^{n \times n}$	... Nullspace stiffness matrix.
$\boldsymbol{D}_d \in \mathbb{R}^{m \times m}$	... Matrix of desired damping.
$\boldsymbol{D}_n \in \mathbb{R}^{r \times r}$	... Nullspace damping matrix.
$\boldsymbol{\Lambda}_d \in \mathbb{R}^{m \times m}$	... Matrix of desired inertia.
$\boldsymbol{M}(\boldsymbol{q}) \in \mathbb{R}^{n \times n}$	... Link side inertia matrix of the manipulator.
$\boldsymbol{C}(\boldsymbol{q}, \dot{\boldsymbol{q}}) \in \mathbb{R}^{n \times n}$	... Link side Coriolis/centrifugal matrix of the manipulator.
$\boldsymbol{g}(\boldsymbol{q}) \in \mathbb{R}^{n}$	... Gravity torques.
$\boldsymbol{\Lambda}(\boldsymbol{x})$	... Cartesian inertia matrix.
$\boldsymbol{\mu}(\boldsymbol{x}, \dot{\boldsymbol{x}})$	... Cartesian Coriolis/centrifugal matrix.

Springer Tracts in Advanced Robotics

Edited by B. Siciliano, O. Khatib and F. Groen

Further volumes of this series can be found on our homepage: springer.com

Vol. 49: Ott, C.
Cartesian Impedance Control of Redundant and Flexible-Joint Robots
190 p. 2008 [978-3-540-69253-9]

Vol. 48: Wolter, D.
Spatial Representation and Reasoning for Robot Mapping
185 p. 2008 [978-3-540-69011-5]

Vol. 47: Akella, S.; Amato, N.; Huang, W.; Mishra, B.; (Eds.)
Algorithmic Foundation of Robotics VII
524 p. 2008 [978-3-540-68404-6]

Vol. 46: Bessière, P.; Laugier, C.; Siegwart R. (Eds.)
Probabilistic Reasoning and Decision Making in Sensory-Motor Systems
375 p. 2008 [978-3-540-79006-8]

Vol. 45: Bicchi, A.; Buss, M.; Ernst, M.O.; Peer A. (Eds.)
The Sense of Touch and Its Rendering
281 p. 2008 [978-3-540-79034-1]

Vol. 44: Bruyninckx, H.; Přeučil, L.; Kulich, M. (Eds.)
European Robotics Symposium 2008
356 p. 2008 [978-3-540-78315-2]

Vol. 43: Lamon, P.
3D-Position Tracking and Control for All-Terrain Robots
105 p. 2008 [978-3-540-78286-5]

Vol. 42: Laugier, C.; Siegwart, R. (Eds.)
Field and Service Robotics
597 p. 2008 [978-3-540-75403-9]

Vol. 41: Milford, M.J.
Robot Navigation from Nature
194 p. 2008 [978-3-540-77519-5]

Vol. 40: Birglen, L.; Laliberté, T.; Gosselin, C.
Underactuated Robotic Hands
241 p. 2008 [978-3-540-77458-7]

Vol. 39: Khatib, O.; Kumar, V.; Rus, D. (Eds.)
Experimental Robotics
563 p. 2008 [978-3-540-77456-3]

Vol. 38: Jefferies, M.E.; Yeap, W.-K. (Eds.)
Robotics and Cognitive Approaches to Spatial Mapping
328 p. 2008 [978-3-540-75386-5]

Vol. 37: Ollero, A.; Maza, I. (Eds.)
Multiple Heterogeneous Unmanned Aerial Vehicles
233 p. 2007 [978-3-540-73957-9]

Vol. 36: Buehler, M.; Iagnemma, K.; Singh, S. (Eds.)
The 2005 DARPA Grand Challenge – The Great Robot Race
520 p. 2007 [978-3-540-73428-4]

Vol. 35: Laugier, C.; Chatila, R. (Eds.)
Autonomous Navigation in Dynamic Environments
169 p. 2007 [978-3-540-73421-5]

Vol. 34: Wisse, M.; van der Linde, R.Q.
Delft Pneumatic Bipeds
136 p. 2007 [978-3-540-72807-8]

Vol. 33: Kong, X.; Gosselin, C.
Type Synthesis of Parallel Mechanisms
272 p. 2007 [978-3-540-71989-2]

Vol. 32: Milutinović, D.; Lima, P.
Cells and Robots – Modeling and Control of Large-Size Agent Populations
130 p. 2007 [978-3-540-71981-6]

Vol. 31: Ferre, M.; Buss, M.; Aracil, R.; Melchiorri, C.; Balaguer C. (Eds.)
Advances in Telerobotics
500 p. 2007 [978-3-540-71363-0]

Vol. 30: Brugali, D. (Ed.)
Software Engineering for Experimental Robotics
490 p. 2007 [978-3-540-68949-2]

Vol. 29: Secchi, C.; Stramigioli, S.; Fantuzzi, C.
Control of Interactive Robotic Interfaces – A Port-Hamiltonian Approach
225 p. 2007 [978-3-540-49712-7]

Vol. 28: Thrun, S.; Brooks, R.; Durrant-Whyte, H. (Eds.)
Robotics Research – Results of the 12th International Symposium ISRR
602 p. 2007 [978-3-540-48110-2]

Vol. 27: Montemerlo, M.; Thrun, S.
FastSLAM – A Scalable Method for the Simultaneous Localization and Mapping Problem in Robotics
120 p. 2007 [978-3-540-46399-3]

Vol. 26: Taylor, G.; Kleeman, L.
Visual Perception and Robotic Manipulation – 3D Object Recognition, Tracking and Hand-Eye Coordination
218 p. 2007 [978-3-540-33454-5]

Vol. 25: Corke, P.; Sukkarieh, S. (Eds.)
Field and Service Robotics – Results of the 5th International Conference
580 p. 2006 [978-3-540-33452-1]

Vol. 24: Yuta, S.; Asama, H.; Thrun, S.; Prassler, E.; Tsubouchi, T. (Eds.)
Field and Service Robotics – Recent Advances in Research and Applications
550 p. 2006 [978-3-540-32801-8]

Vol. 23: Andrade-Cetto, J,; Sanfeliu, A.
Environment Learning for Indoor Mobile Robots – A Stochastic State Estimation Approach to Simultaneous Localization and Map Building
130 p. 2006 [978-3-540-32795-0]

Vol. 22: Christensen, H.I. (Ed.)
European Robotics Symposium 2006
209 p. 2006 [978-3-540-32688-5]

Vol. 21: Ang Jr., H.; Khatib, O. (Eds.)
Experimental Robotics IX – The 9th International Symposium on Experimental Robotics
618 p. 2006 [978-3-540-28816-9]

Vol. 20: Xu, Y.; Ou, Y.
Control of Single Wheel Robots
188 p. 2005 [978-3-540-28184-9]

Vol. 19: Lefebvre, T.; Bruyninckx, H.; De Schutter, J. Nonlinear Kalman Filtering for Force-Controlled Robot Tasks
280 p. 2005 [978-3-540-28023-1]

Vol. 18: Barbagli, F.; Prattichizzo, D.; Salisbury, K. (Eds.)
Multi-point Interaction with Real and Virtual Objects
281 p. 2005 [978-3-540-26036-3]

Vol. 17: Erdmann, M.; Hsu, D.; Overmars, M.; van der Stappen, F.A (Eds.)
Algorithmic Foundations of Robotics VI
472 p. 2005 [978-3-540-25728-8]

Vol. 16: Cuesta, F.; Ollero, A.
Intelligent Mobile Robot Navigation
224 p. 2005 [978-3-540-23956-7]

Vol. 15: Dario, P.; Chatila R. (Eds.)
Robotics Research – The Eleventh International Symposium
595 p. 2005 [978-3-540-23214-8]

Vol. 14: Prassler, E.; Lawitzky, G.; Stopp, A.; Grunwald, G.; Hägele, M.; Dillmann, R.; Iossifidis. I. (Eds.)
Advances in Human-Robot Interaction
414 p. 2005 [978-3-540-23211-7]

Vol. 13: Chung, W.
Nonholonomic Manipulators
115 p. 2004 [978-3-540-22108-1]

Vol. 12: Iagnemma K.; Dubowsky, S.
Mobile Robots in Rough Terrain – Estimation, Motion Planning, and Control with Application to Planetary Rovers
123 p. 2004 [978-3-540-21968-2]

Vol. 11: Kim, J.-H.; Kim, D.-H.; Kim, Y.-J.; Seow, K.-T.
Soccer Robotics
353 p. 2004 [978-3-540-21859-3]

Vol. 10: Siciliano, B.; De Luca, A.; Melchiorri, C.; Casalino, G. (Eds.)
Advances in Control of Articulated and Mobile Robots
259 p. 2004 [978-3-540-20783-2]

Vol. 9: Yamane, K.
Simulating and Generating Motions of Human Figures
176 p. 2004 [978-3-540-20317-9]

Vol. 8: Baeten, J.; De Schutter, J.
Integrated Visual Servoing and Force Control – The Task Frame Approach
198 p. 2004 [978-3-540-40475-0]

Vol. 7: Boissonnat, J.-D.; Burdick, J.; Goldberg, K.; Hutchinson, S. (Eds.)
Algorithmic Foundations of Robotics V
577 p. 2004 [978-3-540-40476-7]

Vol. 6: Jarvis, R.A.; Zelinsky, A. (Eds.)
Robotics Research – The Tenth International Symposium
580 p. 2003 [978-3-540-00550-6]

MIX
Papier aus verantwortungsvollen Quellen
Paper from responsible sources
FSC® C105338

If you have any concerns about our products,
you can contact us on
ProductSafety@springernature.com

In case Publisher is established outside the EU,
the EU authorized representative is:
Springer Nature Customer Service Center GmbH
Europaplatz 3, 69115 Heidelberg, Germany

Printed by Libri Plureos GmbH
in Hamburg, Germany